Monte Verità Proceedings of the Centro Stefano Franscini Ascona

Edited by K. Osterwalder, ETH Zürich

Automatic Extraction of Man-Made Objects from Aerial and Space Images

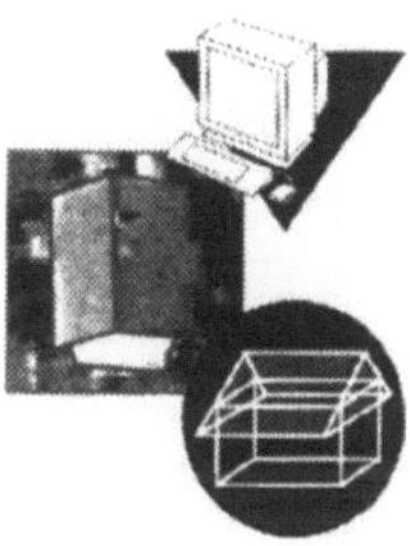

Edited by
A. Gruen
O. Kuebler
P. Agouris

1995

Birkhäuser Verlag
Basel · Boston · Berlin

Editors:

Prof. Dr. Armin Gruen
Institute of Geodesy & Photogrammetry
ETH-Hoenggerberg
CH-8093 Zürich

Prof. Dr. Olaf Kuebler
Communications Technology Laboratory
ETH-Zentrum
CH-8092 Zürich

Dr. Peggy Agouris
Institute of Geodesy & Photogrammetry
ETH-Hoenggerberg
CH-8093 Zürich

e-mail: peggy@p.igp.ethz.ch

A CIP catalogue record for this book is available from the Library of Congress, Washington D.C., USA

Deutsche Bibliothek Cataloging-in-Publication Data

Automatic extraction of man-made objects from aerial and space images / ed. by A. Gruen ... – Basel ; Boston ; Berlin : Birkhäuser, 1995
(Monte Verità)
ISBN-13: 978-3-0348-9958-1 e-ISBN-13: 978-3-0348-9242-1
DOI: 10.1007/978-3-0348-9242-1

NE: Gruen, Armin [Hrsg.]

e-mail: agruen@p.igp.ethz.ch

Printed on acid-free paper produced from chlorine-free pulp. ∞

Softcover reprint of the hardcover 1st edition 1995

9 8 7 6 5 4 3 2 1

Preface

The idea for this Workshop originated in discussions we had during the early stages of our joint research project AMOBE (Automation of Digital Terrain Model Generation and Man-Made Object Extraction from Aerial Images). One of the objectives of this project is to combine the expertise of our photogrammetry and computer vision groups, which in essence translates to supporting computer vision activities with the quantitative approaches of photogrammetry, including precise calibration methods and camera models and a general 3-D world and analysis concept. In return, it was anticipated that photogrammetric "know-how" would be amended by the most recent image analysis algorithms provided by the computer vision community. It seemed to us that this was a most appropriate strategy to tackle the difficult problem of automation in Digital Terrain Model generation and building/road extraction.

It soon became clear to us that it would be highly beneficial, in order to get first-hand information about the state-of-the-art of this quickly developing area, to bring together the worldwide leading experts of this field. With Monte Verità in Ascona as the location and Centro Stefano Franscini as the administrative unit, the Federal Institute of Technology (ETH) Zurich has at its disposal ideal facilities to conduct a workshop aiming at critical introspection without completely losing contact with the lovely outside surroundings. Seeking to provide a forum to outline the advantages and limitations of existing and proposed approaches as well as to furnish guidelines for future research, we invited the principal groups worldwide which, according to our information, had proven and significant experience and expertise in building and road extraction.

In retrospect, we feel that our Workshop was truly successful. First reactions of the participants indicate that the aims of the meeting were fully reached. Concentrated and thematically focused scientific-technical presentations and discussions were supported and intensified by the pleasure of working and living in a scenic environment and the opportunity to get personally acquainted with colleagues in a relaxing and cozy atmosphere.

It is our hope and desire that something of this unique atmosphere and high quality of our Workshop will be preserved by these Proceedings. We have structured the contents of this volume into four major themes, namely General Strategies, Building Extraction, Road Extraction, Map-Based Extraction, as they reflect the key issues of the Workshop.

We would like to thank all those individuals who contributed to the organization and administration of this Workshop, in particular Dr. Emmanuel Baltsavias, Dr. Scott Mason, and Mr. Haihong Li. We also acknowledge the excellent support provided by the Centro Franscini through Mrs. Katia Bastianelli. Above all, however, we are grateful for the enduring work and patient attitude of Dr. Peggy Agouris, our Workshop scientific manager.

Armin Gruen
Institute of Geodesy and Photogrammetry

Olaf Kuebler
Communications Technology Laboratory

Table of Contents

General Strategies

Building Extraction

Road Extraction

Map-Based Extraction

General Strategies

Using Context to Control Computer Vision Algorithms

Thomas M. Strat

Artificial Intelligence Center
SRI International
333 Ravenswood Avenue
Menlo Park, California 94025, USA
strat@ai.sri.com

Abstract

We are investigating the design of an architecture that can be used as the basis for controlling the invocation of image understanding algorithms for cartographic feature extraction. The key research question is whether sufficient contextual constraints are available to choose algorithms and their parameters for aerial photo analysis. Our approach has been to apply the context-based architecture incorporated in CONDOR, an SRI system for automatically constructing scene models of natural terrain from ground-level views. The semiautomated nature of cartographic feature extraction allows access to additional sources of contextual constraints that were not available to CONDOR. [1]

1 Introduction

Thirty years of research in image understanding (IU) have produced an enormous number of computer vision algorithms, many of which have demonstrated reliable performance at solving particular tasks in restricted domains. However, the development of computer vision systems that are reliable in more general circumstances has proved elusive [JB91]. It is in response to this situation that a framework is offered within which computer vision algorithms of specialized competence can be used and integrated with other such algorithms to produce a reliable vision system that operates effectively in a broader context than any of its individual component algorithms can. Other authors have also stressed the use of context to aid image interpretation [Alo90, CGHS89, FHS94].

The IU system that we envision comprises a large number of computer vision algorithms, each tailored to accomplish a particular task under a particular set of circumstances. The goals of these algorithms may overlap or even duplicate each other's goals, but the assumptions that they make and the data with which they operate will often differ.

The strategy for integrating such a collection of computer vision algorithms is to represent explicitly the assumptions made by each algorithm, and to use the context of the present task to select the most appropriate algorithms for solving that task. By doing

[1] The research reported here was supported by the Advanced Research Projects Agency and the U.S. Army Topographic Engineering Center under Contract DACA 76-92-C-0034.

so, we seek to avoid the source of many failures of computer vision techniques — the employment of an algorithm outside the bounds of its intended domain of competence.

2 The CONDOR Architecture

In the context-based vision architecture the development of general-purpose visual capabilities is attempted by assembling large numbers of special-purpose algorithms. Their invocation and the interpretation of results are mediated by the consideration of contextual information.

The image understanding system CONDOR was designed along these lines to serve as the perceptual architecture for a hypothetical outdoor robot [SF91b]. Given an image and a possibly extensive spatial database, a world model describing the robot's environment, the system is to analyze the image and to augment its world model. CONDOR's recognition vocabulary consists mainly of natural objects such as trees, bushes, trails, and rocks. Because of the difficulty of recognizing such objects individually, CONDOR accepts an interpretation only if it is consistent with its world model. CONDOR recognizes entire contexts, rather than individual objects [SF91a, Str92].

CONDOR associates a data structure called a *context set* with each IU algorithm. The context set identifies those conditions that must be true for that algorithm to be applicable. Efficient and effective visual recognition can be achieved only by invoking IU algorithms in those contexts in which they are likely to succeed.

While CONDOR has demonstrated a significant capability for recognizing natural objects in ground-level outdoor imagery, perhaps its more enduring contribution lies in its context-based architecture (CBA), which offers a design methodology with broad applicability. For example, CBA is an attractive framework for organizing a cartographic feature extraction system using many algorithms that extract different features under different circumstances. The remainder of this paper describes the use of CBA in a 3–D cartographic modeling system.

2.1 Context Sets

Context sets are used to specify the conditions that must be met for a given algorithm to be applicable. The context set can also specify the conditions that must be met for a given parameter setting to be useful. For example,

MBO(closed-curve, rectangular-corners, manual-entry, gradient-descent)

specifies the parameters for a model-based optimization (MBO) algorithm[2] that could be used to extract roof boundaries under some circumstances. The following context set encodes conditions that must be met for the successful employment of that algorithm to extract roofs.

{ image-is-bw, image-resolution$\leq$ 3.0, interactivity-is-semiautomated }

This context set gives the requirements that must exist for the MBO algorithm to be applicable.

In practice, a large number of context sets governing the application of MBO algorithms as well as other algorithms could be constructed and used to implement a cartographic

[2]Snakes are a simple, well-known example of an MBO algorithm.

feature-extraction system suitable for site-model construction. It is clear that such a collection could be unwieldy and difficult to maintain. A more structured representation of the context-set concept is needed.

2.2 Context Tables

One alternative representation for context sets is the *context table* — a data structure that tabulates the context elements in a more structured fashion [Str93]. An IU algorithm is associated with each row in the table; each column represents one context element (see Table 1).

Table 1: An Example of a Context Table

	feature	interactivity	resolution	geography	algorithm
1	roof	semiautomatic	≤ 3 m	—	MBO(topology=closed-curve, obj-fn=rectangular-corners, init=manual-entry, opt=gradient-descent)
2	roof	manual	≤ 10 m	—	CME(primitive=closed-curve)
3	road	semiautomatic	≤ 1 m	hilly	MBO(topology=ribbon-curve, obj-fn=(smoothness(0.5),continuous, parallel), init=manual-entry, opt=gradient-descent)
5	road	semiautomatic	≤ 1 m	flat	MBO(topology=ribbon-curve, obj-fn=(smoothness(0.8),continuous, parallel), init=manual-entry, opt=gradient-descent)
7	road	manual	≤ 10 m	—	CME(primitive=open-curve)
8	road	manual	≤ 1 m	—	CME(primitive=ribbon-curve)
9	road	semiautomatic	≤ 2 m	—	ROAD-TRACKER (control=bidirectional-search, init=manual-entry)

The context table is equivalent to a collection of context sets. Conceptually, it provides a more coherent view of the contextual requirements of related algorithms. Applicable algorithms are selected by finding rows for which all conditions are met. One drawback to the table representation is its potentially large size. Each algorithm may require many rows to capture the contextual constraints of its various parameter combinations. Its chief value is its organization of contextual information for knowledge-base construction.

2.3 Context Rules

A third alternative for representing context sets is to encode them as production rules whose antecedent is the context set, and whose consequent is the applicable algorithm. For example,

$$\{\ \text{image-is-bw, image-resolution} \leq 3.0\text{, interactivity-is-semiautomated}\ \} \implies$$
$$\text{MBO(closed-curve, rectangular-corners, manual-entry, gradient-descent).}$$

One advantage of encoding the constraints as rules in a logic program is that using the logic program interpreter eliminates the need to devise special machinery to test satisfaction

```
(:algorithm ?name ?parms)
    ?name:  The name of a lisp function
    ?parms:  A list of the parameters to the function

(:task ?name ?feature ?given)
    ?name:  A generic task name, such as :extract, :refine
    ?feature:  A feature type (eg, :building, :road, :functional-area)
    ?given:  An object to be used, an image, a region, or other designation
          of a location where the task is to be accomplished
```

Figure 1: Documentation of compound terms.

of context sets. Unification provides a generic mechanism for matching the constraints embodied in a rule to the current context. Context rules can be more compact than the equivalent context table because additional predicates can be introduced to capture common context elements that appear in multiple rows of the table.

Whatever representation is chosen, it is clear that context sets can encode the information necessary to make intelligent choices of IU algorithms to solve particular tasks. Furthermore, they can be employed in either direction. In the forward direction, the context sets are used to find applicable algorithms. In the opposite direction, the sets can be used for several purposes, including the selection of images on which to invoke a given algorithm.

3 Context-Based Architecture Specification

We have chosen to use logic programming to implement the context-based architecture because it provides a natural declarative language for expressing the constraints, because unification provides a powerful mechanism for matching the contextual constraints (as encoded in the rules) to the current context, and because the logical backchaining of a rule-based system provides the ability to search the rule base for algorithms that are applicable.

In this paper we illustrate the concepts using a version of Prolog [SS86] that is defined in the book *On Lisp* [Gra94]. A synopsis of the syntax of this Lisp-based Prolog is included as Appendix A. It should be noted that the choice of logic programming (and Prolog in particular) is primarily a matter of implementation convenience — other schemes could be employed as effectively.

3.1 Terms

Before describing the context-based architecture in detail, we introduce two concepts that are central to its operation: algorithms and tasks (Figure 1).

An *algorithm*, represented by a logical term, designates a piece of code that implements a computer vision technique. Each algorithm term indicates the name of the procedure to be invoked and a list of the parameters it expects. The parameters will typically include a list of the images to be employed, any initial conditions to be provided, and the typical arguments to the function. For example,

```
(:algorithm road-tracker (fh-image-2 3d-ribbon-curve (942 1516)))
```

(a)
```
; (invoke ?alg)
;     ?alg is an algorithm that could be invoked to solve the given task
;      in the current context.
(<- (invoke ?alg)
        (desired-task ?task)
        (solves ?alg ?task)
        (desired-interactivity ?int)
        (interactivity ?alg ?int)
        (desired-accuracy ?acc)
        (accuracy ?alg ?acc) )
```
(b)
```
; (solves ?alg ?task)
;      An algorithm that solves a task using an image,
;      also solves the task for the site of the image.
(<- (solves ?alg (:task ?name ?feature ?site) )
        (image-site ?image ?site)
        (solves ?alg (:task ?name ?feature ?image)) )
```

Figure 2: Documentation of some of the top-level rules.

indicates that the road tracker algorithm is to be invoked on the image named `fh-image-2`, creating an object of class `3d-ribbon-curve` starting at image coordinate (942, 1516).

A *task* is a term that indicates an operation that could be performed. It includes the name of the task, the type of feature that is desired, and a list of the data to be used in accomplishing the task. An example task is

```
(:task extract wide-road world-point(-5407.3 5732.2)).
```

Here, the task is to extract a wide road starting from the specified point whose georeferenced coordinates are given.

3.2 Predicates

The two most important predicates used in the context-based architecture are `:solves` and `:invoke`. The clause `(:solves ?alg ?task)` is true if the indicated algorithm can be used to solve the indicated task. The specification of the algorithm includes the specification of the imagery, parameters, and other given information to be used. The task provides a complete specification of the goal to be accomplished and the constraints to be exploited.

The clause `(:invoke ?alg)` is true if the indicated algorithm is applicable to solving the current task. It is the truth of this predicate, and the associated binding of `?alg`, that the context-based architecture is expected to compute.

3.3 Rules

The primary rule governing the operation of the CBA is shown in Figure 2a. It specifies the conditions that must be satisfied in order for an appropriate algorithm to be found to solve the currently desired task in the current context. In words, this rule states that an algorithm to be invoked must be capable of solving the desired task, and that its level of interactivity must be what is desired and its accuracy must meet the desired accuracy. Assertions of the clauses `(desired-task ?task)`, `(desired-interactivity ?int)`, and `(desired-accuracy ?acc)` are expected to be provided in advance by the user.

Several auxiliary rules, such as the one shown in Figure 2b, add versatility to the control structure. This rule states that an algorithm that solves a task using a particular image also solves that task for the site captured by the image. For example, the task

```
(:task :extract :building fort-hood)
```

can be reduced to

```
(:task :extract :building fh-image1)
```

if `fh-image1` actually portrays the site, Fort Hood.

Posing the query `(invoke ?alg)` triggers the search of a binding for `?alg` that solves the desired task, `?task`. If successful, the binding of `?alg` that is produced is typically a compound term specifying an algorithm, the imagery to be used, and any parameters it needs. If no applicable algorithm can be found, the search terminates and the failure is reported.

3.4 Rule Packets

Introduction of a new algorithm to the system requires specification of a packet of rules that delimit its scope of applicability, its properties, and its assumptions.

For example, Lynn Quam's Road Tracker algorithm [QS91] seeks to follow the extent of a road by correlating successive cross sections starting from an initial seed point. It requires that the width of the road occupy at least 4 pixels. The rule packet governing the application of this algorithm is given in Figure 3. The first rule states that the algorithm can be used to extract any feature that is a subclass of type road, that the image it uses must have a ground-sample-distance of less than 1.5 meters in the vicinity of the road (i.e., a 6-meter-wide road must span at least 4 pixels), and that the algorithm will use

```
; (road-tracker ?image ?class)
; The top-level lisp function to invoke the SRI road tracker algorithm.
;   ?image:  The image to be used
;   ?class:  The RCDE class to be instantiated
(<- (solves (:algorithm 'road-tracker ?image ?class ?feature )
            (:task :extract ?feature ?image) )
        (image ?image)
        (subtype :road ?feature)
        (rcde-class-of-feature-type ?class ?feature)
        (gsd ?image ?x)
        (lisp (and (numberp ?x) (< ?x 1.5)) ))

(<- (interactivity (:algorithm 'road-tracker ?image ?class ?feat) :semi))

(<- (accuracy (:algorithm 'road-tracker ?image ?class ?feature) :coarse) )
```

Figure 3: The rule packet for the Road Tracker algorithm.

an instance of a RADIUS Common Development Environment (RCDE) class suitable for modeling the desired feature type.

The remaining two rules specify that the road tracker algorithm is inherently semiautomatic (because it requires the user to provide an initial seed point), and that the delineation accuracy that can be expected is qualitatively designated as "coarse."

4 Examples

We have implemented a prototype of the context-based architecture using Prolog [SS86]. This software is presently being tested and extended to provide more flexibility and more powerful constructs for use in the algorithm-specific rule packets.

Each algorithm that is to be controlled by the CBA is represented by a packet of rules. Our implemented system contains rule packets for the following algorithms:

- Mouse input of points and lines
- Interactive RCDE object-modeling primitives [MWQ+92]
- A subset of the model-based optimization algorithms [FH91]
- The Quam Road Tracker [QS91]
- A cookie-cutter algorithm for automatically cloning structures

The rule packets for these algorithms illustrate the basic operation of the CBA. As implemented, the CBA performs only some simple selections of algorithms and parameters. Additional feature extraction algorithms are being contemplated for integration [CDD+93, CZD+94, SM93, LHN94]. More advanced concepts are being devised and implemented to allow more precise specification of when an algorithm should be used and how its parameters are to be determined.

To illustrate the operation of the CBA, consider the goal of semiautomatically extracting a model of a road, using the rule packets for the algorithms listed before. The complete logic program used by the CBA contains a collection of facts about the current context, including a list of images that are available and their properties. In practice, these are computed at runtime through procedural attachment within the logic program.

The user asserts additional facts as follows to specify the task and the constraints on that task.

```
(desired-task (:task :extract two-lane-road fort-hood))
(desired-accuracy :coarse)
(desired-interactivity :semiautomatic)
```

The CBA is then activated by posing the query to the Prolog interpreter:

```
(invoke ?alg)
```

to which the only acceptable binding for **?alg** turns out to be

```
(:algorithm road-tracker fh-image4 3d-ribbon-curve)
```

This example makes use of the rule in Figure 2b to find images of the Fort Hood site for further consideration. Only **fh-image4** has sufficient resolution for use by the **road-tracker**

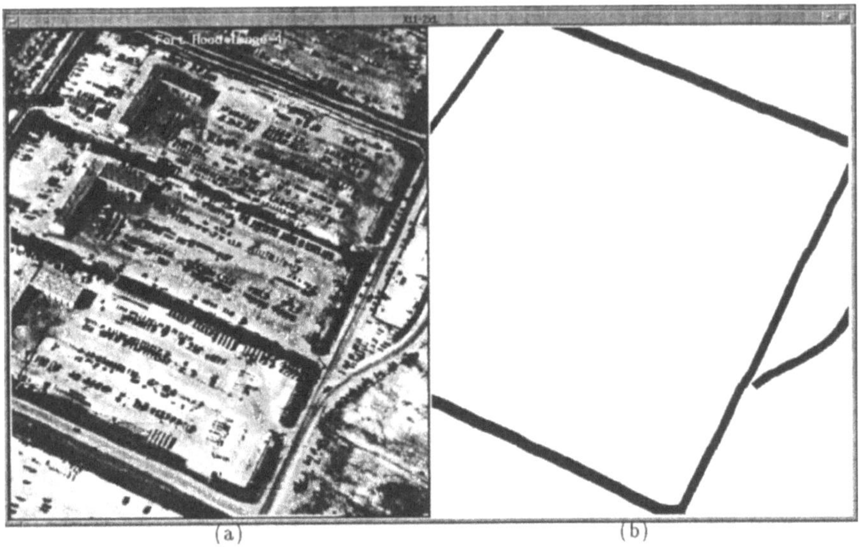

Figure 4: Sample task: (a) An image of Fort Hood (b) Result of invoking the road tracker algorithm.

algorithm. The example illustrates how the CBA has identified a suitable algorithm, its parameters, and an image, given only the specification of the task to extract a road from Fort Hood. The MBO algorithms were ruled out because they require an initial estimate of the road boundaries, which were not available in this case. The interactive modeling techniques of the RCDE were eliminated from consideration because the user specified his desire for a semiautomatic process. If more than one binding of `?alg` would have solved the query, the Prolog interpreter would compute each one of them. Additional machinery must be developed to allow the CBA to choose the best algorithm, or best parameter settings, when more than one solution exists.

5 Discussion

The context-based architecture described was designed to enable the construction of large, reliable image understanding systems by integrating a collection of reliable, but specialized algorithms. Initial experiments indicate that progress has been made in this direction, but formal testing of even larger assemblages of algorithms is necessary to validate this belief.

Some of the known limitations of our current implementation are

- There is no notion of likelihood of success — all algorithms are characterized as being applicable or not. Incorporating machinery to manipulate probabilities or other uncertainty measures is doable, but ascertaining the likelihood of success of IU algorithms under many and varied circumstances could be difficult.

- Incorporating additional algorithms requires hand crafting the rule packets that govern their employment. Automated knowledge acquisition to automate this process is needed to alleviate the burden.
- The CBA already has the ability to find sequences of algorithms to solve a given task. However, it has no procedure to check the validity of intermediate results at runtime.

It is clear that the performance of an IU system employing the context-based architecture could also be attained by integrating the same computer vision algorithms via more traditional methods. However, the explicit representation of contextual constraints affords a number of additional benefits that would be lost to a purely functional integration. These benefits include

Task specification: The context-based architecture allows the user to specify the task to be accomplished, leaving the selection of specific algorithms to be decided by the system. For example, the user can state that he would like the system to construct a 3-D model of a building, and the system would decide which of several building extraction algorithms would be most appropriate given the currently available imagery and auxiliary data. The user can make effective use of the IU system while possessing little knowledge of the capabilities and limitations of the individual computer vision algorithms.

Incremental integration: When building large systems in an evolutionary fashion, it can be difficult to add new capabilities without jeopardizing the integrity of existing capabilities. The modular decomposition of the context rules allows the developer to integrate a new algorithm by adding a packet of rules governing its use, without modifying existing code.

Our work on the context-based vision paradigm has led to the development of an architecture for integrating independently developed IU algorithms within a single system. The architecture provides the additional benefit of allowing the photoanalyst to specify a desired result, rather than a specific procedure to be followed, in carrying out a cartographic feature extraction task.

References

[Alo90] Yiannis Aloimonos. Purposive and Qualitative Action Vision. In *Proc. DARPA Image Understanding Workshop*, pages 816–828, Pittsburgh, Pennsylvania, September 1990.

[CDD+93] R. Chellappa, L.S. Davis, D. DeMenthon, A. Rosenfeld, and Q. Zheng. Site-Model-Based Change Detection and Image Registration. In *ARPA Image Understanding Workshop*, April 1993.

[CGHS89] V. Clément, G. Giraudon, S. Houzelle, and F. Sandakly. Interpretation of Remotely Sensed Images in a Context of Multisensor Fusion using a Multispecialist Architecture. *IEEE Transactions on Geoscience and Remote Sensing*, 31(4), July 1989.

[CZD+94] R. Chellappa, Q. Zheng, L.S. Davis, C.L. Lin, X. Zhang, C. Rodriguez, A. Rosenfeld, and T. Moore. Site-Model-Based Monitoring of Aerial Images. In *DARPA Workshop on Image Understanding*, November 1994.

[FH91] P. Fua and A.J. Hanson. An Optimization Framework for Feature Extraction. *Machine Vision and Applications*, 4(2):59–87, Spring 1991.

[FHS94] Daniel D. Fu, Kristian J. Hammond, and Michael J. Swain. Vision and Navigation in Man-Made Environments: Looking for Syrup in all the Right Places. In *Proceedings of the Workshop on Visual Behaviors*, pages 20–26, Seattle, WA, June 1994.

[Gra94] Paul Graham. *On Lisp: Advanced Techniques for Common Lisp.* Prentice Hall, Englewood Cliffs, NJ, 1994.

[JB91] Ramesh C. Jain and Thomas O. Binford. Ignorance, Myopia, and Naiveté in Computer Vision Systems. *CVGIP: Image Understanding*, 53(1):112–117, January 1991.

[LHN94] C. Lin, A. Huertas, and R. Nevatia. Detection of Buildings using Perceptual Grouping and Shadows. In *Proc. IEEE CVPR-94*, Seattle, June 1994.

[MWQ+92] J.L. Mundy, R. Welty, L. Quam, T. Strat, W. Bremmer, M. Horwedel, D. Hackett, and A. Hoogs. The RADIUS Common Development Environment. In *DARPA Workshop on Image Understanding*, pages 215–226, 1992.

[QS91] Lynn H. Quam and Thoms M. Strat. SRI Image Understanding Research in Cartographic Feature Extraction. In *Proc. ISPRS2*, Munich, September 1991.

[SF91a] T. M. Strat and M. A. Fischler. Context-Based Vision: Recognizing Objects Using Both 2D and 3D Imagery. *IEEE Trans. on Pattern Analysis and Machine Intelligence*, 13(10):1050–1065, October 1991.

[SF91b] T.M. Strat and M.A. Fischler. Natural Object Recognition: A Theoretical Framework and Its Implementation. *Proc. IJCAI-91*, August 1991.

[SM93] J.A. Shufelt and D.M. McKeown. Fusion of Monocular Cues to Detect Man-Made Structures in Aerial Imagery. *CVGIP: Image Understanding*, 57(3):307–330, May 1993.

[SS86] Leon Sterling and Ehud Shapiro. *The Art of Prolog.* The MIT Press, Cambridge, Mass., 1986.

[Str92] Thomas M. Strat. *Natural Object Recognition.* Springer-Verlag, New York, 1992.

[Str93] Thomas M. Strat. Employing Contextual Information in Computer Vision. In *DARPA Workshop on Image Understanding*, 1993.

A Syntax of OnLisp Prolog

The syntax used for the implementation of the context-based architecture is the same as standard Prolog with the following exceptions [Gra94]:

1. Variables are represented by symbols beginning with question marks instead of capital letters.

2. [ ] becomes ( ); $[x \mid y]$ becomes $(x \,.\, y)$; and $[x,\ y,\ ...]$ becomes $(x\ y\ ...)$.

3. Predicates are moved inside parentheses, and no commas separate arguments: *pred*$(x,\ y,\ ...)$ becomes (*pred* $x\ y$ ...).

4. A rule is specified in prefix notation: *head* :- *clause*, *clause* ... becomes (<- *head clause clause* ...). A rule with a head and no clauses is a fact: (<- *head*).

Automatic Extraction of Man-Made Objects
from Aerial and Space Images
Monte Verità, © 1995 Birkhäuser Verlag Basel

Inferring Homogeneous Regions from Rich Image Attributes*

Olof Henricsson

Communication Technology Laboratory,
Swiss Federal Institute of Technology
ETH – Zentrum, CH-8092 Zürich, Switzerland
Ph.: +41 (0)1 632 5281, Fax: +41 (0)1 632 1199
email: ohenri@vision.ethz.ch

Abstract

Image segmentation is an important part in any computer vision framework. However, the transition from local low-level representations to useful structures and relations in the intermediate levels has turned out to be a truly difficult problem. This paper addresses the difficult transition from low-level into intermediate-level vision, where the latter deals with producing a description of image and scene attributes in which more global relations are made explicit. We propose to combine a rich attributed contour representation with very general geometric contour relations. The implemented geometric relations, which are proximity, curvilinearity, parallelism and corner-like relations, allow to handle general man-made objects whose projected surfaces can be described by combinations of the defined relations. The combination of rich image attributes and geometric relations allows to discriminate between strong and weak contour relations. Strong relations require that not only the geometrical constraints are met but also that the contour attributes (e.g. photometric) are in agreement. We describe the approach and show some preliminary results.

1 Introduction

It has been argued that neither edge- nor region-based segmentation algorithms *alone* suffice to correctly analyze complex environments. The key problem with these segmentation techniques is that they operate locally on the image intensities and do not make practical use of global information. Model based (high-level) techniques usually rely on *a priori knowledge* of the objects in the scene, and while impressive results can be obtained for restricted domains these approaches usually fail on complex scenes. However, interesting approaches have been made with *generic models* (Fua and Hanson, 1988) instead of strictly passive techniques like template matching.

An alternative, more flexible, approach is that of *perceptual organization* which consists of organizing the fragmented low-level into higher level descriptions. Previous

*The research described in this paper has been supported by the Swiss National Science Foundation, Grant no. 20-36431.92.

work based on perceptual organization, e.g. (Lowe, 1985; Mohan and Nevatia, 1992; Havaldar *et al.*, 1994), relies solely on the geometrical arrangements of binary edge-points and their collations to produce higher level descriptions, hence not taking into account important information contained in low-level maps and attributes. Low-level properties, e.g. intensity, color, or local edge/line classification impose strong constraints on the segmentation process, and should therefore be incorporated in the process whenever available.

We propose to combine a rich attributed contour representation with geometric contour relations. Contour attributes are information tags assigned to single contours whereas relations are defined between pairs of contours. We distinguish between standard contour attributes and flanking region attributes. The former characterize important properties along the pixels constituting the contour, e.g. integrated contrast, integrated local edge/line classification (Freeman, 1992), and contour shape (straight/curved). The latter, the flanking region attributes, capture properties in the surround of the contour, e.g. intensity, color (Sumanaweera *et al.*, 1988), material or texture homogeneity. A weak geometrical relationship can be leveraged into a strong one if the rich contour attributes are in agreement.

The implemented geometric relations, which are proximity, curvilinearity, parallelism and corner-like relations, allow to handle general man-made objects whose projected surfaces can be described by combinations of the defined relations, hence they are not directly applicable to amorphous objects such as bushes and clouds. By enforcing the geometric relations to also have similar region attributes we restrict ourselves to handle man-made objects whose projected surfaces are fairly homogeneous.

2 Low-Level Scheme

For the low-level segmentation we use the energy filters described in (Heitger *et al.*, 1992) in conjunction with a novel contour detector (Heitger, 1995). The goal is to obtain a representation of image curves in terms of an edge/line dichotomy. The scheme produces a more accurate representation of edges and lines in images of natural scenes than traditional edge detectors since it can better handle featural interferences between edges (and lines) for example at sharp corners or at junctions. Contours are usually defined as aggregates of local edge evidences that are contiguous in the image. We use the contour aggregation scheme described in (Henricsson and Heitger, 1994), a scheme which exploits all information available from regular edge-detectors, without making explicit assumptions how this information was acquired or pre-processed. Especially, the aggregation scheme does *not* require a fully thinned binary edge map to produce a coherent contour graph.

Two dimensional image features, such as junctions, corners and line-ends represent another important class of image information that can serve the definition of object boundaries. These points reflect prominent events in the course of a boundary that allow to divide it into "natural" parts and to extract meaningful shape decompositions. We propose a combination of two complementary methods, which allows a subdivision of object boundaries at positions where the geometry or the contrast or both change considerably. The "contrast" approach is described in (Henricsson and Heitger, 1994), where an independent representation of *key-points* is computed and

where the contours converge upon the key-points during aggregation. The "geometry" approach consists of detecting and localizing points of strong curvature along the already established contours. We used a slightly modified version of the recursive split-and-merge algorithm for linear features described in (Grimson, 1990).
The result is a graph-like representation of contours and vertices. The contours are labeled either *straight*, *curved* or *closed*. By construction, each contour is delimited by two end-points at pixel precision. To make the straight contours less sensitive to residuals in the end-point coordinates we calculate their sub-pixel positions by using standard least squares methods. The contour is approximated with a straight line and the integer end-points are then projected back onto the approximated straight line. As a result, each straight contour will have standard contour attributes defining its straight line equation $ax + yb + c = 0$ and its sub-pixel end-point coordinates. In figure 3B is a typical example of an attributed contour graph. For lack of space we omit a more detailed description, which can be found in (Henricsson, 1995) and (Heitger, 1995).

3 Rich Image Attributes

The graph structure contains only the basic information of the contours and vertices and how these are connected. To make the graph structure more useful, certain rich attributes are assigned to each contour or vertex. Contour attributes either reflect properties on (along) the actual contour or reflect region properties on either side. Region properties can for example be color, intensity or material homogeneity, or texture uniformity or any other relation that bears a grain of empirical validity. To our knowledge, the only work where flanking regions have been used for segmentation is (Sumanaweera *et al.*, 1988), where the regions are defined by sampling the contours (or edgels) on each side along the perpendicular direction. Due to lack of space only the flanking region attributes will be described.

3.1 Construction

The construction of the flanking regions is straightforward, see Fig. 1A. A coordinate system is chosen such that the origin is placed on one of the end-points of the contour and where one axis (denoted $\vec{t}$) goes through the other end-point. The second axis ($\vec{n}$) is chosen in the normal direction to the first axis. The region is defined by a discrete translation of the original contour along the $\vec{n}$-axis. The translation is discretized along the $\vec{n}$-axis with the Bresenham line algorithm. Of course, a corresponding region is defined on the other side of the contour.
The construction of the flanking region, by a discrete translation of the original contour along the $\vec{n}$-axis, allows to represent it in terms of rows and columns over the pixel grid. Along each row we have the different iso-distance lines, see Fig. 1B-F where each iso-distance line is encoded in a gray-value representing the distance from the source contour. The columns in the region representation can be visualized as rays leaving the corresponding positions on the contour in the direction of the $\vec{n}$-axis.
The two parameters *offset* and *width* define the extent of the region. In (Sumanaweera *et al.*, 1988) the authors notice that "...[a] guard size of two pixels was found to be reasonable ..." to reduce the blurring effects occurring at and around the actual contour. An experiment in (Henricsson, 1995) corroborates an offset of two [pixels]

for the flanking regions. The *width* must be chosen such that the region contains enough data samples to make a robust statistical analysis of the region properties. Of course, if the intensity properties in the defined region do not reflect the "true" properties of the object surface, we will have little use of the computed region attributes. As a compromise, we set the *offset* to two and the *width* to five [pixels] in all our experiments.

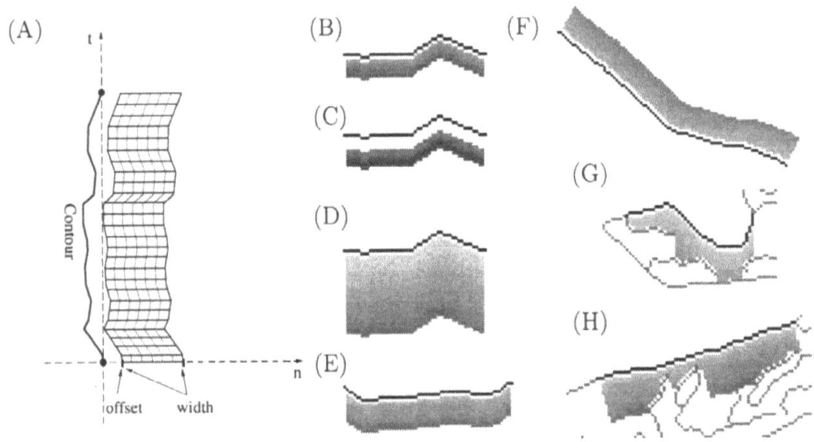

Figure 1: The source contour is black, the interfering contours are gray and the region is encoded in gray-values indicating the distance from the source contour. **(A)** schematically shows the construction of a flanking region, **(B)** a flanking region with *width* = 5, *offset* = 2, **(C)** same contour as in (B) but with *offset* = 5, **(D)** same contour as in (B) but with *width* = 25, **(E) and (F)** two other regions with *width* = 10, *offset* = 2, **(G) and (H)** two examples of contour interferences. Interfering contours truncate the flanking regions (see section 3.2 for truncating mechanism). In the case where the ray does not interfere with any other contours the column stays intact.

3.2 Propagation and Interferences

The above construction of the flanking regions works perfectly well if there are no other contours in the image. In images of natural scenes containing noise, texture and neighboring contours the flanking regions must be modified (truncated) to reduce the effect of interfering contours. A typical example is when two contours meet at an acute angle, see Fig 1G and H.

This interference problem is solved by analyzing each "ray" (column) of the region. Starting from the source contour, we search for the first position where the ray collides with another contour. All positions in the given column beyond the interfering position are deleted, i.e the "ray" is absorbed by the interfering contour. In the case where the ray does not interfere with any other contours the column stays intact. With the propagation and interference procedure we achieve the results in Fig. 1G and H. Notice, with the above "ray" construction it is impossible for the flanking region to be defined "behind" another contour as would be the case for true oriented dilation techniques as in (Dubuc and Zucker, 1994).

3.3 Statistical Analysis

A robust statistical evaluation of the intensity properties of the flanking regions is crucial for a successful combination of geometric and attribute relations. Assuming the data to be normally distributed, the average and the standard deviation can be calculated. However, as we are dealing with images of natural scenes, disturbances like chimneys, bushes, shadows etc. are likely to be within the defined regions. Anomalies of this kind have been successfully located and discounted in related work. For example in (Fua and Hanson, 1988) the authors use deviation histograms and plane approximations to account for these anomalies in an MDL framework. If such anomalies (or artifacts) are contained in the region, but constitute no more than a small fraction of the data samples, we can treat the stated problem with robust estimates. We exchange the average value with the median value and standard deviation with the interquartile range (IQR). The IQR is formed by subtracting the 25th percentile of the data from the 75th percentile of the data. The IQR is a robust estimate of the spread of the data since changes in the upper and lower 25% of the data do not affect it. For examples, see (Henricsson, 1995).

4 Geometric Contour Relations

Two straight contours form the basic unit for all contour relations. The curved labeled contours are only used to connect (link) other straight or curved contours. The contour relations will be evaluated in the order of increasing computational complexity: (1) proximity, (2) cocurvilinearity, (3) parallel relations, and finally (4) corner-like relations. Notice that all relations are symmetrical by construction, i.e. it does not matter if the relation is evaluated from $c_1 \rightarrow c_2$ or $c_2 \rightarrow c_1$. This also means that the relation matrix is symmetrical. Depending on the restrictions imposed on the geometric relations (esp. the proximity distance), the relation matrix is often also sparse. Further, all geometric contour relations are defined from static monocular grey-valued images, hence we do *not* consider groupings in 3D space[1].

4.1 Proximity

At the scale of objects, there exists an upper bound on how distant two contours can be from each other and still belong to the same object. We use proximity relation as a criterion in order to reduce the computations in the successive contour relations, hence only those contour pairs that are proximate can also be curvilinear, parallel, or define a corner relation. In determining the minimum distance between two straight contours, eight distances have to be evaluated, four end-point and four projection distances. If the smallest distance of the eight is below a certain threshold (e.g. 60 pixels) the two contours are considered proximate. See (Henricsson, 1995) for a detailed description.

[1]However, using a straight line stereo algorithm on the extracted straight contours, would allow to generate an attributed contour graph where each and every straight contour is defined in the 3D space. With a 3D attributed contour graph it is then possible to define 3D geometrical relations, e.g. curvilinearity, parallelism and corner-like relations in 3D space.

4.2 Curvilinearity

Cocurvilinearity is a relationship where contours are grouped into larger smooth contours. In (Mohan and Nevatia, 1992) the authors state that "cocurvilinearity of curves can be broken into two components: continuity and proximity of similarly oriented curves". We have implemented an algorithm in which continuity and proximity are merged into one grouping process. For details, see (Henricsson, 1995).

4.3 Axial Symmetry of Straight Contours

Symmetrical structures are common in man-made objects, thus the existence of symmetrical properties should explicitly be used for object detection, see (Ylä-Jääski and Ade, 1995). The symmetry axis of a pair of straight contours lies on the angle bisector between the lines fitting the contours. By using the sub-pixel end-point coordinates we make the symmetry axis less sensitive to residuals in the end-point pixel coordinates. Following Fig. 2A, the generating straight contours are labeled $C1$ and $C2$, the angle bisector is denoted Ls and the actual *symmetry axis* Ss (the dark painted part of Ls). The point Pc on the angle bisector is the point of intersection, i.e. the point where $C1$ and $C2$ intersect. The opening angle is denoted Oa.
To each symmetry axis there are *geometric* as well as *symbolic* attributes. The geometric attributes define the position, length, width, opening angle and symmetry axis of the pair of contours. The symbolic attributes indicate a first coarse classification of the symmetry relation between the contours. By using symbolic attributes, pairs of contours can be selected or rejected by applying simple logical rules, e.g. overlapping, covered or separate in Fig. 2.

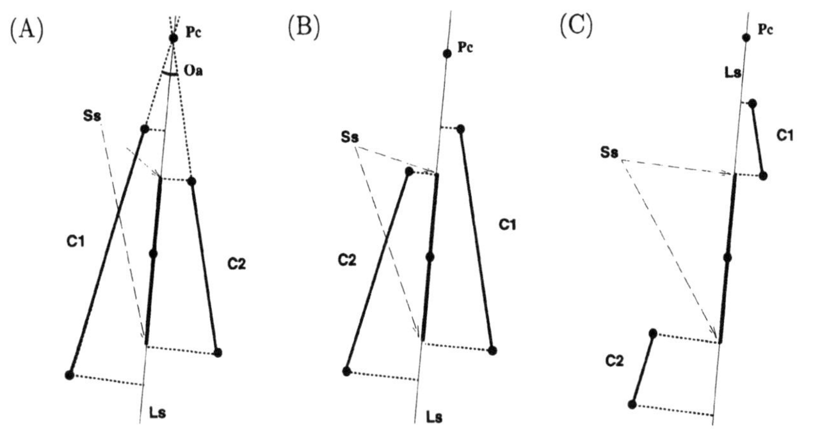

Figure 2: Three symmetry axes with different classes of overlapping, (A) the projection of one side completely covers the other → *covered*, (B) the projections of the sides overlap and no one covers the other → *overlapped*, and (C) the projections of the sides do not overlap each other → *separated*.

Parallel Symmetry. Two straight contours define a parallel relation if the constructed symmetry axis is of type *covered* or *overlapped*, and if the opening angle between the two contours is smaller than some threshold ($O_a < 15°$). The minimum distance from the end-points to the symmetry axis must further be within a certain user defined range ($2 < d_{min} < 60$). There is no restriction on the length of the actual symmetry axis Ss.

Corner-like Symmetry. Two straight contours form a corner-like relation if they define a symmetry axis that is *covered* or *overlapped*, and if the opening angle between the two contours is within a certain user defined range ($15° < O_a < 140°$). A further very important criterion is that the distance from the closest end-point of each contour to the *point of intersection* is smaller than some user defined distance (7 pixels). This construction allows to infer corners even if the contours do not join completely.

5 Attributed Contour Relations

The combination of the rich region attributes in section 3 and the geometric relations from section 4 allows to discriminate between strong and weak contour relations. Strong relations require that not only the geometrical constraints are met but also that the flanking region attributes are in agreement. If the absolute difference between the median values (50th percentiles) of the *opposing* regions is smaller than some threshold and if the interquartile range (IQR) of *one* of the *opposing* regions is below a second threshold the two contours define an attributed or strong relation. Which region attributes are *opposing* is unambiguously defined by the "inside" relation of the symmetry axis. Of course, the two thresholds must be selected according to the estimated level of image noise. Both thresholds were set to 10 [pixels] for the results in Fig. 4.

6 Results

The presented framework has been tested on a large amount of images, including the distributed data sets for the Ascona Workshop and for the ISPRS Working Group III/3. Due to lack of space we can not present all the results; we refer to (Henricsson, 1995) for more examples. In general, the novel contour detection scheme, presented in section 2, took most of the computing time. The contour aggregation, assignment of contour attributes and the calculation of contour relations required only a small fraction of the total computation time.
The figure 3A, a city scene from Basel, was processed with the programs and algorithms described in sections 2 to 5. The resulting attributed graph with all its contours and vertices is depicted in Fig. 3B. In figure 3C the back projection of the flanking regions (attributes) together with the contours are shown. To demonstrate the usefulness of combining geometric relations and rich contour attributes we manually selected a few contours by clicking with the mouse. We then let the computer extract all geometric relations and visualize them as either strong or weak.
In Figs. 4A, B and C the selected contours show a large discrimination between weak and strong relations. In figure 4D the extracted contour relations (with one single exception) all describe strong relations. The weak contours bounding the same window shutter as the selected contour in Fig. 4C are weak (*non*-attributed) due to the IQR-threshold. The IQR of the flanking regions for those contours bounding the textured window shutters are between 20 and 25 [pixels], values clearly exceeding the set threshold of 10 [pixels]. The IQR values of those contours bounding the homogeneous areas in the image are around 2-3 [pixels].

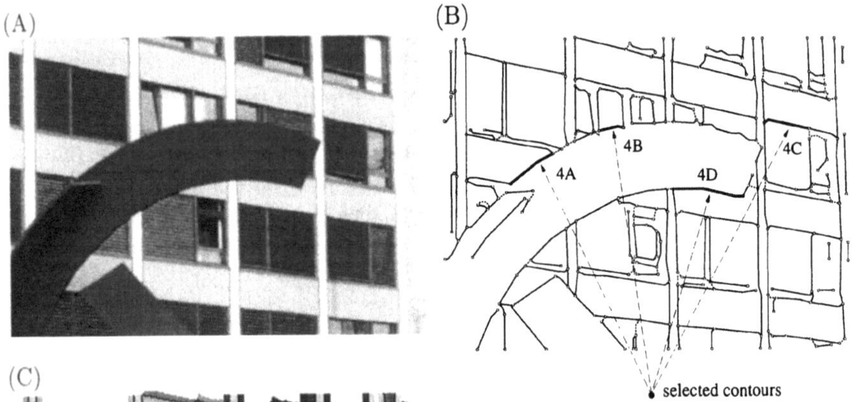

Figure 3: (A) a cut-out 311 × 215 from the original city scene from Basel (Switzerland), (B) The resulting attributed graph with all its contours and vertices. The weak and strong relations calculated from each of the four selected contours (indicated with arrows) are depicted in Fig. 4. (C) the back projection of the flanking regions (attributes) together with the generating contours.

The flanking region attributes do not only make sense in a larger framework where contour relations are initially established and where successive groupings of these related contours allow to infer closed regions and finally object surfaces. The region attributes also have a perceptually appealing effect. We venture to say that humans tend to perceive the scene in Fig. 3C with greater ease than Fig. 3B. The binding problem, i.e. which contours belong to the same object surfaces, seems much easier to solve with the region attributes projected back into the image than without them.

7 Conclusions and Future Work

In this work we have presented a scheme where we combine an attributed contour representation with very general geometric contour relations. The combination of the two allows to discriminate between strong and weak contour relations. In a strong contour relation not only the geometrical arrangement is instantiated but the contour attributes are also in agreement.
Future work will follow two main paths. First of all, our approach should facilitate top-down generation of suitable generic models that comply with the implemented relation repertoire. This will include the construction of general grouping algorithms that allow to infer larger coherent structures in the image. The hypothesized structures will most certainly still be incomplete and a transition to more global methods, e.g. discrete global optimization, will therefore be required. The flexibility of the presented system should then be used to analyse which relations and attributes are particularly important for the final segmentation.

Secondly, the presented methods are all defined in 2D. However a transition from 2D to 3D representation should facilitate the task of detecting and describing man-made objects in aerial imagery. An attributed contour representation in 3D would allow to define new and very powerful contour attributes and geometrical relations in 3D, e.g. coplanarity of straight lines.

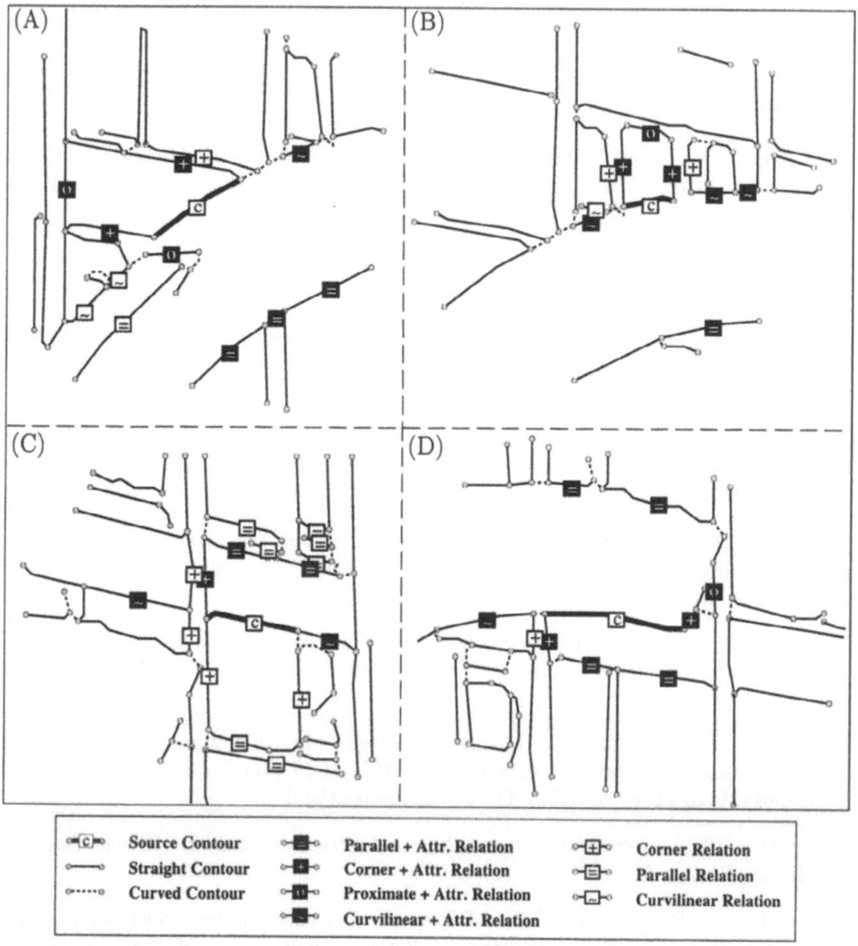

Figure 4: Contour relations from selected contours in Fig. 3. The generating contours (source contours) were selected manually by clicking with the mouse. All relations of proximate contour pairs, where one of the contours is the generating contour, are shown. The weak proximity relations do not have a marker. (A) and (B) Relations from a generating contour positioned on the upper boundary of the occluding sculpture, (C) Contour relations from a generating contour situated on the upper side of a window shutter, (D) Relations from a generating contour positioned on the lower boundary of the occluding sculpture.

Acknowledgements

I am grateful to Friedrich Heitger for helpful discussions. I would further like to thank Markus Stricker, Olaf Kübler, Antti Ylä-Jääski and Frank Ade for their help.

References

Dubuc B. and S.W. Zucker (1994). Curve-like Sets, Normal Complexity, and Representation. In *International Conference on Pattern Recognition*, pp. 216–221.

Freeman W.T. (1992). *Steerable Filters and Local Analysis of Image Structure.* PhD thesis, MIT, Media Laboratory, Cambridge MA.

Fua P. and A.J. Hanson (1988). Extracting Generic Shapes Using Model-Driven Optimization. In *DARPA Image Understanding Workshop*, pp. 994–1004, Cambridge, Massachusetts.

Grimson W.E.L (1990). *Object Recognition by Computer: The Role of Geometric Constraints.* The MIT Press, Cambridge, Massachusetts.

Havaldar P., G. Medioni, and F. Stein (1994). Extraction of Groups for Recognition. In J.O. Eklundh, editor, *Computer Vision - ECCV'94*, volume I, pp. 251–261. Springer Verlag, Berlin.

Heitger F., L. Rosenthaler, R. von der Heydt, E. Peterhans, and O. Kübler (1992). Simulation of neural contour mechanisms: From simple to end-stopped cells. *Vision Research*, Vol. 32,pp. 963–981.

Heitger F. (1995). Feature Detection using Suppression and Enhancement. Technical Report BIWI-TR-160, Institute for Communications Technology, Image Science Laboratory, ETH, Zürich, Switzerland.

Henricsson O. and F. Heitger (1994). The Role of Key-Points in Finding Contours. In J.O. Eklundh, editor, *Computer Vision - ECCV'94*, volume II, pp. 371–383. Springer Verlag, Berlin.

Henricsson O. (1995). Inferring Homogeneous Regions by Combining Contour Relations and Rich Image Attributes. Technical Report BIWI-TR-162, Institute for Communications Technology, Image Science Lab, ETH, Zürich, Switzerland.

Lowe D.G. (1985). *Perceptual Organization and Visual Recognition.* Kluwer Academic Publishers, Boston, MA.

Mohan R. and R. Nevatia (1992). Perceptual Organization for Scene Segmentation and Description. *IEEE Transactions on Pattern Analysis and Machine Intelligence*, Vol. 14,No. 6,pp. 616–635.

Sumanaweera T.S., G. Healey, B. Lee, T.O. Binford, and J. Ponce (1988). Image Segmentation Using Geometric and Physical Constraints. In *DARPA Image Understanding Workshop*, pp. 1091–1099, Cambridge, Massachusetts.

Ylä-Jääski A. and F. Ade (1995). Grouping Symmetrical Structures for Object Segmentation and Description. *Computer Vision, Graphics, and Image Processing: Image Understanding*, Accepted.

Automatic Extraction of Man-Made Objects
from Aerial and Space Images
Monte Verità, © 1995 Birkhäuser Verlag Basel

Stereo Reconstruction from Multiple Views

P. Fua (fua@ai.sri.com) *

SRI International
333 Ravenswood Avenue, Menlo Park, CA 94025, USA

Abstract

We present a unified framework for 3–D shape reconstruction that allows us to combine multiple information sources—such as stereo and shape-from-shading—derived from images whose vantage points may be very different. A formal integration framework is critical because, in recovering complicated surfaces, the information from a single source is often insufficient to provide a unique answer.
We describe two complementary implementations of this paradigm that both rely on deforming a generic object-centered 3–D representation of the surface so as to minimize an objective function. The first implementation models the surface as a triangulated mesh. The second models it as a set of particles that interact with each other through forces that tend to align them.

1 Introduction

The recovering of surface shape from image cues, the so-called "shape from X" problem, has received tremendous attention in the computer vision community. But no single source of information "X," be it stereo, shading, texture, geometric constraints or any other, has proved to be sufficient across a reasonable sampling of images. To get good reconstructions of a surface, it is necessary to use as many different kinds of cues with as many views of the surface as possible. In this paper, we present and demonstrate a working framework for surface reconstruction that combines image cues, such as stereo and shape-from-shading, obtained from multiple images whose vantage points may be very different.
Clearly, a single image-centered representation is inadequate for our purpose. Instead, object-centered surface representations are required. Here, we advocate the use of two such representations that have proved effective under different circumstances:

- **Triangulated Meshes** A regular 3–D triangulation is an example of a surface representation that meets the criteria stated above. Our approach to recovering surface shape and reflectance properties from multiple images is to deform a 3–D representation of the surface so as to minimize an objective function. The free variables of this objective function are the coordinates of the vertices of the triangulation, and the process is started with an initial surface estimate.

  We have successfully used these meshes to model surfaces whose topology is known *a priori*, such as human faces or relatively low-resolution terrain.

*Support for this research was provided by contracts from the Advanced Research Projects Agency.

- **Sets of 3–D Particles** Real-world scenes often contain several objects whose topology may not be known in advance: some surfaces are best modeled as sheets, while others are topological spheres or contain holes. One cannot typically assume that there is only one object and one surface of interest.

  To deal with these complex issues, we replace the triangulations by a set of connected surface patches or "oriented particles" as defined by Szeliski and Tonnesen (1992). These particles are instantiated by fitting local surfaces to traditional stereo data, and their positions are refined by minimizing an objective function analogous to the one used for triangulations.

We view the contribution of this paper as providing both the framework that allows us to combine diverse sources of information in a unified and computationally effective manner, and the specific details of how these diverse sources of information are derived from the images.

2 Using Meshes to Model Single Surfaces

We represent a surface $\mathcal{S}$ by a hexagonally connected set of vertices called a *mesh*. Such a mesh is shown in Figure 1(a). The position of a vertex v_j is specified by its Cartesian coordinates (x_j, y_j, z_j).

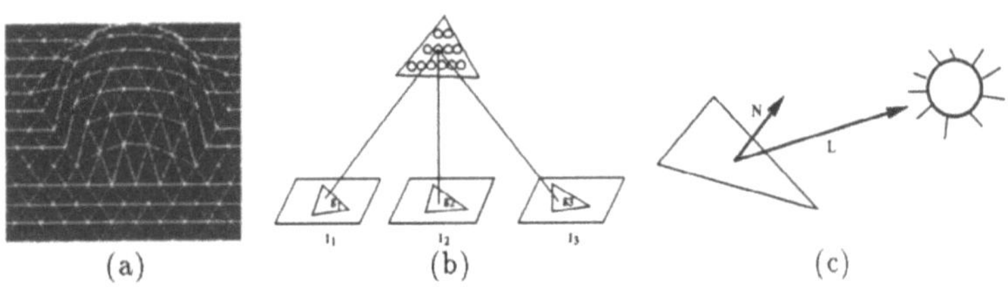

Figure 1: Mesh representation and computation of the objective function's image terms. (a) Wireframe representation of a mesh. (b) Facets are sampled at regular intervals; the circles represent the sample points. The stereo term is computed by summing the variance of the gray level of the projections of these sample points, the g_is. (c) Each facet's albedo is estimated using its normal N, the light source direction L, and the average gray level of the projection of the facet into the images. The shading term is the sum of the squared differences in estimated albedo across neighboring facets.

For each input image, we generate a "Facet-ID" image by encoding the index i of each facet f_i as a unique color, and projecting the surface into the image plane, using a standard hidden-surface algorithm. We use it to determine which surface points are occluded in a given view.

2.1 Objective Function and Optimization Procedure

The objective function $\mathcal{E}(\mathcal{S})$ that we use to recover the surface is a sum of terms that take into account the image-based constraints—stereo and shape-from-shading—that are brought to bear on the surface. To minimize $\mathcal{E}(\mathcal{S})$, we use an optimization method that is inspired by the heuristic technique known as a continuation method (Terzopoulos, 1986) in which we add a regularization term to the objective function and progressively reduce its influence. We define the total energy of the mesh, $\mathcal{E}_T(\mathcal{S})$, as

$$\mathcal{E}_T(\mathcal{S}) = \lambda_D \mathcal{E}_D(\mathcal{S}) + \mathcal{E}(\mathcal{S}) \text{ with } \mathcal{E}_(\mathcal{S}) = \sum_i \lambda_i \mathcal{E}_i(\mathcal{S}) \ . \tag{1}$$

The $\mathcal{E}_i(\mathcal{S})$ represent the image-based constraints discussed below, and the λ_i their relative weights. $\mathcal{E}_D(\mathcal{S})$, the regularization term, serves a dual purpose. First, we define it as a quadratic function of the vertex coordinates, so that it "convexifies" the energy landscape when λ_D is large and improves the convergence properties of the optimization procedure. Second, in the presence of noise, some amount of smoothing is required to prevent the mesh from overfitting the data, and wrinkling the surface excessively (Fua and Leclerc, 1994).
In our implementation, we take $\mathcal{E}_D$ to be a measure of the curvature or local deviation from a plane at every vertex. Using finite differences, $\mathcal{E}_D$ can be expressed as a quadratic form (Fua and Leclerc, 1994)

$$\mathcal{E}_D(\mathcal{S}) = 1/2(X^T KX + Y^T KY + Z^T KZ) \ , \tag{2}$$

where X, Y, and Z are the vectors of the x, y and z coordinates of the vertices, and K is a sparse and banded matrix. Because $\mathcal{E}_D$ is quadratic and decouples the three spatial coordinates, our energy term is amenable to a "snake-like" optimization technique (Kass *et al.*, 1988).
To speed the computation and prevent the mesh from becoming stuck in undesirable local minima, we typically use several levels of mesh size to perform the computation. We start with a relatively coarse mesh that we optimize. We then refine it by splitting every facet into four smaller ones and reoptimizing. Finally, we repeat the split and optimization processes one more time.

2.2 Combining the Components

The total energy of Equation 1 is a sum of terms whose magnitudes are image- or geometry-dependent and, as a result, not necessarily commensurate. One therefore needs to scale them appropriately, that is to define the λ weights so as to make the magnitude of their contributions commensurate and independent of the specific radiometry or geometry of the scene under consideration. Since the dynamics of the optimization are controlled by the gradient of the objective function, an effective way to normalize the contributions is to introduce a set of weights λ'_i such that $\lambda'_D = 1 - \sum_{1 \leq i \leq n} \lambda'_i > 0$. The λs are taken to be

$$\lambda_i = \frac{\lambda'_i}{\| \vec{\nabla} \mathcal{E}_i(\mathcal{S}^0) \|} \ , \ \lambda_D = \frac{\lambda'_D}{\| \vec{\nabla} \mathcal{E}_D(\mathcal{S}^0) \|} \ , \tag{3}$$

where $\mathcal{S}^0$ is the surface estimate at the start of each optimization step. In practice we have found that, because the normalization makes the influence of the various terms comparable irrespective of actual radiometry or dimensions, the user-specified λ'_i weights are context-specific but not image-specific. In other words, we use one set of parameters for images of faces when combining stereo and shape-from-shading, and another when dealing with aerial images using stereo alone, but we do not have to change them for different faces or different landscapes.

2.3 Image Constraints

In this work, we use two complementary image-based constraints: stereo and shape-from-shading.

The stereo component of the objective function is derived by comparing the gray levels of the points in all the images for which the projection of a given point on the surface is visible, as determined using the Facet-ID image. As shown in Figure 1(b), this comparison is done for a uniform sampling of the surface. More specifically, we take the stereo component of our objective function to be the variance in gray-level intensity of the projections in the various images of a given sample point on a particle, summed over all sample points in a facet, summed over all facets. This method allows us to deal with arbitrarily slanted regions and to discount occluded areas of the surface. This term can be viewed as a 3–D generalization to the purely image-based correlation techniques that use deformable windows (Nishihara, 1984; Baltsavias, 1991; Devernay and Faugeras, 1994).

The shading component of the objective function is computed using a method that does not invoke the traditional constant albedo assumption. Instead, it attempts to minimize the variation in albedo across the surface, and can therefore deal with surfaces whose albedo varies slowly. This term is depicted by Figure 1(c).

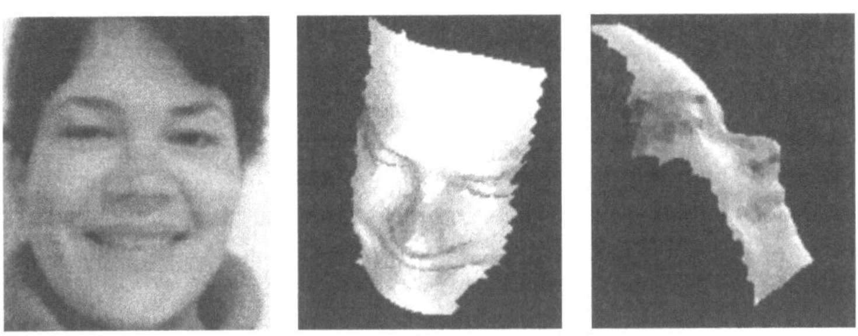

(a) (b) (c)

Figure 2: Combining stereo and shape-from-shading. (a) First image of a triplet (courtesy of INRIA). (b,c) Shaded views of the reconstructed surface.

Stereo information is very robust in textured regions but potentially unreliable elsewhere. We therefore use it mainly in textured areas by weighting the stereo component most strongly for facets of the triangulation that project into textured image areas. Conversely, the shading information is more reliable where there is little texture and is weighted accordingly. For more details, we refer the interested reader to an earlier publication (Fua and Leclerc, 1994).

3 Using Particles to Model Multiple Surfaces

Our approach to recovering multiple surfaces is to model them as sets of local surface elements that interact with one another. Following Szeliski and Tonnesen (1992), we refer to the surface patches as "oriented particles." The forces that bind them can be understood as "breakable springs" that tend to align the particles with each other but may break for particles that are too far out of alignment.

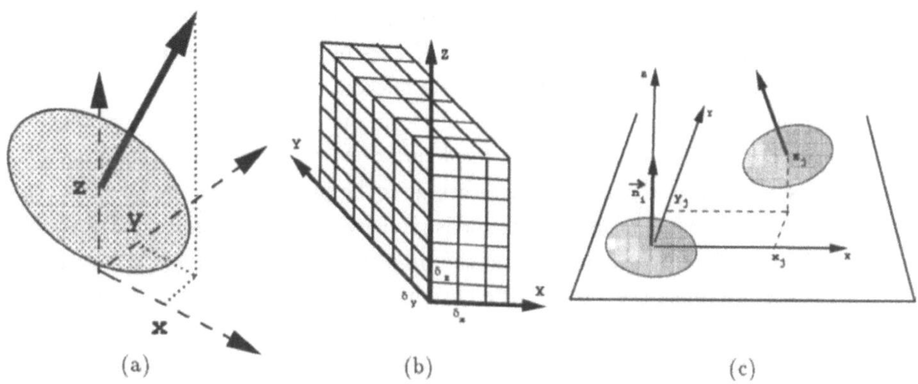

Figure 3: Data structures and metric. (a) A particle is a disk to which we associate a local referential. We allow the center of gravity to shift along the z axis and parametrize the orientation using the projections of the normal vector on the x and y axes. (b) The input 3–D points are stored in a cube-shaped set of voxels and we instantiate a particle in each voxel containing enough such points. (c) The "distance" between two particles is primarily a function of the distance of the center of gravity of one particle from the tangent plane of another.

Our surface elements are disks whose geometry is defined by the positions of their centers, the orientations of their normals and their radii. In theory, these disks have six degrees of freedom. However, in the modeling of a global surface in terms of such disks, translations along the tangent plane of the surface can be ignored as long as the disks remain roughly equidistant from one another and the radius can be chosen so that the disks approximately cover the surface. Therefore, in practice, we deal with only three degrees of freedom.

As shown in Figure 3, to achieve an orientation-independent implementation, we assign to each particle a local referential. We define a particle's position by the translation of the center along the local vertical and its orientation by the x and y projections of the normal on the local x and y axes.

Our procedure involves three steps: (1) instantiating a set of particles from raw stereo data, (2) refining their position and orientation by minimizing an image-based objective function, and (3) eliminating spurious particles and clustering those that appear to belong to the same global surfaces.

3.1 Initialization

We typically start with a set of stereo pairs or triplets, and corresponding disparity maps, of a given scene as our input data. We then turn each valid disparity into a 3–D point. These points typically form an extremely noisy and irregular sampling of the underlying global 3–D surfaces.

To generate a set of regularly spaced particles from such data, we pick spatial step sizes δ_x,δ_y, and δ_z along the X, Y and Z axes of an absolute referential. We use them to define a cube-shaped set of 3–D buckets, such as the one of Figure 3(b). We then store the 3–D points computed from our initial correlation data into the appropriate buckets. By fitting a local surface to every bucket containing enough points, we generate particles whose center is the projection of the bucket's center onto the surface and whose orientation is given by the surface's normal at that point. In the presence of very noisy data, the projection may fall outside the bucket. In this case, we reject the particle, thereby ensuring that there is only one particle per bucket and that the particles are regularly distributed.

For the initialization phase to be successful, it is important both to choose the right kind of surface model and to use a robust method to perform the fitting. We have used both planar and quadric models. The quadrics, even though they involve more computation, have proven very effective because they allow the use of larger sets of points than planes without introducing any appreciable bias. We use not only the points in the bucket under consideration but also in the buckets that are its immediate neighbors. This method allows us to robustly fit local surfaces of arbitrary orientation using a relatively large set of 3–D points, and tends to enforce consistency of orientation among neighboring particles.

3.2 Clustering

To cluster the isolated particles into more global entities, we define a simple "same surface" relationship $\mathcal{R}$ between particles P_i and P_j as follows:

$$P_i \mathcal{R} P_j \iff d_{\text{part}}(P_i, P_j) < \max_d \ , \tag{4}$$

where d_{part} is a distance function and $\max_d$ a distance threshold. We could take d_{part} to be the Euclidean distance between particle centers. However, such a distance would not be discriminating enough for our purposes because it does not take the particles'orientation into account. It has proved much more effective to define a distance function that penalizes more heavily the distance of one particle's center from the tangent plane of the other than the distance along the tangent plane. The simplest way to achieve this result is to define d_{part} as follows:

$$\begin{aligned} d_j &= kz_j^2 + (1-k)(x_j^2 + y_j^2) \ , \\ d_i &= kz_i^2 + (1-k)(x_i^2 + y_i^2) \ , \\ d_{\mathrm{part}}(P_i, P_j) &= max(d_i, d_j) \end{aligned} \tag{5}$$

where x_j,y_j and z_j are the coordinates of the center of P_j in a referential whose Z direction is the normal of P_i and whose origin is the center of P_i, as shown in Figure 3(c), and k is a constant larger than 0.5. In this paper, we take $k = 0.9$; x_i, y_i, and z_i are defined symmetrically.

In essence, the threshold $\max_d$ on d_{part} limits the curvature of the common underlying surface to which particles may belong. As such it is domain-dependent.

The data set equipped with the relationship $\mathcal{R}$ can now be viewed as a graph whose connected components are the surfaces we are looking for. In practice, there will usually be spurious particles that are weakly linked to legitimate clusters. In such cases, we have found that removing all points that do not have a minimum number of neighbors allows us to throw away the gross errors and generate meaningful clusters.

To demonstrate the effectiveness of our initialization and clustering methods given relatively clean stereo data, we use the sequence of forty 512x512 images depicted by Figure 4. The images were acquired with a video camera over a period of a few seconds by turning around the subject who was trying to stand still. Camera models were later computed using standard photogrammetric techniques at the Institute for Geodesy and Photogrammetry, ETH-Zürich.

We ran a correlation-based algorithm—once for each consecutive pair of images in the sequence—stored the resulting 3–D points in a 80x80x80 set of voxels, and instantiated particles in all voxels containing at least 200 points. The results are shown in the second row of Figure 4. Because we use a large number of images, the main features of the head—including the nose, mouth, chin, ears, and even the boundary of the skullcap—are clearly captured by our representation. However, because the correlation-based algorithm produced erroneous, but not random, disparities around occlusion boundaries, we also find a number of spurious particles around the nose, chin, and back of the head. To get rid of them we computed the interparticle distance described above and eliminated all particles not having at least four neighbors within 1.2 times the median distance between neighbors, as shown in the third row of Figure 4.

(a)

(b)

(c)

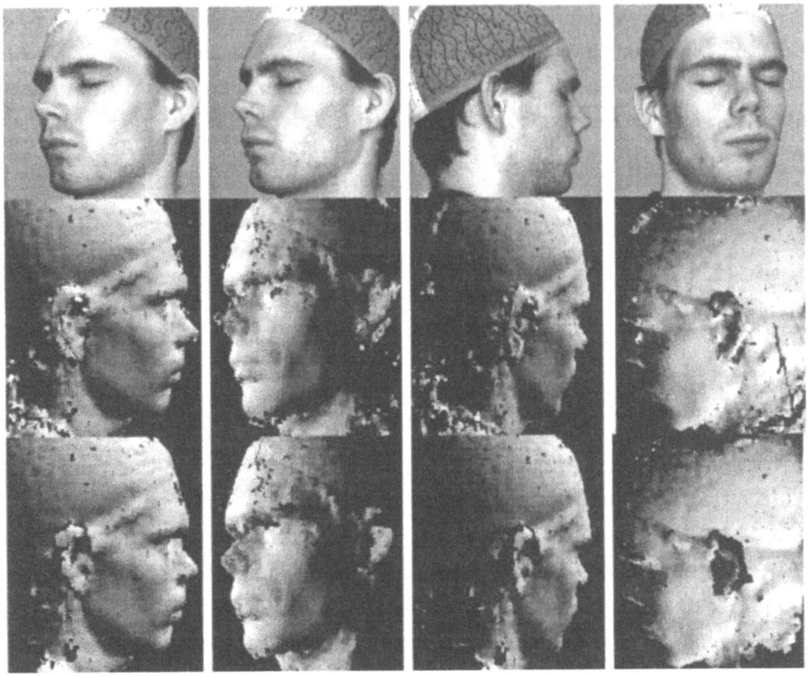

Figure 4: Modeling a complete head. Row (a): A stereo pair of a person's head and left images of two additional stereo pairs taken from different viewpoints (courtesy of ETH-Zürich). Row (b): Four shaded views of the particles instantiated by fitting local surfaces to the 3–D points derived by correlating the images of Figure 4. Row (c): Similar views of the subset of particles that belong to the same global surface. Erroneous ones have been eliminated.

3.3 Refining

Because it is extremely difficult to design a stereo algorithm that never produces correlated artifacts, we cannot expect any robust fitting technique to exclude all erroneous 3–D points. Furthermore, fitting local surfaces to the initial data amounts to smoothing and may result in spurious particles that appear to line up with legitimate ones and become very hard to eliminate. To resolve such problems, it is necessary to return to the original images and assess the quality of each particle. For each disk-shaped particle, we define a stereo term by projecting the 3–D disks into 2–D elliptical patches in each image and measuring how well these patches correlate. This term is equivalent to the one defined in Section 2.3, except that it is now a disk as opposed to a triangular facet that is sampled at regular intervals

(Fua, 1995). It is a function of the three degrees of freedom of each particle and can therefore be used to perform optimization. We allow the particles to interact with one another and to rearrange themselves to minimize an energy term that is the sum of the multi-image intensity correlation term and of a deformation energy term that tends to enforce consistency between neighboring particles (Szeliski and Tonnesen, 1992). While optimizing the energy term, the particles that actually correspond to the same underlying global surfaces will "stick together" and the ones that do not will tend to move in separate directions, stop lining up with each other, and be easily eliminated by the clustering technique of Section 3.2.
Formally, we write the total energy of a set of particles $\mathcal{E}_T$ as

$$\mathcal{E}_T = \mathcal{E}_{St} + \lambda_D \mathcal{E}_D \ , \tag{6}$$

where $\mathcal{E}_{St}$ is the multi-image intensity correlation term, $\mathcal{E}_D$ the deformation energy term, and λ_D a weighting coefficient that is dynamically adjusted using the method of Section 2.2.
To define $\mathcal{E}_D$, we follow Szeliski and Tonnesen (1992). $\mathcal{E}_D$ as the sum of "conormality" and "coplanarity" potentials that control the surface's resistance to bending. The total energy, $\mathcal{E}_{St}$ is a function of $3n$ state variables, where n is the number of particles and is optimized using conjugate-gradient.

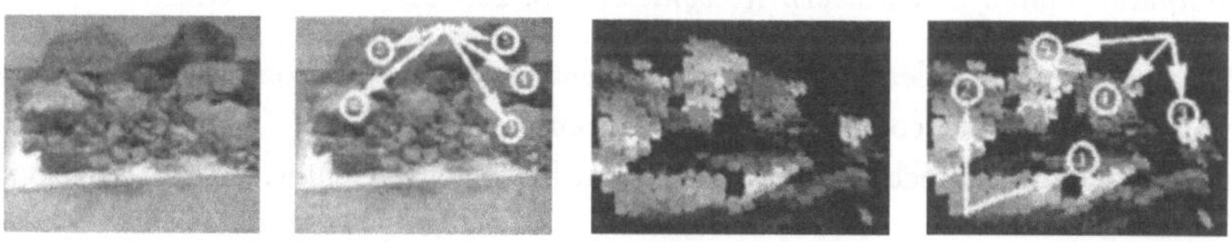

(a) (b) (c) (d)

Figure 5: Modeling a pile of rocks. (a) The first image of a triplet (courtesy of INRIA). (b) The same image with the five largest rocks labeled from 1 to 5. (c) Shaded view of the particle set after refinement and clustering seen from a viewpoint located on the left side of the rock pile. (d) The same view with the five largest rocks labeled as in (b). Note that the overhang of rock number 1 is well recovered, a result that would be difficult to achieve using a 2-1/2-D representation.

In our final example, shown in Figure 5, we used all three steps of our surface reconstruction procedure to reconstruct a ground-level scene using three triplets of images acquired by the INRIA mobile robot and the precomputed robot's ego-motion. Note that the five main rocks in the scene, including one that overhangs, appear in the reconstruction.

4 Conclusion

We have presented a surface reconstruction framework that uses object-centered representations to recover geometry and reflectance properties from multiple images. When the surface's topology is known *a priori*, we represent it as a triangulated mesh and are able to handle self-occlusions while merging information from several viewpoints, thereby allowing us to eliminate blindspots and make the reconstruction more robust where more than one view is available. The reconstruction process relies on both monocular shading cues and stereoscopic cues. We use these cues to drive an optimization procedure that takes advantage of their respective strengths while eliminating some of their weaknesses.
When the surfaces's topology is unknown or there may be more than one surface in the scene, we replace the triangulation by a set of oriented particles that allow us to achieve similar results, but with a higher computational cost.

References

E. P. Baltsavias (December 1991). *Multiphoto Geometrically Constrained Matching.* PhD thesis, Institute for Geodesy and Photgrammetry, ETH Zurich.

F. Devernay and O. D. Faugeras (June 1994). Computing Differential Properties of 3–D Shapes from Stereoscopic Images without 3–D Models. In *Conference on Computer Vision and Pattern Recognition*, pp. 208–213, Seattle, WA.

P. Fua and Y. G. Leclerc (1994). Object-Centered Surface Reconstruction: Combining Multi-Image Stereo and Shading. *International Journal of Computer Vision.* In press, available as Tech Note 535, Artificial Intelligence Center, SRI International.

P. Fua (June 1995). Reconstructing Complex Surfaces from Multiple Stereo Views. In *International Conference on Computer Vision*, Boston, MA. Also available as Tech Note 550, Artificial Intelligence Center, SRI International.

M. Kass, A. Witkin, and D. Terzopoulos (1988). Snakes: Active Contour Models. *International Journal of Computer Vision*, Vol. 1,No. 4,pp. 321–331.

H.K. Nishihara (1984). Practical Real-Time Imaging Stereo Matcher. *Optical Engineering*, Vol. 23,No. 5.

R. Szeliski and D. Tonnesen (July 1992). Surface Modeling with Oriented Particle Systems. In *Computer Graphics (SIGGRAPH)*, volume 26, pp. 185–194.

D. Terzopoulos (1986). Regularization of inverse visual problems involving discontinuities. *IEEE Transactions on Pattern Analysis and Machine Intelligence*, Vol. 8,pp. 413–424.

Model Registration and Validation*

Andres Huertas, Mathias Bejanin and Ramakant Nevatia

Institute for Robotics and Intelligent Systems
University of Southern California
Los Angeles, California 90089-0273, U.S.A.
Ph.:(213) 740-6428, Fax: (213) 740-7877
e-mail: {huertas,nevatia}@iris.usc.edu

Abstract

An important application of machine vision is to provide a means to monitor a scene over a period of time and report changes in the content of the scene. We have developed a validation mechanism that implements the first step towards a system for detecting changes in images of aerial scenes. Validation seeks to confirm of the presence of model objects in the image. Our system uses a 3-D site model of the scene as a basis for model validation, and eventually for detecting changes and to update the site model. The validation process is implemented in three steps: resection, fine registration of the image to the model by matching of model features to image features, and validation of the objects in the model. Our system, if necessary, is aided by shadows to help validate the model. The system has been tested using a hand-generated site model and several images of a 500:1 scale model of the site, acquired form several viewpoints.

1 Introduction

Change detection is an important task in the process of photo-interpretation. It is also a tedious task as it requires careful comparison of images (and their models) taken at different times under possibly vastly varying conditions. We believe that even partial automation of this task will greatly increase human productivity and possibly also enhance the reliability of the results.

The task of change detection consists of finding significant differences between the new data and the model derived from the older data. The significance of the differences may be task specific though in most cases man-made changes are more important than those caused by natural factors such as seasonal changes. We are only interested in those changes in the image that come from some changes in the site rather than from changes in imaging conditions.

Previous work on change detection relies on image differencing. The work described by Lillestrand (1972) and Ulstad (1973) follows this approach. A new image to be compared with a reference image is first corrected spatially to bring them into registration. This is done at each point by finding the best conjugate point using a correlation-based criterion. Global intensity corrections are made so that the first two moments of the image match, allowing for

* This research was supported, in part, by the Advanced Research Projects Agency of the United States Department of Defense and was monitored by the U.S. Army Topographic Engineering Center under Contract No. DACA76-93-C-0014. Mathias Bejanin was a Visiting Researcher at USC, supported in part by Matra Cap Systemes of France.

any differences in luminosity or film sensitivity. Substraction of the two images then reveals small scale changes between the two views. Although this algorithm works very well for this particular problem of detecting changes between two images, it is unable to separate the effects of changes resulting from different viewing conditions (such as different viewing positions, different illumination and seasonal changes) from important structural changes.

Another approach is to use general classification algorithms after spatial registration to assign each pixel to a limited number of classes, thus transforming the intensity range of a pixel into a reduced number of levels. This assumes that preferably more than one band of image data is available, as in Landsat's multiple band images, for example. Depending on a pixel's position in the n-dimensional data space, and relying on the statistical description of each class (which is user-defined), a pixel is assigned a class. The classes could represent water, forest, roads, urban areas, and so forth. Once the classification is done for each image, the detection of changes becomes straightforward.

Change detection involves comparing a *new* image of a site (or a collection of images) to the information associated with that site. These information may consist of a site model and one or more previous images and results of previous analyses on these images. We assume that in all cases, a site model of suitable resolution and complexity is available.

In this paper we address the first step of our work in change detection, *model validation*, the process of confirming the presence of model objects in images. This task requires that we first register the new image to the site model and second that we validate the model objects. In our work we have chosen to work with features extracted from model information and from the new image, rather than developing pixel based techniques.

Lastly, we show results, applied to building structures, using a site model and associated imagery supplied to us by RADIUS, a U.S. Government sponsored research program.

2 Overview of the Change Detection Task

Fig. 1 shows a flowchart of the change detection system. The process clearly requires a comparison of new imagery with the old imagery (and the models constructed from the old imagery). The techniques of simple image differencing are inadequate for this task. We follow a four step approach, the first two, the subject of this paper:

- *Registering site model to image:* The first step in change detection is to register the new image(s) to the model(s) contained in the site folder.
- *Model validation:* After a coarse registration between the image and the model, we apply a matching process to fine-register the model and the image. Match results are used to validate model objects. Model features that are not present in the image represent likely changes. Some missing features are due simply to viewing conditions. Most of these, however, can be predicted and explained from the site model itself. The task of finding objects in the image that are not in the model is more difficult since the model is no longer as useful in directing the processing. We plan to do this in the future by the next two steps.
- *Focus of attention*: A collection of techniques that will draw attention to significant structures in the image that are not explained by the existing model. To separate the significant changes from the insignificant ones, a *perceptual grouping* operation that organizes lower level features into higher level structures will be necessary (see Lin et al., 1994).
- *Detailed analysis*: In this step, we analyze in detail the structures indicated by the focus

of attention processes. This step requires the development of automated or semi-automated site modeling techniques.

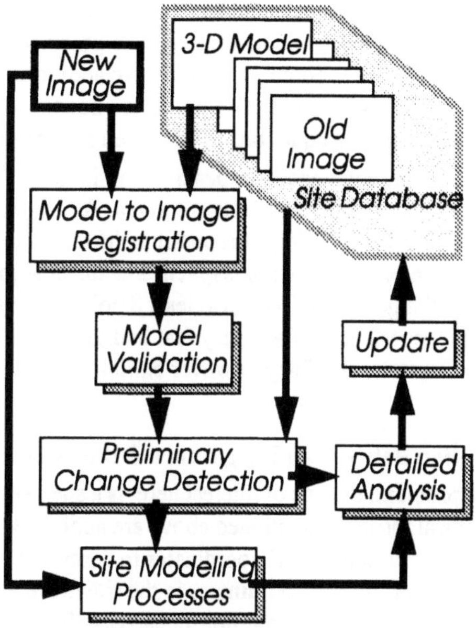

Fig. 1 : Flowchart of change detection system.

3 Validation of Buildings

We are given a 3-D site model (consisting of a set of objects such as buildings, houses, and storage tanks, describing a site) and an image of this site, taken at a later time. The site model consists of geometric information only: we do not have any data about albedo or color for any of the objects. Specifically, we apply our technique to validating the *buildings* in the model. Extension to most objects of similar structure and geometry is straightforward.

Model validation involves three steps, with details given below:

- *Coarse registration* of the model and the image. This is equivalent to finding the correct position and orientation of the camera at the time the image was taken.
- *Matching* model to image. We project model features (line segments) onto image coordinates, and match them to line segments extracted from the image. Matching features represent hypotheses of the presence -or the absence- of a model feature in the image.
- *Validation* of hypotheses. If necessary, we verify the hypotheses using shadow information extracted from the image.

3.1 Registration

The orientation and position (the external parameters) of the camera are usually known approximately for a given image. Otherwise, if the intrinsic parameters of the camera -focal length and position of the principal point - are known, then it is possible to derive this information. This problem is known as *relative* orientation for the case of image-to-image registration, and *exterior* orientation for the image-to-model registration case.

3.1.1 Relative Orientation Between Two Images

Most algorithms proposed in the past use a set of conjugate points given by the user. In theory, given three parameters for rotation, three for translation, and allowing for an overall scale factor, there are five constraints. A pair of conjugate points give three equations (denoting the transformation of the three coordinates) plus two additional unknowns (the depth, in both coordinate systems). Therefore, only five points are needed to compute the transformation (Horn, 1986). In this case the equations to be solved are non-linear.

These methods are usually computationally expensive and in general, a unique solution is not possible when the minimal number of points are given. An example of a powerful nonlinear solution can be found in Horn (1991). It is possible, however, to solve the problem using only linear equations, if more information is given. The linear algorithms (Longuet-Higgins, 1981, Hartley, 1992, Faugeras, 1992) need at least 8 points correspondences. The main advantage of linear methods is that they are very fast and guarantee the uniqueness of the solution, except for degenerate cases. They however exhibit sensitive behavior in the presence of real, noisy data.

3.1.2 Exterior Orientation Between an Image and a 3-D Model

The exterior orientation problem, the one of interest here, is a special case of the relative orientation problem, thus the algorithms mentioned above are applicable with only slight modifications. Instead of giving two sets of 2-D coordinates to describe the conjugate points, we will give for a 'conjugate point' its 3-D coordinates in the site model and its 2-D coordinates in the image. In our experiments we use the non-linear USGS Space Resection algorithm provided as part of the RADIUS Common Development Environment (RCDE), the environment used for ARPA sponsored RADIUS research work and obtained good results, within a few pixels.

3.2 Matching

Our matching technique consists of two steps: First, we match model features to image features. We have chosen to match line segments. The visible model line segments are constructed from the description of the model objects. The line segments are extracted from the image using our LINEAR feature extraction system (Nevatia, Babu, 1980). Secondly, each matching pair (model segment, image segment) denotes a hypothesis that the object the model feature is part of, has a corresponding object in the image.

The Matching Algorithm

Our matching algorithm assumes translational errors only and described in detail in Medioni et al. (1991). We have a *candidate* pair (S_m, S_i) of matching segments, (model segment, image segment) when S_m and S_i overlap (a segment ends project inside the other.) The segments also must lie within an certain "distance" d of each other. If S_m and S_i intersect, then $d = 0$. Otherwise, d is the smallest projected distance of each segment end point on the other segment, if it falls inside that segment. In Fig. 2, $d = min\ (d2, d3)$.

Each pair of candidate segments produces a *vote* as a function of the distance between the two segments and the differences of segment orientation and length.

We compute votes for candidate pairs and add their contribution into an accumulator array.

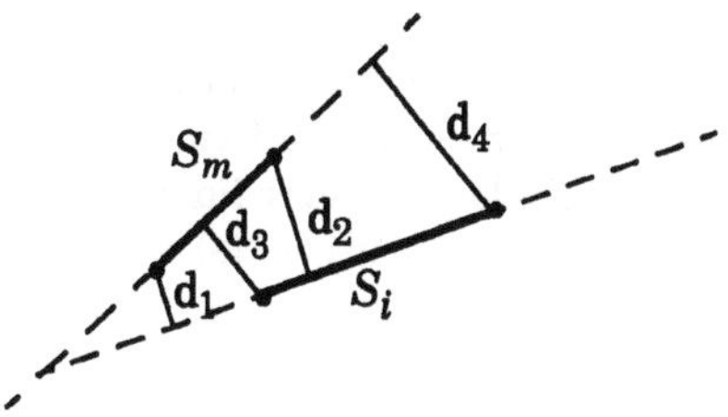

Fig. 2 : Distance between S_m and S_i.

The array axes denote translation. For noise effect reduction, the vote is cast not only at the point corresponding to the translation between the two segments (the translation between the two segment centers), but in a rectangular region around these points.

A peak in the accumulator array gives the position of the best translation between the two sets of segments. A second pass of the algorithm collects the matching pairs that contribute to the peak. Matching pairs thus corresponds model segments to image segments.

After matching, validation continues at the object level. For each building object in the model, we now have a record of the matching of its component edges. We define a *strong* match when at least four of the possible nine visible segments in the simplest model object (a cube) have a corresponding match among the segments from the image. Otherwise the match is *weak*. While objects having a strong match are considered validated (Fig. 3), buildings having weak (or no) matches, require *verification* before they are considered validated.

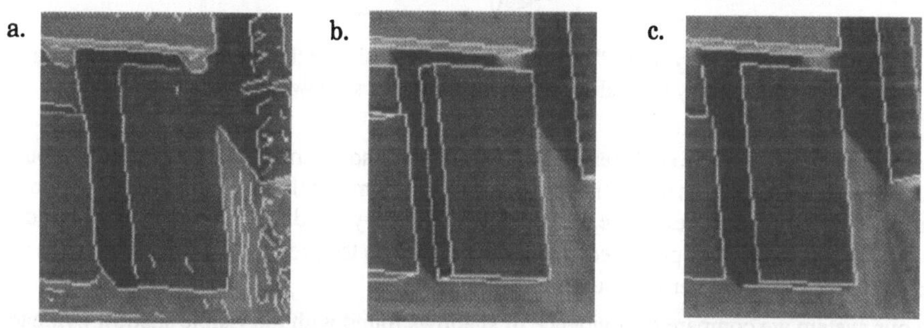

Fig. 3 : Example of segment matching on building 36. (a) Image segments. (b) Model segments. (c) segments of the model that were matched. This match is considered strong.

3.3 Verification of Objects

Weak matches can be the result of several conditions, such as inaccurate registration, poor contrast or occlusion by other objects. Our current system uses shadow clues for verification if these are available. Shadows are detected from the image using the techniques described in Lin et al. (1994). We assume that the sun angles are known *a-priori*. To verify model objects we generate the shadows they are supposed to cast and match these to the shadow evidence found in the image (see below.) We have considered the ground plane to be locally planar in our experiments (Fig. 4.)

Shadow Detection

We label image segments as potential shadow segments by noting the consistency of the "dark" side of the segment with respect to the direction of illumination. Segments oriented parallel to the direction of illumination also correspond to possible shadow lines cast by vertical object edges. Similarly, we detect shadow junctions. The L-junctions formed (allowing for gaps) by potential shadow lines are labeled potential shadow junctions. For more details on the shadow labeling of segments and junctions see Huertas, Nevatia (1988) and Lin et al. (1994).

Validation of Hypotheses

The 3-D position of the objects having weak or no matches, is known from the model. We can predict their position in the image. Thus, the process consists of calculating the shadows cast in 3-D space and predict their location in the image. Then we search around the predicted locations for evidence of shadows among the potential shadows extracted from the image. If we find a sufficient evidence of shadows, then the presence of the building is confirmed

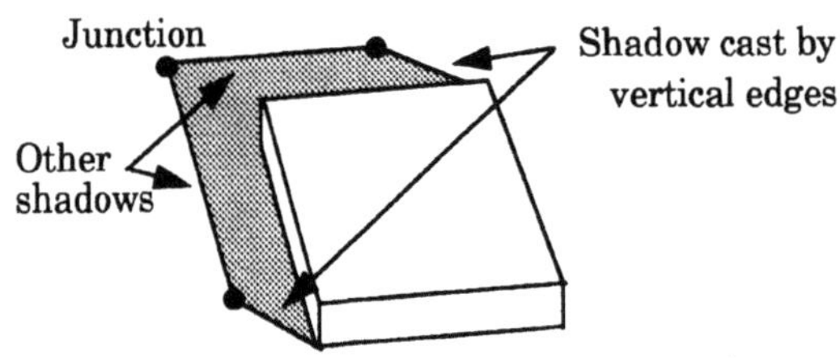

Fig. 4 : Shadows cast by "cubic" building. Evidence consists of three junctions, two shadows cast by vertical segments and two other shadow segments.

Some objects may not be sufficiently isolated and its shadows may not be cast on the ground. Also, shadows may be cast on other objects and sometimes will not even be visible from the viewpoint. In Fig. 5, for example, the shadow 1-3 cast by building C is occluded by building B, and the shadow 1-2 is projected on building A, not on the ground. We cannot take this into account until we have verified building B or C.

In our system we compare the evidence of shadows found with the visible shadow evidence. Visible shadows are determined by the knowledge of which objects have already been verified. In the "counting" (or accumulation) of shadow evidence, we also give greater importance to shadows cast by vertical edges and to shadow junctions that correspond geometrically to shadow casting object structures. We consider this very strong and reliable evidence.

Not all weak matches are verified in a single pass due to poor contrast, occlusion, self-occlusion, lack of a DTM, and errors in the model itself. We however use the knowledge of previously verified buildings. This knowledge also allows us to predict more accurately the position of shadows on a subsequent pass. We repeat the process until no further verification of weakly matched objects using shadows is possible.

The image location of buildings that have *no* matched segments can still be predicted in relation to the location of already validated buildings. We proceed in the same manner described to verify these. In this case however the absence of shadows confirms the absence of the building but the presence of shadows does not guarantee the presence of a building. In this case attention is focused at this image location for further processing: We extract *all* the

a. 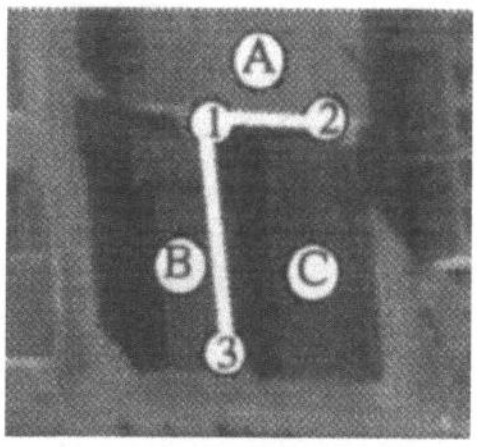

b. 

Fig. 5 : Example of a situation where building verification is made difficult, for buildings C and E, because of the presence of surrounding buildings.

possible edges in the neighborhood of the object and use the translation computer earlier to obtain a richer set of matching segments. The previous analysis is applied to these windows, including shadow processing, to help validate the remaining model objects.

4 Experimental Results

The validation system has been tested with a 3-D site model and images denoted as the Model Board 1 data set, provided to us by the RADIUS program. The site model describes geometrically a scene containing several buildings and other objects, constructed at the approximate scale of 1:500. The images are 1320x1035 in size with a ground sampling distance between 18.5 and 32.5 inches. A typical image is shown in Fig. 6. The features in these images consist of buildings (of cubic shape), houses (simple buildings with a gable) and storage tanks and chimneys (cylindrical shapes). The site model (Fig. 6a) contains 60 cubes and houses. We have tested our model validation system on some 40 images, RADIUS data sets J1-J8 and K1-K10, a set of 8 images from a real site near Denver, and two images from Fort Hood (Texas). The system software is written in Common Lisp and runs under the RCDE on Unix workstations.

The matching process gives excellent estimates of misregistration. We use an accumulator array of size 50 x 50 allowing a 25-pixel misregistration offset in any direction. Initial coarse registration using the USGS resection technique was in the order of a few pixels. An example of an accumulator array is shown in Fig. 6c. The peak is very sharp, allowing unambiguous determination of the translation error.

We summarize our validation results in the tables 1, 2 and 3 below. The entries in the table denote the number of visible/verified objects, and the corresponding percentage of success, for each image at the end of each of the steps in the process. All the images were processed at one-half of the available resolution (660x517 pixels). Although our results are very encouraging, further testing is required to deal with possible false matches, that is, validation of model objects that are not in the image. There is also room for improvement at the registration level and at the verification level. Inaccuracies in the model should be tolerated to a good extent at this stage, for example. These improvements will also help processing images that lack shadow definition or shadows altogether.

A typical intermediate result of our system is shown in Fig. 7. The model objects are shown superimposed on the scene (J2 in this example) after registration and matching, but before shadow validation of weak matches. The objects shown in dashed lines are not visible in the image. The objects shown in bold lines were not validated by matching only. All others were

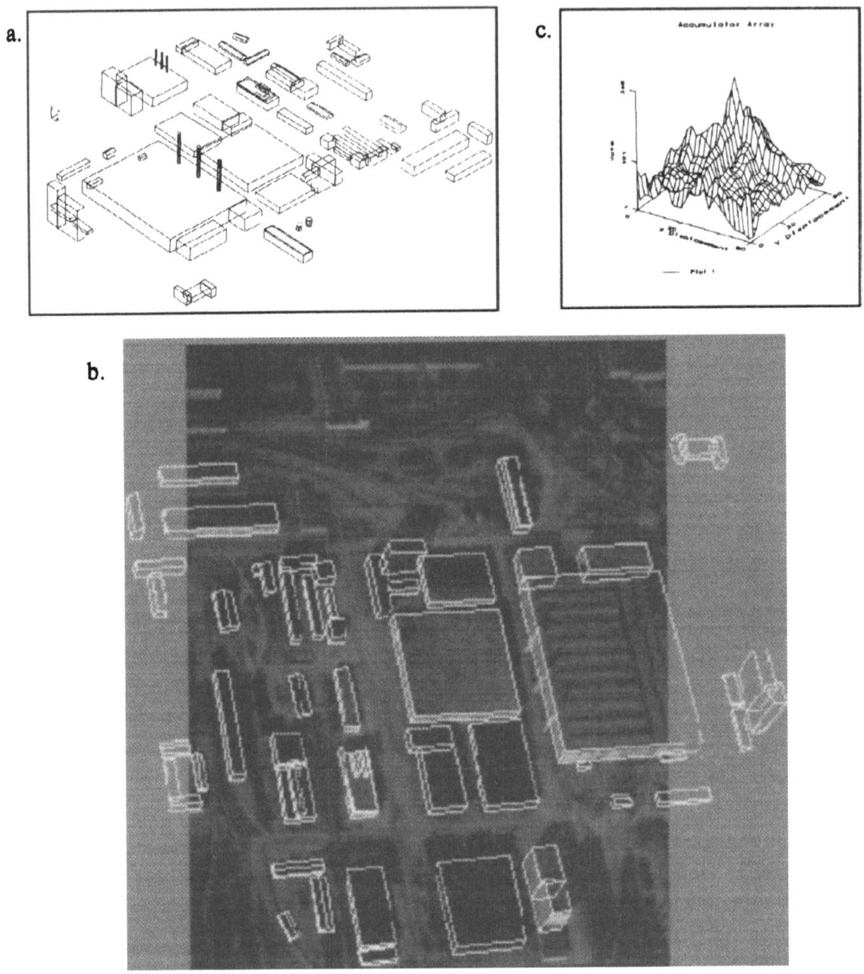

Fig. 6 : (a) The model. (b) The model registered to image J1. c) The accumulator array

validated. As table 1 shows the non-validated objects are later validated either by shadows or by reprocessing.

5 Conclusion

We have presented a model-based system capable of verifying the presence in an image of objects described in a 3-D site model. This task, called model validation, is the first step towards the construction of a full *change detection* system. We perform the model validation in three steps: first we register the image on the model, then we perform segment matching and build hypothesis to represent the possible presence of each model object in the image, and finally we verify those hypotheses using the shadow information extracted from the image. In the future we plan to use available DTMs for improved shadow processing.

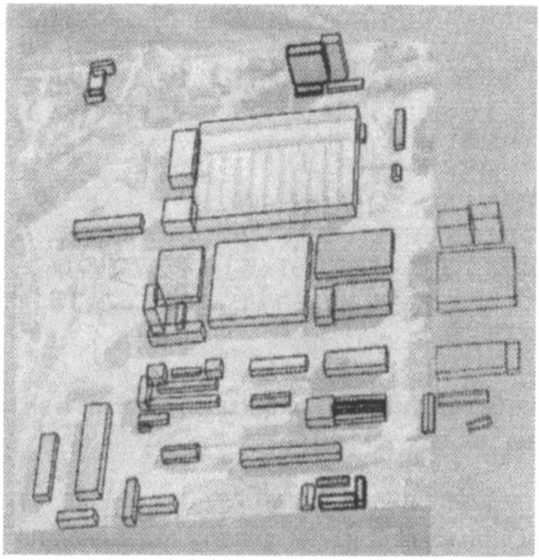

Fig. 7 : Validation result for image J2 after initial match (strong matches). Objects outside the field of view are shown with dashed lines. Non-validated objects are shown in bold lines.

Image	After Match	After Verification	After Reprocess	Final Result
J1	37/40 (92.5%)	38/40(95.0%)	40/40(100%)	40/40 (100%)
J2	41/47(87.2%)	43/47(91.5%)	47/47(100%)	47/47 (100%)
J3	37/38(97.4%)	38/38(100%)	*	38/38 (100%)
J4	42/46(91.3%)	43/46(93.5%)	43/46(93.5%)	43/46 (93.5%)
J5	40/52(76%9%)	45/52(86.5%)	52/52(100%)	52/52 (100%)
J6	33/41(80.5%)	38/41(92.7%)	40/41(97.6%)	40/41 (97.6%)
J7	38/40(95.0%)	40/40(100%)	*	40/40 (100%)
J8	44/49(89.8%)	46/49(93.9%)	49/49(100%)	49/49 (100%)

Table 1: Results for Modelboard J set

Image	After Match	After Verification	After Reprocess	Final Result
K1	18/19 (94.7%)	18/19 (94.7%)	19/19 (100%)	19/19 (100%)
K2	20/25 (80.0%)	23/25 (92.0%)	25/25 (100%)	25/25 (100%)
K3	30/39 (76.9%)	31/39 (79.5%)	36/39(92.3%)	36/39 (92.3%)
K4	17/17 (100%)	*	*	17/17 (100%)
K5	22/28 (78.6%)	27/28 (96.4%)	28/28(100%)	28/28 (100%)
K6	35/42 (83.3%)	39/42 (92.9%)	41/42 (97.6%)	41/42 (97.6%)
K7	17/17 (100%)	*	*	17/17 (100%)
K8	25/32 (78.1%)	26/32 (81.2%)	29/32 (90.6%)	29/32 (90.6%)
K9	29/37 (78.4%)	32/37 (86.5%)	37/37 (100%)	37/37 (100%)
K10	27/36 (75.0%)	33/36 (91.7%)	36/36 (100%)	36/36 (100%)

Table 2: Results for Modelboard K set

Image	After Match	After Verification	After Reprocess	Final Result
D8	2/2 (100%)	*	*	2/2 (100%)
D10	13/14 (92.9%)	14/14 (100%)	*	14/14 (100%)
D51	11/14 (78.6%)	12/14 (85.7%)	13/14 (92.9%)	13/14 (92.9%)
D70	2/2 (100%)	*	*	2/2 (100%)
D71	2/2 (100%)	*	*	2/2 (100%)
D74	14/15 (93.3%)	15/15 (100%)	*	15/15 (100%)
D34	27/27 (100%)	*	*	27/27 (100%)
D35	24/27 (88.9%)	25/27 (92.6%)	25/27 (92.6%)	25/27 (92.6%)

Table 3: Results for Denver set

Acknowledgment

The authors appreciate Gerard Medioni's contribution of his ideas and discussions on matching images using linear features.

References

Faugeras O., Luong Q, Maybank S. (1992) *Camera Self-Calibration: Theory and Experiments*, in Proceedings of the 2nd European Conference on Computer Vision, pp 321-334, Santa Margharita, Ligure, Italy.

Hartley R, Gupta R, Chang T. (1992) *Stereo from Uncalibrated Cameras*, in Proceedings of the IEEE conf. on Computer Vision and Pattern Recognition, pp 761-764, Champaign, Illinois.

Horn B. (1986) *Robot Vision*, Mc Graw Hill, New York.

Horn B. (1991) *Relative Orientation Revisited*, in Journal of the Optical Society of America, Vol. 8, No. 10, pp 1630-1638.

Huertas A., Nevatia R. (1988) *Detecting Buildings in Aerial Images*, Computer Vision, Graphics and Image Processing Vol. 41, pp 131-152.

Lillestrand R, (1972) *Techniques for Change Detection*, IEEE Trans. on Computers, Vol. C-21, Number 7, pp 654-659.

Lin C., Huertas A., Nevatia R. (1994) *Detection of Buildings using Perceptual Grouping and Shadows*. Proceedings of the IEEE conf. on Computer Vision and Pattern Recognition, pp 62-69, Seattle.

Longuet-Higgins H. (1981) *A Computer Algorithm for Reconstructing a Scene from two Projections*, in Nature, Vol. 293, pp 133-135.

Médioni G., Huertas A.,Wilson M. (1991) *Automatic Registration of Color Separation Films*, Machine Vision and Applications, Vol. 4, pp 33-51.

Nevatia R., Babu R. (1980) *Linear Feature Extraction and Description*, Computer Vision, Graphics and Image Processing, Vol 13, 257-269.

Ulstad M. (1973) *An Algorithm for Estimating Small Scale Differences Between two Digital Images*, Pattern Recognition, Vol. 5, pp 323-333.

Automatic Extraction of Man-Made Objects
from Aerial and Space Images
Monte Verità, © 1995 Birkhäuser Verlag Basel

Automatic Matching of 3-D Models to Imagery

Walter Mueller and James Olson

GDE Systems Inc.
P.O. Box 509008
San Diego, CA 92150-9008
Ph: 619-592-5381 (Mueller) 619-592-5380 (Olson)
Fax: 619-592-5420
email: mueller@gdesystems.com olson@gdesystems.com

Abstract

This paper presents a technique for automatically locating known control features in uncontrolled imagery. The automatic matching process exploits the available image support data and information concerning the imaging process, including sensor operation, environmental factors, and control feature information. One key feature of the algorithm is the radiometric modeling which is used to apply sensor characteristics, object reflectance, and illumination conditions. The importance of radiometric modeling is demonstrated.

1. Introduction

Photogrammetry is used to obtain accurate positional information of an object. Prior to projecting image coordinates to ground coordinates, the image needs to be controlled via both interior orientation and exterior orientation. Techniques for automating interior orientation exist and are applied in today's image scanner products. Exterior orientation however is still a labor intensive process of accurately locating control points in imagery. Photogrammetric adjustment of the camera model can be performed after control points are located.

For sites which are frequently imaged, locating control can be automated using a site model of the area. After the image is controlled, model supported exploitation techniques as being developed on the RADIUS program (Gerson and Wood, 1994) can be applied to the image.

Several techniques have been developed which locate objects in imagery (Brown, 1992). Techniques range from direct image-to-image correlation (Chellappa, 1994) to feature extraction and feature edge matching (Beveridge, 1993). Image-to-image correlation techniques require storing multiple patches of imagery for each control feature under various imaging geometries and illumination conditions. Large storage capacity would be required for such a system. On the other hand, feature extracting and edge matching techniques can suffer from combinatorial complexity in computing an optimal solution from matching the model to a set of line features extracted from an image. Beveridge (1993) tries to alleviate this problem by using random starting positions and iterating to find a local maximum fit (which may not be the correct answer). By running the algorithm several times from several random starting points, the best fit location of among all runs is assumed the right answer but again it may not be correct. This paper discusses a model-based technique for locating predefined features in imagery using both the geometry of the feature and the predicted radiometry of the feature. This approach attempts to exploit all known information about the feature to be located and the given imaging process.

For this approach, features are modeled geometrically and radiometrically. Monochromatic images contain two dimensions of geometry (rows and columns) and one dimension of radiometry (pixel intensity). Prediction of object appearance in an image requires more than just a geometric description of an object. Object color variations, varying camera exposures, and differing illumination conditions all manifest themselves in image pixel intensities. The digital image acquisition process is assumed known and constant. The illumination from ambient light and direct sunlight is predicted.

The radiometry of the feature is the reflectance of the object as collected by the image sensing system sensor. Ambient light conditions and the direct illumination contribution of the sun are modeled to predict the pixel intensity of the feature in the image. A direct comparison of the predicted intensity and the actual intensity found in the image is then performed to measure the match. Correlation techniques are weaker in this regard because they cannot be used to directly compare a single pair of pixels and all linear relations between pixels produce equivalent results. If the absolute intensity can be accurately predicted, it should be exploited by differencing rather than correlating.

In section 2, the basic algorithm for developing the model and then using the model to locate its position in an input image is presented. In Section 3, results are given applying the algorithms to the given AMOBE images. Two matching approaches are described and compared. Section 4 discusses observations and recommendations for further research.

2. Model-Based Location Process

There are two basic processes: an interactive model building process which populates a model database and an automatic model location process which locates the image position of the model reference point in an image. An interactive modeling process is used to build the object models. Object models are constructed in order to capture an object-space description which can be projected into an image. Descriptions of both object geometry and radiometry are used to define not only which pixels correspond to which faces, but also what reflectances are expected on each face as captured by the imaging system. The model building process has as inputs imagery, image support data and control points for controlling the imagery. In the geometric modeling function, stereo imagery is controlled and the 3D geometry of the target object is extracted. The object is defined geometrically by a group of planar facets. After geometric modeling is completed, the radiometric modeling function defines the radiometry of each facet as collected by the imaging system. The radiometry is defined in terms of the observable object reflectance for the given sensor. The resultant feature model (geometric and radiometric) is stored in a model database for later retrieval by the automatic model locating function.

The automatic object location process uses the image and image support data to search for the object model in a given image. It is assumed that the image support data has enough information to compute the sun angle. The output of the automatic object location process is the image location with the best match to the model and the resultant figure of merit (FOM) of the best match. The entire image is searched in the process. A minimum FOM is required for accepting the match to eliminate false alarms due to occlusions or objects not in the image.

2.1 Interactive Model Building

Geometric Modeling. A three-dimensional geometric model is required of the model-based algorithm. Photogrammetric techniques are used to measure the size, shape and orientation of each facet. Stereo imagery is first loaded into a stereo photogrammetric workstation such as Helava's SOCET SET® and controlled. The stereo imagery can then be used to derive a precise feature point in object space and precise shape and orientation information to define the planar facets which comprise the object.

Radiometric Modeling. The radiometric modeling process assumes that the radiometric characteristics of the imaging system are constant. The radiometry of the facets of the objects are modeled so they can be predicted in the automatic model locating function. The radiometric function assumes ambient light and direct sunlight are the only sources of illumination for the various object facets. Only the reflectivity as quantified by the given imaging system is stored in a model database for each facet in the model (for this application it is assumed each facet has a homogenous radiometric characteristic). Image exposure, object reflectance and illumination conditions are the independent variables which define the objects radiometry for a given facet.

The camera's image exposure parameters define the amount of a facet's reflected light is captured by the camera's sensor. Most imaging systems use exposure controls to ensure effective utilization of the exposed media under various illumination conditions. Common exposure controls include exposure duration and aperture size control. These controls change the amount of light from the image subject which is required to produce a given amount of light in the captured image. Determining the amount of light sensed is then a matter of compensating for the amount of exposure allowed by the image sensor. Typically, image exposure compensation cannot be performed on digital images due to the lack of radiometric support data. This is the case regarding the images presented in this paper. However, if the exposure setting remains constant, there is not need to calibrate for exposure settings.

The object reflectance defines the amount of incident light on a facet that is reflected. Objects are visible due to their ability to reflect light. Darker objects reflect less light than lighter objects. Monochromatic imaging systems typically have a fixed spectral response and object reflectance can be characterized for that response under given illumination conditions. Object reflectance is characterized by measuring the calibrated pixel intensity (sensed radiance) under nominal ambient illumination (typical shadow). As long as consistency is maintained among various objects, any lighting condition can be used to measure object reflectance.

Illumination defines the amount of light incident on the reference objects. For typical outdoor optical images, the sun is the primary source of illumination, but not only a direct source. In a typical scene, reflections from the atmosphere and neighboring objects contribute towards ambient illumination. For typical outdoor scenes, the amount of illumination can be constrained to eliminate unlikely possible illuminations.

2.2 Automatic Model Locating

The automatic object location process is shown in Figure 1. There are two functions in the process: the Predict Object Image Geometry and Radiometry function and the Search for Object function. The Predict Object Image Geometry and Radiometry function reads in the image support data and the absolute location of the control feature in object space to photogrammetrically project the feature into the predicted image space. Sun angle is also determined from the support data and used to compute the contribution of direct sunlight to the total illumination of the facet. The Search for Object function will search the entire image for the feature and return the image position of the model reference point of the best match in the image. The Search for Object function has three input parameters: homogeneity, ambient light range, and direct light range. The homogeneity parameter describes the amount of deviation allowed from the predicted pixel intensity. The light ranges are the ranges searched to maximize the amount of correctly predicted intensities for the modeled facets. The output of the Search for Object function is the object image position with the best match, the figure of merit (FOM) for the match, and the resultant value for the determined ambient and direct light parameters which gave the best match. The FOM is returned to reject points which are below a level of certainty (due to clouds, changes in feature appearance, etc.).

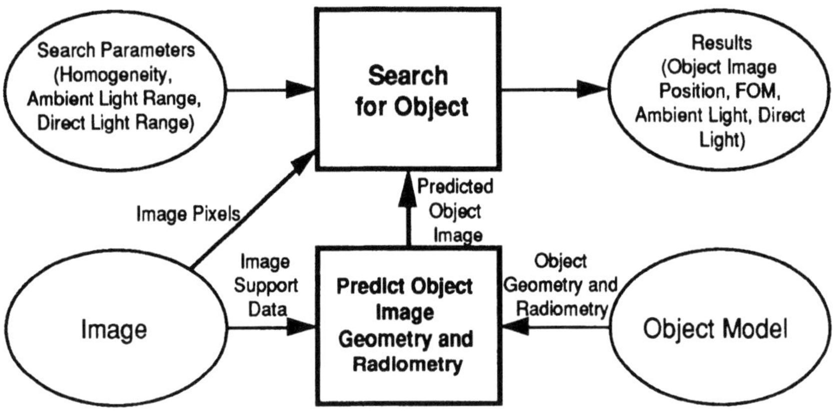

Fig. 1: The automatic object location process.

Pixel Intensity Prediction. This section describes the pixel intensity prediction of a facet in the model. To ensure that only pixels which include intensity from a single object face are compared, pixels intensities are only predicted when all four corners of a given pixel position are contained within a given face. Pixel areas which straddle facets are not included because effective prediction of these pixels is impossible without precise characterization of the imaging system pixel impulse response. For each facet, the pixel intensity of all pixels interior to the facet is predicted by:

$$p_i = r_i (a + d\, s_i) \tag{1}$$

where

p_i is the predicted pixel intensity for the face in the image
r_i is the reflectance of the facet
a is the ambient lighting factor
d is the direct solar illumination factor
s_i is the exposure of the facet plane to the sun (computed as the cosine of the angle between the surface normal and illumination direction vectors; zero for facets which face away from the sun)

For each image location, the algorithm cycles through various values of a and d in (1) and applies to all facets in predicting the pixel intensity.

Measure of Match. For each pixel in each facet and for a given value of a and d in (1), the predicted pixel is compared to the actual image pixel. The number of pixels in a all facets within a given percentage of the predicted value is counted. The percentage of visible pixels predicted within the given range is used as the measure of match.

3. Results

The gray level AMOBE images were used to evaluate the model matching process. One building each from the industrial and residential scenes were modeled. Of the four images given for each scene, a stereo pair was used to create the geometry of the model, a single image was used to create the radiometric model, and all four were used to test how well the algorithm located the model in the image. The models were also searched in images where the models were not located (i.e. the industrial building was searched in the four residential images and the residential building was searched in the four industrial images).

When searching an image, the entire image was searched to identify the best match to the model prediction. To measure the necessity for radiometric prediction, each test was conducted with and without radiometric prediction. For the test without radiometric prediction, it was assumed only that each facet was homogenous in intensity.

3.1 Making the Models

Our current modeling capabilities include convex three-dimensional objects with planar facets which are diffuse reflectors. Stereo imagery containing the object was controlled using SOCET SET to subsequently measure the geometry of the objects facets. The radiometry of the faces of the object (1) was derived as follows:

1. Identify an area on the image which is partially shadowed and has a homogeneous texture.
2. Measure the intensity ratio of the direct sunlight texture and the shadowed texture.
3. Assume the ambient illumination factor is 1.0 and estimate d as one less than the value measured in step 2, divided by the exposure of the area to the sun, s.
4. Identify the location of the object being modeled and project the object geometry into an image.
5. Measure the average pixel intensity for each visible facet.
6. Estimate the surface reflectance for each facet by solving equation (1) while using the a and d estimated in step 3.

The building models from both the industrial site and the residential site are given in Figure 2. For the industrial scene (Figure 2a), we modeled the gable-roofed building in the lower-left corner of the scene. The model included two facets for the roof, four wall facets, one door, and a portion of ground near the building. The two sides of the building which were not visible in the scene used to develop object radiometry were modeled without doors or windows (even though examination of the images revealed a significant presence of doors and windows). The stereo pair from scenes 5882 and 5883 were used to measure the geometry of the facets of the chosen building object. Image 5882 was used to derive the object radiometry.

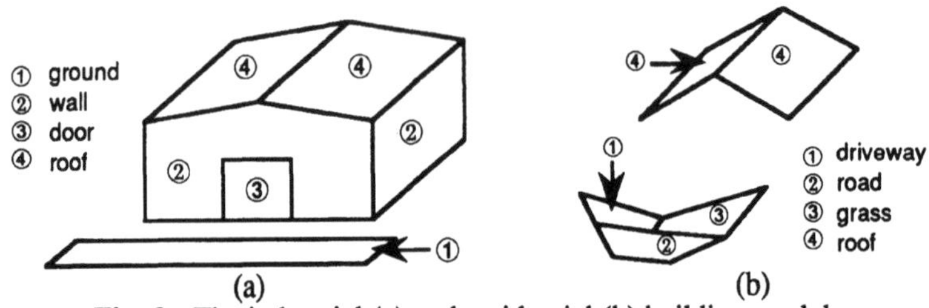

Fig. 2: The industrial (a) and residential (b) building models.

The residential scene model (Figure 2b) was constructed to match the house with the dark gabled roof and driveway on the road which passes through the middle of the scene. This model included two roof facets and three facets on the ground. The ground facets cover areas of road, driveway, and grass. The building was geometrically modeled using the pair 5888 and 5889. Object radiometry was derived using only image 5888.

3.2 Results of Search

Both models were applied to all 8 gray level AMOBE images and results follow. The sun angle for each image was computed from the given support data (time, date, and image geoposition). Figure 3 shows the results of searching for the industrial building model on industrial image 5874 and the results of searching for the residential building on image 5897. Figure 3a shows the original industrial image 5874. Figure 3b shows the resultant figure of merit (FOM) at each location in the image as measured as a percentage of pixels in all facets falling within 10% of the predicted value (brighter values for higher percentages). Figure 3c shows the model overlaid on the image where the highest FOM was found. Figure 3d shows the original residential image 5897. Figure 3e shows the FOM image and Figure 3f shows the model overlaid on the image where the highest FOM was found. As shown in these examples, the position of the highest FOM is clearly dominant and accurately indicates the position of the actual object.

In Figure 4, the FOM output for each pixel location in the image is compared to the FOM output of a model which just assumes homogeneity in each of the modeled facets. This was done to test the need for radiometric modeling of the facets over just modeling the facets as a homogeneous texture. Figure 4a is the original industrial image 5883. Figure 4b is an image of the FOMs derived when the feature was located using a model of the feature assuming the facets were homogenous. Figure 4c is an image of the FOMs derived when the feature was located predicting the radiometry of the facets. Figures 4d and 4e outlines the location of where the object was located with the highest FOM in the image using the homogeneity measure only and using the radiometric prediction respectively. As shown, using only the homogeneity information in the model causes many false alarms and the highest FOM does not correctly locate the feature. Using the radiometric model however produces a strong FOM only on the feature and the other areas of the image have much lower FOMs.

Figure 5 is a graph of the results of applying both modeled features on all images. All images were used to see what kind of FOM was obtained in an area were it was known the feature did not exist. For the radiometric modeled feature, all features were located correctly in all the images containing the feature. The second local maximum FOM was well below the percentage of pixels predicted over the primary maximum FOM. For the homogeneity model, all winners (image positions with the highest FOM) in the industrial images were incorrect and all winners in the residential scene were correct. In industrial

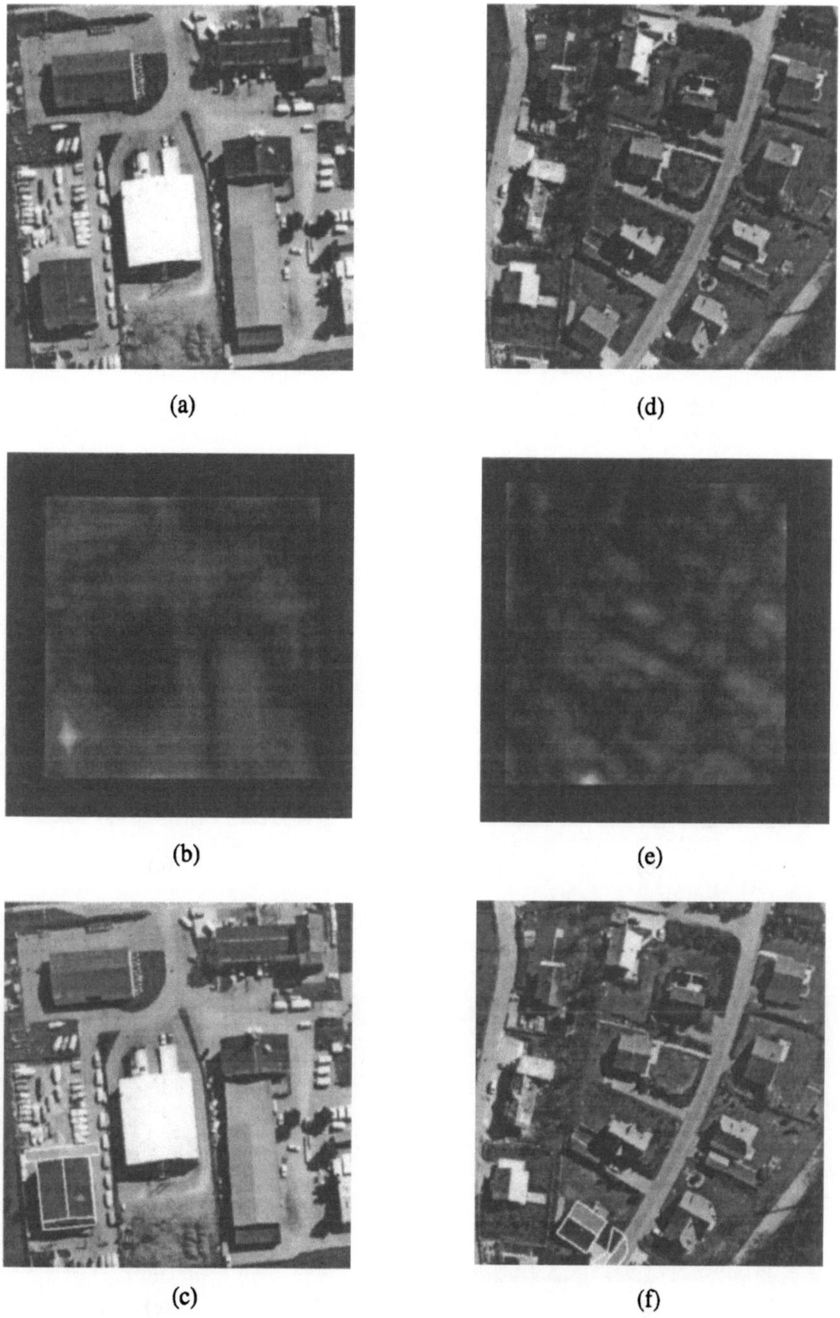

Fig. 3: Results of Comparing Predicted Model to Images

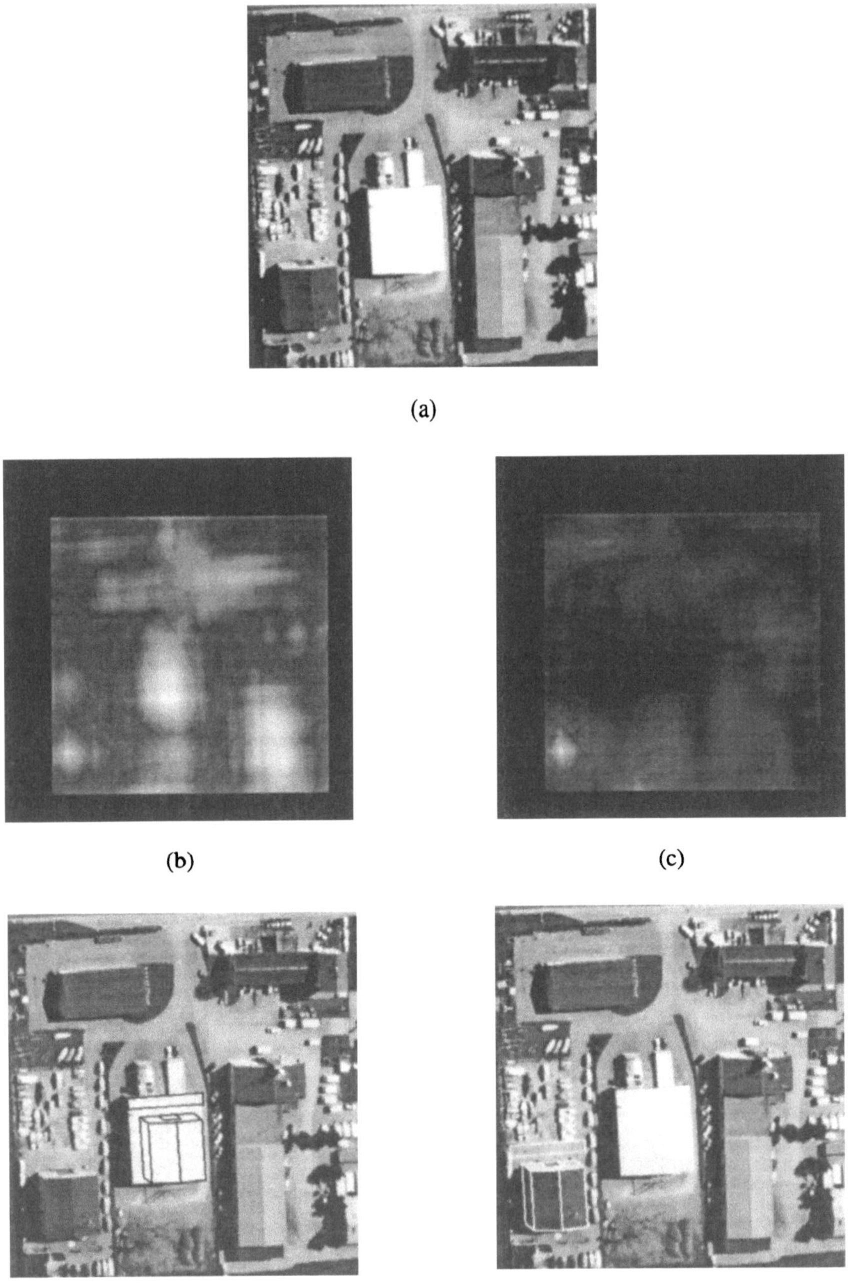

Fig. 4: Comparing Predicted Radiometry vs. Homogeneity Only

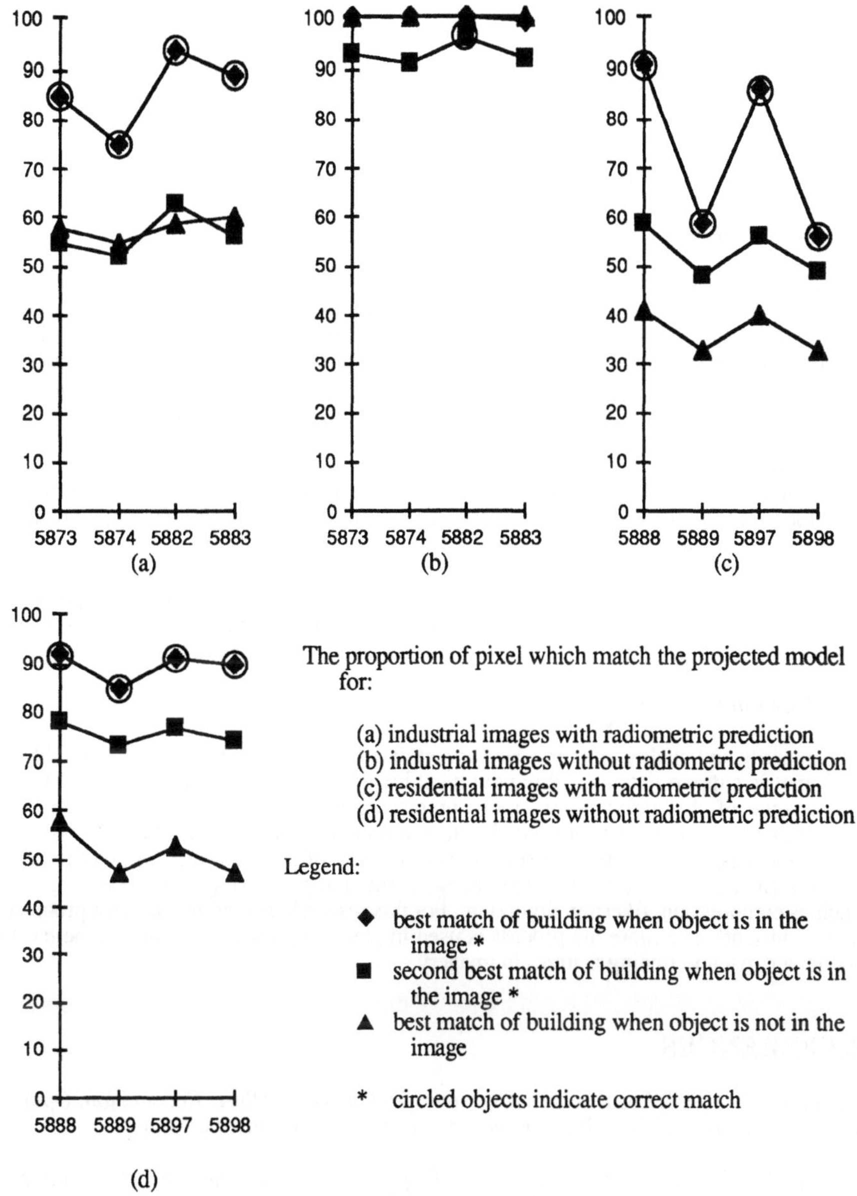

Fig. 5: The proportion of model pixels matched for the industrial scene and the residential scene both using and ignoring model radiometry.

image 5882, the second best FOM using the homogeneity model was the correct answer and for the other industrial images (5873, 5874 and 5883), the correct answer came in as the third best FOM or lower. Two images of the residential scene (5889 and 5898) had predicted coverage well below the other two images using the radiometric models. It was noticed that these two images were considerably darker than the other two which probably occurred during the digitizing process. However, the correct answers were obtained nonetheless but with less confidence than in the other images.

4. Conclusions and Recommendations

4.1 Conclusions

The radiometric prediction approach to predicting and locating an object in the scene was shown to reliably locate modeled objects. The radiometric model prediction approach was clearly superior to the approach relying on simply homogeneity of regions. Evidence is shown which suggests that absolute percentages of predicted pixels found in the image can be used to differentiate between finding an object in the image and finding false alarms.

One of the drawbacks to any approach for finding given objects in a scene is that changes to the objects cannot be predicted and will result in reduced reliability of the automatic location process. For the purpose of controlling imagery, identifying features which do not change much over time is desired.

Also, while the geometry of the feature is sensor independent, the object-model's radiometry as measured in this process is image system dependent. Different sensors can share geometry but require different values for radiometry in the model.

4.2 Recommendations

The quantity and quality of geometric support data far exceeds the quantity and quality of radiometric support data available with most images. With additional sensitivity to the importance of this information, more precise predictions can be made to improve algorithms such as we have presented. Most important in this regard is the processing of the imagery between the time of sensing and the time of pixel production. Examples of this information include intermediate image formation processing and processing steps such as compression, filtering, digitizing. For this research, it is assumed these processes were constant. The more the process is used in predicting the pixel value the better the model approach to finding features in imagery.

REFERENCES

Beveridge, R. (1993) *Local Search Algorithms for Geometric Object Recognition: Optimal Correspondence and Pose,* PhD thesis, University of Massachusetts, Amherst.

Brown, L.G.(1992) *A Survey of Image Registration Techniques*, ACM Computing Surveys, Vol. 24, No. 4, pp. 325-376.

Chellappa R., Q. Zheng, L.S. Davis, C.L. Lin, X. Zhang, C. Rodriguez, A. Rosenfeld, T. Moore (1994) *Site-Model-Based Monitoring of Aerial Images*, Image Understanding Workshop Proceedings, pp. 295-318.

Gerson D., S. Wood (1994) *RADIUS Phase II, The RADIUS Testbed System,* Image Understanding Workshop Proceedings, pp. 231-237.

Automatic Extraction of Man-Made Objects
from Aerial and Space Images
Monte Verità, © 1995 Birkhäuser Verlag Basel

Structural 3D-Analysis of Aerial Images with a Blackboard-based Production System

Uwe Stilla, Eckart Michaelsen, Karl Lütjen

Research Institute for Image Processing and Pattern Recognition (FIM/FGAN)
Eisenstockstr. 12, D-76275 Ettlingen, Germany
Ph.: +49 7243 99252, Fax: +49 7243 99229
e-mail: usti@gate.fim.fgan.de

Abstract

A model-based method for the analysis of structures in aerial images is described. The knowledge about object structures is represented by a set of productions. Together with the object concepts these productions form a net. Such production systems are implemented in a blackboard architecture. The database is realized in form of an associative memory. A simple example net is used to explain the 3D-analysis and the detection of a high building. The result of the image analysis and partial results obtained by a derivation graph are displayed. Problems with the associative access are discussed and an extended hardware concept is proposed.

1 Introduction

The presented work is part of the DFG research project *Analysis of aerial and satellite images for automatic determination of the ground sealing of urban areas.* The aim of this project is to attain a context-oriented automatic image analysis, which comprises in addition to multispectral and texture classification an image structure recognition by using two- and three-dimensional models.

In the field of pattern recognition, knowledge-based methods are increasingly used for the analysis and description of aerial imagery (Nagao & Matsuyama, 1980)(McKeown et al., 1985) (Nicolin & Gabler, 1987). A special group are structure-oriented hierarchical methods, which build up structure hierarchies by composing complex structures from less complex structures. Using this approach the analysis proceeds step by step, with constant reference to the patterns being analyzed, producing intermediate results with increasing degrees of abstraction.

A context-oriented recognition of objects may use different information sources, for instance: (i) internal context information e.g. neighbourhood relations to objects of the background (Stilla & Jurkiewicz, 1991), (ii) external context information e.g. a map corresponding to the image (Stilla & Hajdu, 1994), (iii) information from differend images or sensors e.g. stereo vision. Additional knowledge could be used in different ways and on different abstraction levels for the image analysis.

This paper describes the procedure of verifying objects with the aid of at least two images. Our emphasis lies more on the image understanding aspect and evidence fusion than exact measuring of size and coordinates of objects in photogrammetric sense.

2 Blackbord-based Produktion System BPI

For structure analysis of complex scenes the *b*lackboard-based *p*roduction system for *i*mage understanding BPI (Lütjen, 1986) is used.

2.1 Production System

A production system consists of three basic components: a database, a set of production rules and a control unit. A production rule, or production, is a statement in the form: IF *condition* holds, THEN *action* is appropriate. The execution of *action* will result in a change of the data contained in the data base. The control unit controls the overall system activity and has the special task of deciding which production (with satisfied condition part) to fire next.

The process of building up more complex structures from less complex structures, using such productions, can be described by a rewriting system. With reference to formal languages the rewriting system may be determined by a Grammar G. Such a formal grammar is defined by a 4-tuple

$$G = (S, V_n, V_t, P),$$

where S is a set of start symbols (target objects), V_n is a set of non-terminal symbols including S (partial objects), V_t is a set of terminal symbols (primitive objects) and P is a set of rewriting rules (productions). Attributes are assigned to the objects, which represent certain structures. The productions determine how a given set of objects is transferred into a set of more complex objects.

In analogy to string grammars we may say that an image content is parsed (bottom up) by the process of image analysis. Instead of examining concatenation as is done by parsers for string grammars, we examine the topologic or geometric relation of objects in the *condition* part of a production. Therefore, a production rule may be written in the form:

$$\boxed{P_i : X \wedge Y \odot \xrightarrow{i} Z}$$

This means that, if an object of type X and an object of type Y fulfil the relation $\odot$, then an object specific generative function $\xrightarrow{i}$ is carried out which produces an object of type Z. Starting with primitive objects, a target object can be composed step by step using the productions repeatedly. Similar to tabular parsing methods (e.g. in Aho & Ullman, 1972) all intermediate results (partial objects) remain stored in the database during the analysis.

For example a simple rewriting system is determined by

$$G_{Example} = (\{R\}, \{R, U, C\}, \{L\}, \{P_1, P_2, P_3\}).$$

According to $G_{Example}$ and starting with the primitive objects L, the partial objects C, U, R are composed using the productions P_1, P_2, P_3 (see Fig. 1a), object R representing the target object.

P_i :	Objects X,Y			$\odot$	$\xrightarrow{i}$	Object Z
P_1 :	L	$\wedge$	L	L	$\xrightarrow{1}$	C
P_2 :	C	$\wedge$	L	⊏	$\xrightarrow{2}$	U
P_3 :	U	$\wedge$	L	□	$\xrightarrow{3}$	R

L	right-angled
⊏	u-structure
□	rectangular

a

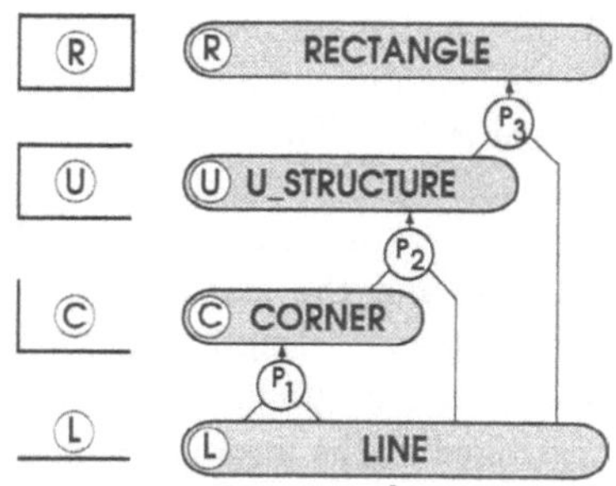

b

Fig. 1: Productions. a) Table of productions, b) Production net

The *general* interaction of productions and the stepwise transfer of objects into objects with a higher abstraction level can be depicted by a production net (Fig. 1b). The compositions for the *actual* objects (instances) are recorded with the aid of pointers and can be illustrated by a derivation graph.

After the analysis, derivation graphs of the target objects can be constructed and used to explain the results. Thus, the subset of primitives, which represents the target objects can be determined. If we compare this subset with the set of primitives, we may say that the production net acts like a filter.

2.2 Blackboard Architecture

In the BPI-System the productions are implemented in a blackboard architecture (Newell, 1962; Nii, 1986). Generally, a blackboard architecture consists of a global database and a set of knowledge sources, which communicate only via the blackboard. In BPI the global database is stored in an associative memory. Knowledge sources are constructed as independent object specific processing modules, which examine a *condition* and execute an *action* of a production.

Systems with a blackboard architecture are essentially ***data driven*** (Velthuijsen, 1992). One or more hypotheses "*part_of* a more complex object" are attached to an object. An object-hypothesis pair (processing element) triggers a processing module to verify the hypothesis. The hypotheses arise from the production net, so that the analysis proceeds in a goal-directed manner. A control unit, containing a priority ordered queue of processing elements, is added to the blackboard system. Further details concerning the dataflow of the BPI-System are described in previous papers (Lütjen, 1986; Stilla & Hajdu, 1994)

2.3 Associative Memory

An associative memory, realized by special hardware, is used to get fast access to object sets (Davis & Lee, 1986). The memory concept which provides an inverted database is demonstrated in Fig. 2.

A bit matrix is used to assign attribute values to the objects. The objects (marked by an object index) are registered in the columns and the attribute values (e.g. length=*31*, length=*32*, length=*33*) are registered in the lines. The bitstring (line) attached to an attribute value indicates to which objects this attribute value is assigned.
A new object is *written* into a free column, placing bits corresponding to the attribute values. In Fig. 2 the four diamond-shaped bits indicate an object of type LINE with length=*31*, orientation=*45°*, and assessment=*average*.

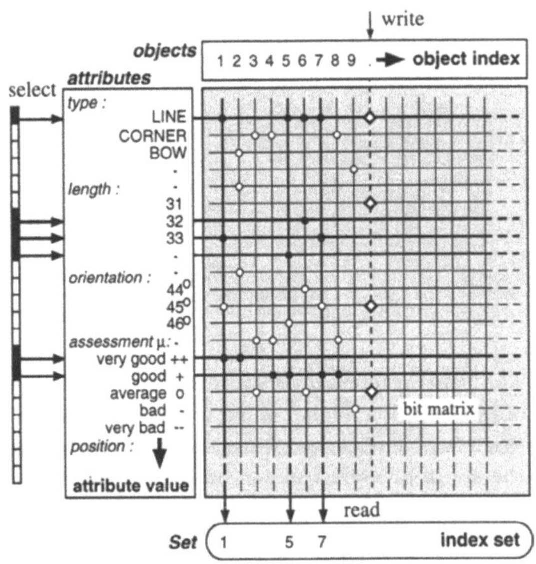

Fig. 2: BPI system: Associative memory

Object sets with certain features are ***selected*** with attribute values, attribute intervals and set operations (e.g. all objects of type LINE with length >*31* and with an assessment of *good (+)* or *very good (++)*). The hardware ***reads*** out an index set containing all objects satisfying the selected attributes.

3 Object Model

The object model defines the object concept and may incorporate a set of model parameters, tolerance parameters and assessment parameters. When a hypothesis is tested for a triggering processing element (object-hypothesis pair), objects are searched which fulfil the model conception with respect to the triggering object and a given tolerance. For this, search ranges R are marked in the attribute space.

Let us suppose a triggering object LINE carries the hypothesis *part_of*_CORNER. In order to test this hypothesis a search area R_A is constructed around an endpoint $e_t=(x_t, y_t)$ of the triggering object and a search interval R_φ for orientation is defined. The search area R_A is bounded by a circle or simply by a square (Fig. 3). Associatively the set S of those objects type LINE is selected, which have a right-angled orientation φ and one endpoint e=(x, y) close to an endpoint of the triggering object LINE.

$$\begin{aligned} R_A &= \{(x,y)\ ;\ (x_t - \Delta x \leq\ x\ \leq x_t + \Delta x)\ \wedge\ (y_t - \Delta y \leq\ y\ \leq y_t + \Delta y)\} \\ R_\varphi &= \{\varphi\ ;\ \varphi_t + 90^o - \Delta\varphi \leq\ \varphi\ \leq \varphi_t + 90^o + \Delta\varphi \mod 180^o\} \\ S &= \{o\ ;\ \text{type}(o) = \text{LINE}\}\ \cap\ \{o\ ;\ e(o) \in R_A\}\ \cap\ \{o\ ;\ \varphi(o) \in R_\varphi\} \end{aligned}$$

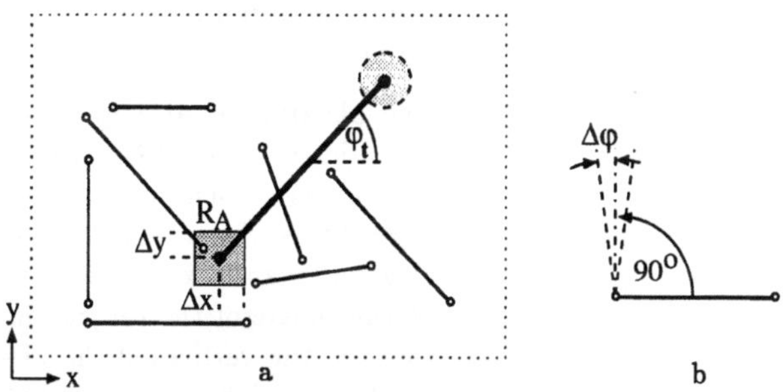

Fig. 3: Search ranges for object CORNER. a) search area, b) search interval

By applying the production P_1 (Fig. 1) repeatedly to a given set of n objects LINE a set of m objects CORNER is produced. In general $n(n-1)/2$ pairs of objects LINE have to be formed and tested with the relation of P_1. If the geometric relations make up a strong constraint m should be much smaller than n. In this case the number of tests can be reduced if only those object pairs are examined which are likely to fulfil the relation. Attaching a hypothesis to each object LINE n hypothesis tests have to be carried out. This means, that the computation effort using the associative access decreases from a quadratic dependency of n to a linear dependency of n.

4 Production Net

To explain the interaction of such model-based productions in a production net, a simple model for the 3D-anlysis of aerial images is chosen. Descriptions of more complex object models can be found in (Stilla & Jurkiewicz, 1991). As an example we select a model for the detection of multi-storied building. Using this model we assume that a multi-storied building can be described as a cuboid and that the cuboid surfaces (the same wall and roof surface) are visible in images from different view-points. These two perpendicularly oriented areas form an edge in space (CUBOID_EDGE). The height of this edge is a typical feature for high buildings and can be used to discriminate this object from other objects.

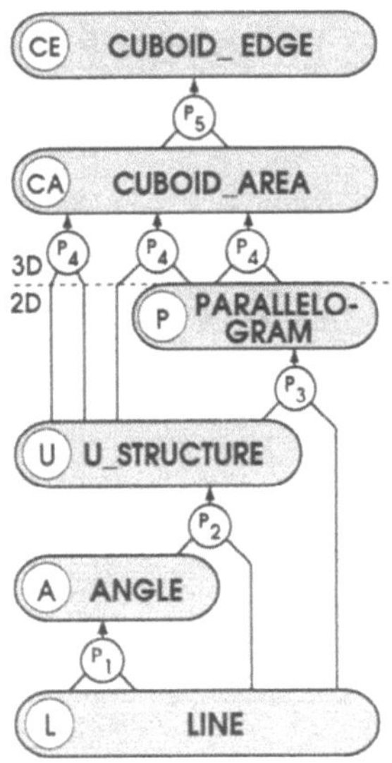

Fig. 4: Production net for object CUBOID_EDGE

A production net for the object CUBOID_EDGE is depicted in Fig. 4. The analysis distinguishes between a 2D- and 3D-scene analysis. At first the 2D-analysis is carried out independently in different images. Starting with the object primitives LINE, the objects ANGLE, U_STRUCTURE and PARALLELOGRAM can be built applying the productions (P_1-P_3). 2D-objects form the primitives of the following 3D-analysis.

The 3D-analysis attempts to find pairs of 2D-objects (U_STRUCTURE or PARALLELOGRAM) which are projections of the same 3D surface. This is done by selecting pairs and examining rays originating at the centre of the projection and passing through the vertices of the 2D-objects. On ideal conditions rays through corresponding object vertices of two images will intersect in the 3D-space.

Due to image noise, processing errors and inaccurate camera parameters the rays may not intersect exactly. Hence, the minimal distance between the rays is calculated. The 2D-objects will be called *not corresponding* if this distance between the rays of pairs of vertices is greater than a given threshold. Additionally, a model-based plausibility check ist carried out (e.g. regarding the position: object points must not be under the earth's surface or behind the camera).

If 2D-objects of different images correspond, the object CUBOID_AREA is generated (P_4). IF objects CUBOID_AREA are oriented in such a way that the surface normals are right-angled and located such that vertices are neighbouring, then a target object CUBOID_EDGE is generated (P_5).

5 Results

Aerial photographs (scale 1:40000, f=153.57 mm) of an urban area (city of Karlsruhe) were used as image data. The transparencies are scanned (B/W) with a resolution of 25 μm per pixel. An exact determination of the camera parameters for each image was carried out by determining matching points manually and calculated iteratively (resection in space).

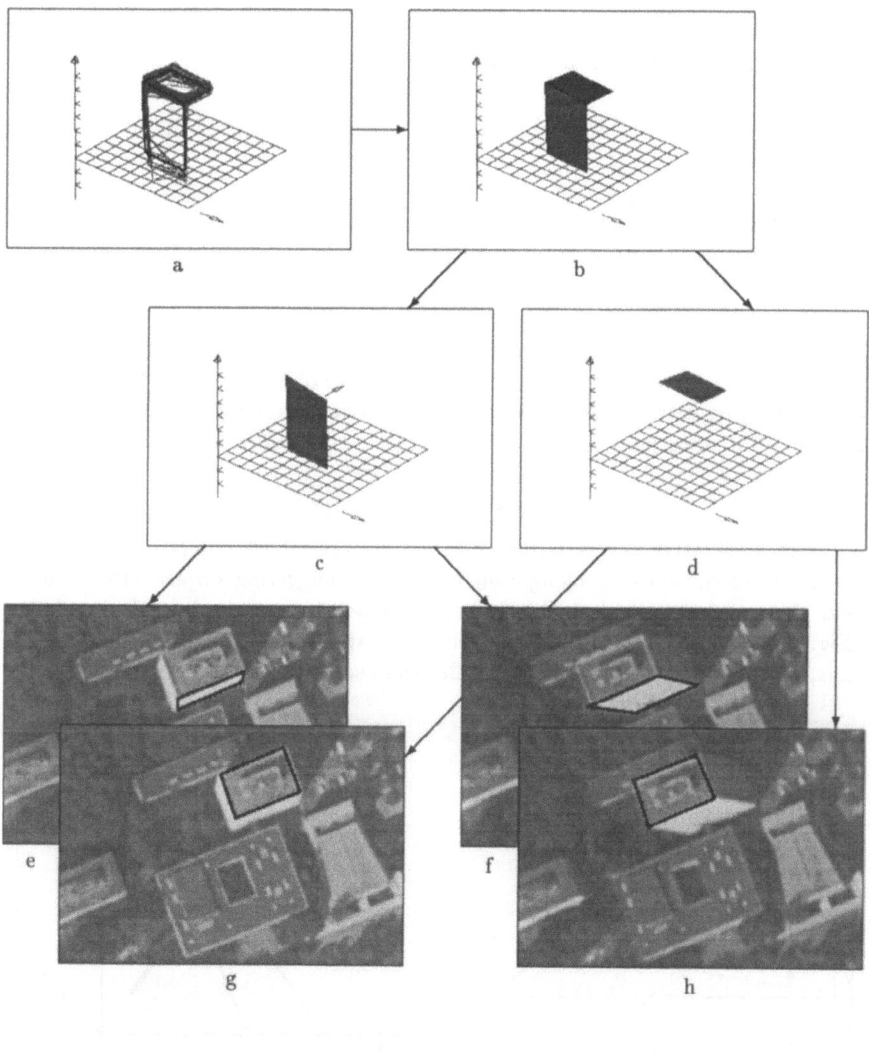

Fig. 5: 3D-analysis. a) set of objects CUBOID_EDGE, b-h) partial results

In the preprocessing stage the scanned image data are transferred into a sequence of binary images, applying a multiple thresholding method. In each binary image the contour lines of the segments are detected and approximated by straight lines using a dynamic split algorithm. These straight lines are stored as primitive objects LINE, attributed with length, orientation, and endpoint coordinates, in the blackboard.

Partial results of a scene analysis are displayed in Fig. 5. The set of objects CUBOID_EDGE, which are found in an expected area is displayed in Fig. 5a. From this set one object is selected (Fig. 5b). The previous objects, which have been obtained by the derivation graph are displayed in Fig. 5c-d (CUBOID_AREA) and in Fig. 5e-h (PARALLELOGRAM).

6 Discussion

Both the development of new models for complex structures and the computationally expensive testing of production nets tends to be quite time consuming. In order to translate the productions into a program code of process modules in an efficient way, software tools are required. On the one hand the programming is supported by a set-oriented programming language for the BPI-system, on the other hand an interactive explanation module (WEEK) is available for debugging and displaying sets and relations of objects. The processing time is reduced by special WPP(word parallel processing)-hardware developed for set-oriented associative access.

In some cases, a geometric condition can not be transformed in such a way, that a direct associative access is possible, for example if all objects LINE have to be determined, whose prolongation intersect a triggering object LINE. To access these objects, the possible intersection points have to be calculated first (Fig. 6a). A similar problem is given in 3D-space when searching for pairs of rays, whose minimum distance is smaller than some threshold (Fig. 6b).

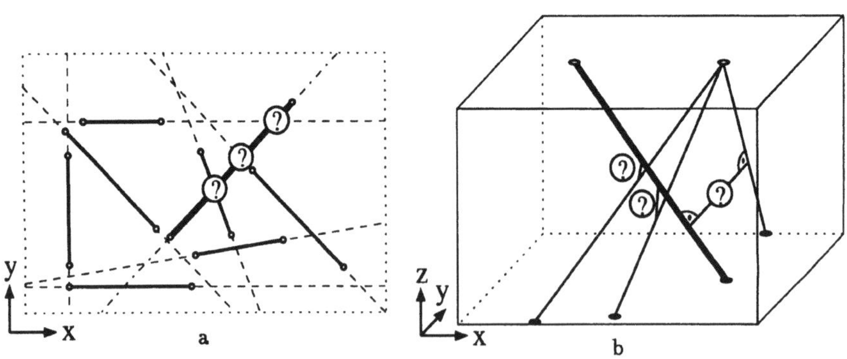

Fig. 6: Access to temporary attributes. a) intersection, b) minimum distance

Hence, the concept for the associative access has to be extended and a new hardware has to be developed. The extended concept for the associative memory ist displayed in Fig. 7. It works in the following way: A function to calculate temporary attributes (e.g. euclidean distance d, intersection (x,y), etc.) has to be selected. The arithmetic operations have to be carried out with a parameter set, which is determined by attribute values of the triggering object. For the calculation, each object has its own processor, which runs with the same program code. For this task SIMD(single instruction multiple data)-processors are suitable. The *selected* objects send the values of specified attributes to their processors and the temporary attributes are calculated for these objects. The results are *written* into the blackboard and available for associative access.

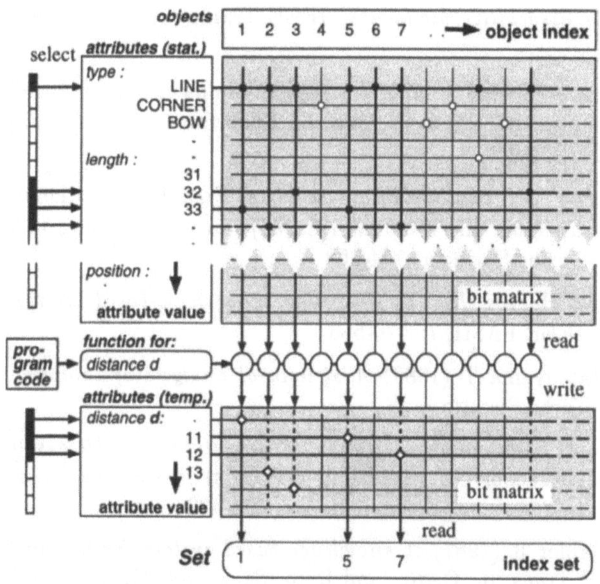

Fig. 7: Extended concept for an associative memory

In the example shown in Fig. 3 objects LINE with neighbouring endpoints to an endpoint of triggering object LINE are searched. Instead of constructing a search area around the endpoint $e_t=(x_t, y_t)$ the distance function $d^2= (x\text{-}x_t)^2+(y\text{-}y_t)^2$ is selected and the SIMD-processors are loaded with the program code. All objects LINE respond to the attribute d^2 with their temporary attribute values. The set of objects LINE for which the condition $d_i^2 < d_{max}^2$ holds is selected associativly. Instead of describing search areas or search volumes by sets of pixels or voxels a more *analytical* description by separating hyperplanes and set operations can be used. For each hypothesis test the extended memory concept permits a fast transformation of attributes in a suitable coordinate system.

Acknowledgements

This project is funded by the Deutsche Forschungsgemeinschaft (DFG) and is carried out in cooperation with the Institut für Photogrammetrie und Fernerkundung, University of Karlsruhe.

References

Aho Av, Ullmann JD (1972) *The theory of parsing, translation, and compiling. Vol I: Parsing.* London: Prentice-Hall Davis WA, Lee DL (1986) *Fast search algorithms for associative memories.* IEEE Transactions on Commputers, C-35: 456-461

Füger H, Stein G, Stilla U (1994) *Multi-populations evolution strategies for structural image analysis.* IEEE Conference on Evolutionary Computation (ICEC'94), Orlando, Vol I, 229-234

Lütjen K (1986) *BPI: Ein Blackboard-basiertes Produktionssystem für die automatische Bildauswertung.* In: Hartmann G (ed) Mustererkennung 1986, 8. DAGM-Symposium. Berlin: Springer, 164-168

McKeown DM, Harvey WA, McDermott (1985) *Rule-based interpretation of aerial imagery.* IEEE PAMI, 7: 570-585

Nagao M, Matsuyama T (1980) *A structural analysis of complex aerial Photographs.* New York: Plenum

Newell A (1962) *Some problems of basic organization in problem-solving programs.* In: Yovits MC, Jacobi GT, Goldstein GD (eds) Proceedings second conference on self-organizing systems, Spartan Books, 393-423

Nicolin B, Gabler R (1987) *A knowledge-based system for the analysis of aerial images.* IEEE Trans. on Geosc. and Rem. Sens., vol. GE-25: 317-329

Nii HP (1986) *Blackboard systems.* AI Magazine, 7: 38-53, 82-106

Stilla U, Jurkiewicz K (1991) *Objektklassifikation mit einem blackboardorientierten Inferenzmechanismus.* Ettlingen: FIM/FGAN, FIM-Bericht Nr. 230

Stilla U, Hajdu A (1994) *Map-aided structural analysis of aerial images.* In: Ebner H, Heipke C, Eder K (eds) Spatial information from digital photogrammetry and computer vision. ISPRS Commission III Symposium, Munich. Bellingham: SPIE, Vol. 30, Part 3/2, 769-775

Stilla U, Quint F, Sties M (1995) *Analyse von Luft- und Satellitenbildern zur automatischen Ermittlung der Bodenversiegelung städtischer Siedlungsbereiche.* Ettlingen: FIM/IPF, DFG-Zwischenbericht II

Velthuijsen H (1992) The nature and applicability of the blackboard architecture. Leidschendam: PTT Research

Semi-Automatic Detection and Extraction of Man-Made Objects in Multispectral Aerial and Satellite Images

Vittala K. Shettigara
Siegfried G. Kempinger
Robert Aitchison

Defence Science and Technology Organisation
PO Box 1500 Salisbury SA 5108 Australia
Ph: 08 259 7176, Fax: 08 259 5619
e-mail: vis@itd.dsto.gov.au

Abstract

Extraction of man-made objects in images is an important problem in computer vision. In this paper a general procedure to extract man-made objects in colour images of terrains using 'Background Discriminant Transformation (BDT) ' is presented. BDT is simple to use, robust and application oriented. The procedure does not require extensive knowledge of the terrain. To illustrate the procedure an example is shown from an airborne colour image. However, the process has been also tested, and reported elsewhere, on spaceborne multispectral images such SPOT images. At this stage the process is semi-automatic.

The procedure is implemented on an Image Processing CAD workstation. The workstation has Image Processing, CAD graphics and GIS modules interacting closely. This integrated system has the capabilities to perform the high-level image processing needed for image interpretation.

1. Introduction

Environmental monitoring, mapping, surveillance and terrain analysis often require extraction of man-made objects in images. Compared to well structured and controlled images that we encounter in industrial computer vision problems, terrain images are significantly more complex. This makes the development of a process for automatic extraction of man-made objects considerably challenging.

Most of the research papers on computer vision deal with monochromatic images. In this paper we deal with colour or multispectral images. Colour images provide simple but effective tools to enhance, segment and extract man-made objects. These stages are vital pre-requisites for high-level image interpretation.

There are many research papers and reports that have endeavoured to separate man-made and natural objects. McKeown (1988) has developed a knowledge based system for detecting and analysing man-made structures in remotely sensed images. Moran and Korvin (1994) have developed a model based approach to detect man-made objects in SAR (Synthetic Aperture Radar) images. Ormsby (1992) has used statistical measures of separation (divergence) between natural and artificial objects to recommend different Landsat TM bands to use for separating different objects. On a similar line Ford and McKeown (1992) have demonstrated the potential of multispectral images for cartographic feature extraction. All of the above procedures require extensive prior knowledge of the area being investigated.

The procedure presented here is simple and more importantly it requires very little prior knowledge of the area. We have implemented the procedures in an Image Processing CAD workstation. In this workstation the Image Processing modules are integrated with CAD graphics and GIS modules. The integrated environment provides capabilities to perform various steps involved in high-level image processing needed for image interpretation.

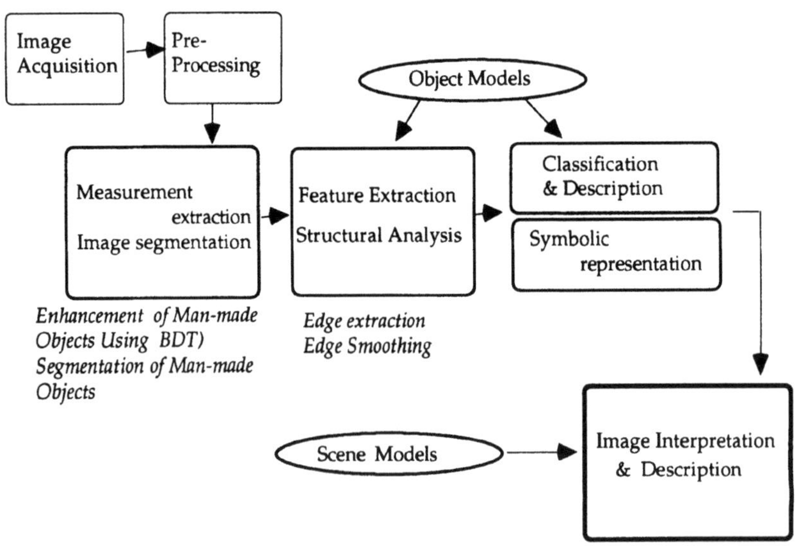

Figure 1: Processing units in a Computer Vision System

Tremendous advances have been made in the last decade in the field of computer vision. However, to a large extent, interpretation of terrestrial images has remained semi-automatic. Figure 1 shows various computational segments involved in automatic image interpretation (adapted from Trivedi, 1990 and Jain, 1989). The areas of contribution of this paper are shown in italics and they will be dealt with below.

2. Enhancement of Man-made Objects in Multispectral Images

The separation of man-made objects from natural objects is achieved using a technique known as Background Discriminant Transformation (BDT; Shettigara, 1991). The terms 'artificial' or 'man-made' refer to objects artificially created using synthetic or natural materials. For example sealed or unsealed roads, asbestos or tile roofed buildings, concrete materials and metallic objects are all artificial objects. Common natural objects we encounter are vegetation, soil and water, at varying percentages of cover.

2.1 Basis of Background Discriminant Transformation (BDT) Technique

The BDT technique is based on the spectral characteristics of objects. As will be shown below, artificial objects are generally separable from natural objects in the spectral domain. As an example, if we consider the spectral bands of SPOT images: green, red, and near infrared, we note the following characteristics. A typical profile of vegetation displays a slight decline from the green to the red band followed by a sharp rise from the red to the near infrared (Figure 2). The sharp rise is due to the chlorophyll pigment in green vegetation. Chlorophyll absorbs the red portion of the spectrum and reflects the green and near-infrared bands in low and high proportions respectively. In contrast, the spectral profile of an artificial object displays a steady decrease in reflectance as the wavelength increases (Wharton 1987). Wheeler (1986) shows a series of profiles contrasting man-made and natural objects in an urban environment. Most man-made objects portray similar profiles but display a distinctly different shape to that of the vegetation profiles. In practice, however, the image pixels are rarely due to pure objects. Backgrounds generally dominate images, making it difficult to enhance artificial objects in relation to natural objects.

Goward and Wharton (1984) found that iron oxides typically absorb the ultraviolet and the visible portion of the spectra. Many artificial objects are iron based and they will display absorption in the shorter wavelengths because of the iron content. The amount of absorption will depend on the proportion of iron oxide in the material. Materials like bitumen display low reflectance through all the Landsat TM bands. These materials are dominated by hydrocarbons which are strong absorbers throughout the TM spectrum (Goward and Wharton, 1984). Thus, they appear dark in all bands.

Crist and Cicone (1984) developed the Tasseled Cap transformation for the Landsat TM images. The Tasseled Cap transformation, in mathematical terms, is a procedure to rotate the data reference axes such that the different physical properties of the terrain are highlighted. The original application of this transformation was for crop monitoring. Crist and Cicone (1984) provided six Tasseled Cap axes for Landsat TM images. Only the first four axes: brightness, greenness, wetness and haze are physically meaningful, and these contain 95% or more of the total data variability. It is interesting to note that in greenness-brightness space, man-made materials consistently stand separate from natural materials. There is no known equivalent to the Tasseled Cap transformation for SPOT images, which are becoming increasingly popular.

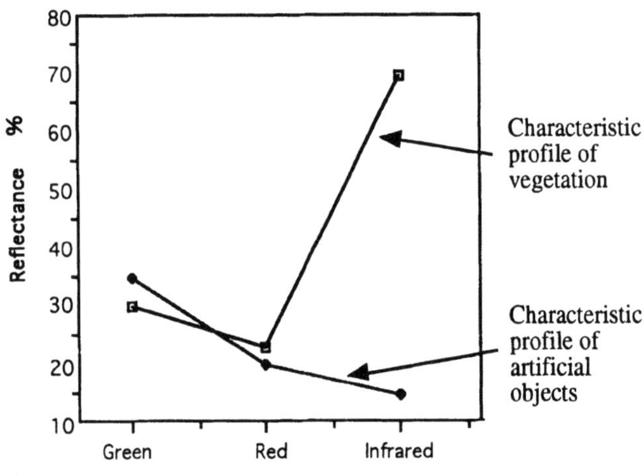

Figure 2: Typical profiles of vegetation and artificial objects

2.2 Background Discriminant Transformation (BDT)

The detection of objects in images is often dependent on the suppression of the background. For example, it is difficult to detect small objects in the presence of background clutter. Numerous filters have been devised to suppress the background in single band images. Boulter (1983) has tested various spatial filters to enhance point targets by reducing background clutter. The Background Discriminant Transformation (BDT) is also a technique to suppress the dominance of the background, but in a multispectral space.

In the BDT technique, the multispectral images are linearly transformed, much like in the principal component transformation. The linear transformation coefficients are computed to maximise the variance (information content) of the non-background objects relative to the background objects (Shettigara, 1991). The technique assumes that an image has two major classes: background and non-background or man-made objects and natural objects.

The method requires the user to identify the background class that needs to be suppressed in order to enhance the useful information. The background class is visually chosen by identifying a few training areas. The method requires mean vectors and covariance matrices of the background class and of the whole image. Our procedure then automatically computes the mean vector and covariance matrix for the non-background class. The theoretical basis of the technique and the algorithm are presented in Shettigara (1991).

The procedure computes the same number of new transformed bands (BDT bands) as the original bands in the multispectral image. In the first band, the ratio of variances of the non-background class and background class is the maximum, indicating the dominance of the man-made objects (non-background) over the natural objects (background). In the last band, the ratio is the minimum, indicating the dominance of the natural objects over man-made objects. In other bands, the ratio of variances will have intermediate values. For example, the second band will have the second best ratio of variances of man-made objects and natural objects.

For visualising an effective separation of man-made objects and natural objects in an image display system the first, the last and the second BDT bands are displayed in red, green and blue colour guns of the monitor. By this arrangement it is possible to *consistently* display man-made objects in reddish/pinkish hues and natural objects in greenish-blue hues. A display of this nature enables the operator to quickly and effectively monitor the man-made activities. Some times it may be necessary to invert one or more bands, depending on the polarity of the transformation vector.

2.3 Advantages of Background Discriminant Transformation

The foremost application of BDT is in enhancing and visualising man-made objects in relation to natural objects. There are many advantages in using BDT compared to other image enhancement procedures such as principal component analysis. In BDT the background is chosen by the user depending on the application. This provides control on the type of enhancement the user requires. The other attraction of the technique is that it is *scale invariant*. This is a very useful property because the gain and offset of the sensor system will not affect the quality of the transformed image. The implication is that images of the same area taken by different sensors or images taken at different times can be processed without significantly affecting the analysis. The same property also enables one to integrate data from dissimilar sensors. For example, SPOT and SAR (synthetic aperture radar) images may be *fused* to enhance man-made objects.

BDT technique is also *robust*. It has been observed that the transformation coefficients derived for one area for a given multispectral image can be used on other similar areas and on similar images. For example, the BDT vectors derived for a SPOT multispectral image of an area in South Australia were used for discriminating man-made objects from an equivalent Landsat TM image (bands 4, 3 and 2) of an area in the northern part of Australia. The results were surprisingly good. The successful adaptation of the transformation coefficients derived for a distinctly different biophysical area indicates the robustness of the technique. The coefficients once derived can be used for a large area, which is a useful property for tasks like monitoring of human activities over a large area and over a period of time. As long as the terrain does not change significantly within that period, like within a season, the transformation coefficients once derived can be used repeatedly.

3. Segmentation of Man-made Objects

After transforming the original colour image using BDT it is displayed as explained above. In the display, the man-made objects appear in pinkish-reddish hue and natural objects display bluish-greenish hue. Using the clear colour separation between man-made and natural objects it is possible to segment the image into a limited number of clusters corresponding to man-made objects and natural objects. For segmentation, standard clustering procedures may be used. For example, in our studies we have used the ISODATA (Hall and Khanna,1977) procedure to cluster the entire image into less than five segments.

In this study, the implementation of ISODATA on an Intergraph ISI-2 image processing system was used. The process is initiated by providing seed values in each distinctly coloured picture areas. In BDT enhanced images the colour variations are limited. This is because BDT assumes the presence of two classes and essentially that is what we see in the enhanced images. Within the major two classes minor variations of colour are seen. Due to this, it is possible to maintain a small number of clusters in an image. The number of classes we choose for man-made objects vary between one and three, depending on the variations of reddish-pinkish hues in BDT enhanced images.

4. Edge Extraction of Man-made Objects

Edges of man-made objects were extracted by mapping the boundaries between the man-made objects and the natural classes. For this purpose a function called RASVEC was used. RASVEC is a part of Intergraph's ISI-2 image processing system. This program locates the edges between various classes derived from the clustering procedure and links them. The edges surrounding each class are put in a separate layer. The edge data are stored as a standard CAD graphics file, and the data can be manipulated like any graphics element. At this level of processing the boundaries of objects follow pixel boundaries and they appear jagged.

After this stage the improvements to object extraction is done manually as follows. Using the CAD tools provided with the ISI-2 image processing system the object boundaries are smoothed using B-splines, arcs of circles or ellipses or simple straight lines. It is possible to reduce complex edges of objects to simple boundaries or symbols as required for the higher level of processing in a computer vision system.

5. Results and Discussion

Extraction of Man-made Objects from Colour Aerial Photographs

In figure 3(a), a part of a scanned aerial photograph of an area north of Adelaide City, Australia, is shown. The original image is a colour image scanned in red, green and blue colours. The pixel size is 3.6m square. The area is an industrial complex. The natural part of the image has open grass fields with varying degrees of exposed soil and some rows of trees. The top right hand corner of the image is a residential area in which houses are often obscured by trees.

The colour aerial image was first enhanced for separating man-made objects from natural objects using BDT. For computing transformation vectors, natural objects such as trees and grass covers were chosen as the background. Using BDT, vectors for new transformed images BDT1, BDT2 and BDT3 were derived. The BDT images were displayed as colour composites of BDT 1, BDT 3 and BDT 2, displayed in red, green and blue respectively. In the BDT image almost all of the man-made objects appear in shades of pink and they stand out clearly from the natural background.

As explained in section 3, the enhanced image was then clustered using the ISODATA procedure. The results of clustering is shown in figure 3b, after applying a 3 by 3 mode filter to the original clustered image. The mode filter was applied to remove "salt and pepper" effects. This has improved the quality of the clusters in the main part of the image, which is covered by the industrial facilities. However, the filter has deteriorated the clusters belonging to houses in the residential area in the upper right hand corner of the image. The point to note is that the mode filter, or any other smoothing filter, has to be applied selectively.

As can be seen in Figure 3b, the process has separated almost all of the man-made objects into one class. This result was achieved without any experimentation with the ISODATA clustering parameters, such as cluster separation or cluster size. One main limitation of our result is that all the man-made objects appear in one cluster and individual objects are not separable as such. We did not concentrate on this problem in the first instance. However, individual components can now be discriminated by clustering only that part of the original image that is separated as man-made objects.

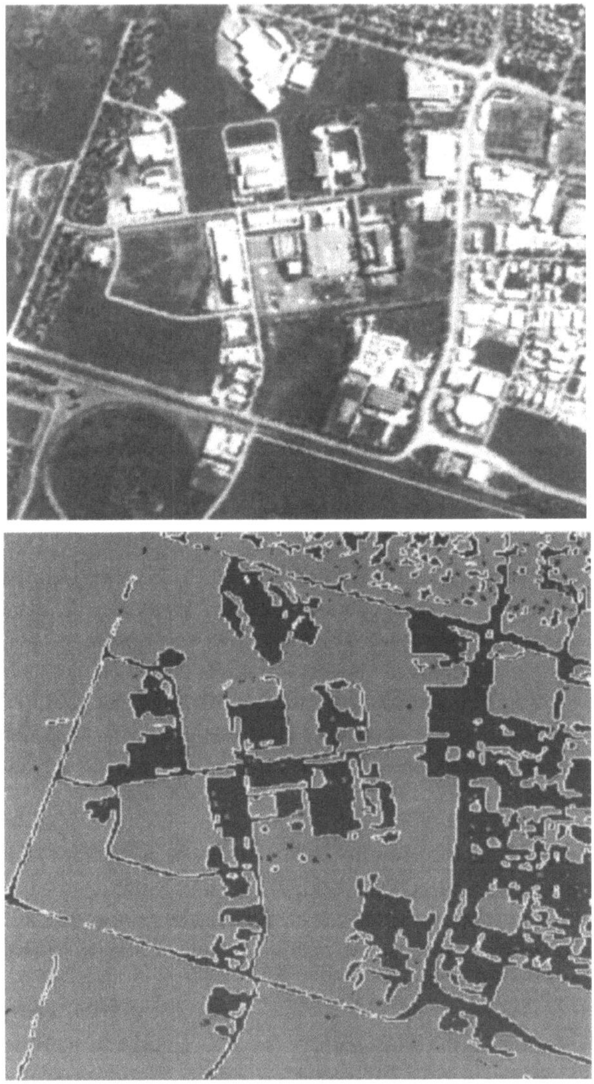

a

b

Figure 3: (a) Red band of the scanned colour aerial photograph; pixel resolution 3.6 m.
(b) Man-made objects, (in dark colour) segmented from BDT enhanced image. White lines are edge-polygons extracted from segmented image.

6. Conclusions

This study has demonstrated a general procedure based on Background Discriminant Transformation (BDT) to extract man-made objects from a colour/multispectral image.

There are theoretical advantages in using the BDT enhancement procedure. The BDT procedure:

- is scale invariant, which enables fusion of different types of data
- is tunable to user requirements by choosing an appropriate background
- requires low level of prior knowledge of the terrain and is simple to use
- is applicable to images with more than three bands
- is robust

Man-made objects can be semi-automatically extracted efficiently after enhancing the images using BDT. It is important to note that the process depends on the spectral separability of man-made objects and natural objects. The detection of man-made objects also depends on the relative dimensions of the objects and the image pixels.

Additional note

The BDT procedure was developed for processing colour (multispectral) images and only colour images convey the potential of the technique fully. For full details and colour illustrations the reader may refer to a report on BDT by the authors (Shettigara and Sumerling, 1994).
Some aspects of this study are the subject of patent applications.

References

Boulter, J.F. (1983) *Spatial Filtering for Enhancing Point Targets in Images from a Space-Based Mosaic IR Detector*, Unclassified No. 4287/83, Research and Development Branch, Department of National Defence Canada.

Crist E.P. and R.C.Cicone (1984) *A physically based transformation of thematic mapper data-the TM Tasseled Cap*, IEEE Transactions on Geoscience and Remote Sensing, GE22, pp. 256-263.

Ford S.J. and D.M.McKeown Jr (1992) *Utilization of Multispectral Imagery for Cartographic Feature Extraction*, Proc. DARPA Image Understanding Workshop, San Diego, Morgan Kaufman Publishers, pp 805-820.

Goward, S.N. and S.W.Wharton (1984) *Use of the TM Tasseled Cap transform for interpretation of spectral contrasts in an urban scene*, 1984 Machine Processing of Remotely Sensed Data Symposium , pp. 84 - 91.

Hall D.J. and D.K.Khanna (1977) *The ISODATA Method Computation for the Relative Perception of Similarities and Differences in Complex and Real Data*, Statistical Methods for Digital Computers (Ed. Enstein K., A.Ralston, H.S.Wilf), Wiley, N.Y., pp340-373.

Jain, A.K. (1989) *Fundamentals of Digital Image Processing*, Prentice Hall, Englewood Cliffs,NJ., p 569.

McKeown Jr, D. M. (1988) *Building knowledge-based systems for detecting man-made structures from remotely sensed imagery*, Philosophical Transactions of the Royal Society of London, 324, pp 423-435.

Moran W. and G. Korvin (1994) *Detection of Man-Made Objects on Synthetic Aperture Radar (SAR) images*, Personal communication.

Ormsby, J. P. (1992) *Evaluation of natural and man-made features using Landsat TM data*, International Journal of Remote Sensing, 13, pp. 303-318.

Shettigara, V.K. and G.M.Sumerling (1994) *Separation of artificial and natural objects and small objects of defence significance in civilian satellite images*, Report ERL-0821-RR, Electronics & Surveillance Research Laboratory, Defence Science and Technology Organisation, Australia, 50p.

Shettigara, V.K. (1991) *Image enhancement using background discriminant transformation*, International Journal of Remote Sensing, 12, pp. 2153-2167.

Trivedi M.M. (1990) *Analysis of High-Resolution Aerial Images, in Image Analysis Applications* (ed. Kasturi, R. and M.M.Trivedi), Marcel Dekker, Inc., NY., pp 281-306.

Wharton, S.W. (1987) *A spectral knowledge based approach for urban land cover discrimination*, IEEE Transactions on Geoscience and Remote Sensing, GE25, pp. 272-282.

Wheeler, D.J. (1986) *Spectral characterisation of urban land covers for Thematic Mapper data*, Symposium on Remote Sensing for Resources Development an Environmental Management, pp. 893 - 898.

Automatic Extraction of Man-Made Objects
from Aerial and Space Images
Monte Verità, © 1995 Birkhäuser Verlag Basel

INFORMATION EXTRACTION FROM DIGITAL IMAGES A KTH APPROACH

Eberhard Gülch

Department of Geodesy and Photogrammetry, KTH
S - 100 44 Stockholm, SWEDEN
Tel: +46-8-790 8925, FAX: +46-8-790 7343, e-mail: ebs@fmi.kth.se

Abstract

The research on Digital Photogrammetry at our Department focuses on automated gathering of cartographic information from multiple digital images. Topics are the description of features and objects to be expected in two-dimensional data. When appropriate, the reconstruction is performed in three-dimensional object space. Objects are buildings in aerial imagery. This approach has potential impact on image coding, image compression and resampling. All major steps of the developed concept are described, experience with alternate approaches are presented and the current state is demonstrated on examples, selected from latest publications of researchers at the Department.

1 INTRODUCTION

Cartographic feature extraction is a major task performed in conventional photogrammetric applications. It is the extraction of man-made objects like roads, buildings, fences, light poles etc., but also vegetation boundaries, forests, fields or hedges. Measurements are performed with highly specialized photogrammetric instruments and a human operator. With the appearance of digitized aerial images and the availability of powerful computers a new research field called Digital Photogrammetry has emerged that has some major fundamental differences compared to conventional photogrammetry: It allows multi-image processing and handling of different data sources and it has the built-in potential for automation of human interpretation and human measurement, i.e. the automated extraction of information from images. At the beginning of research and development in digital photogrammetry major efforts have been put in automating well understood, well defined measuring tasks, like the automated generation of digital elevation models (Gülch, 1994b). Usually algorithmic solutions were sufficient to solve these tasks. Only in recent years research has started in the photogrammetric domain on the automation of more complex measuring tasks, like the extraction of cartographic features. The major challenge here is the automation of the interpretation tasks, currently applied to man-made objects like buildings or roads.

Automated recognition and description of objects present in images would reduce the demand for interaction with an operator to a minimum. The problem of recognition seems to be the more investigated problem of the two. In the field of mapping, the description problem is the more interesting one. To point in an aerial image to identify for instance buildings is a simple task for a human operator. To give a detailed description of all these buildings is extremely time-consuming while at the same time being a well defined task that could be computerized. Highly automated tasks give very few possibilities for user superintendence. Various tests are performed on input and output data. However, properties of data only seldom conform with the requirements of the methods of analysis, which in turn are not infinitely robust. There is a definite need for developing methods for judging the quality of all kinds of image analysis tasks. Only then is it possible to compare different procedures or to use obtained results with any confidence.

A research project on the design of a system for object description and precise localization using logical knowledge on objects and multi-image information has been running at the Division of Photogrammetry during the last years. The project has resulted in an overall concept for description of objects using a procedure that allows quantitative measurements and their evaluation on images which have been identified by a human interpreter. The approach is thus not autonomous, but semi-automated with respect to interpretation, and automated with respect to measurement. The ongoing research work and the results obtained are presented following the major steps of the concept.

2 The Concept For Building Reconstruction

Any two-dimensional data set contains information which is of two kinds, geometrical and semantical. The geometrical information consists of the image having different properties in different regions and is retrieved using a segmentation procedure generating segments with specified borders and internal properties. Any interpretational task, like the reconstruction of buildings involves semantics and must consider a segmentation to find an image representation which is suitable for parsing and then perform the parsing.

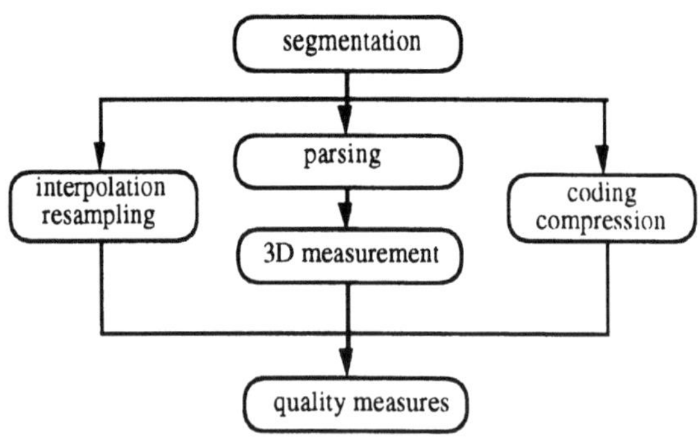

Fig. 1: Strategy for solving different image analysis tasks - A KTH approach. The segmentation is object related and serves different information extraction tasks. They include geometry and semantics for 3-D building reconstruction or just geometry for resampling or compression (figure designed by J. Stokes).

The semantic information consists of 3-D labels for these segments and is obtained using parsers taking the segmentation output as input and using prior knowledge about the possible objects to generate labels. Any task not involving semantics (e.g. resampling or coding) must consider segmentation in order to obtain subsets of data which are homogeneous in the sense that they have known properties.

The topics included for research work within this frame plan (figure 1) are:

- Segmentation of the image to obtain a well defined set of information carriers for future parsing, including a description of the global, regional and local quality of the image as well as the performance of image operators when run on the images used.
- Parsing of the segmentation output is based on expected object types given by a generic object model. Iterative parsing is performed on single and multiple images.
- Given the original images and the output from the parsers the 3-D digital line photogrammetry allows a detailed quantitative description or measurement of the geometry of the objects (point locations, line angles, etc.) including error estimations.
- Generating the geometrical information contained in an image, i.e. the set of segments with their boundaries and internal properties. The set of data describing this information can be used to recover the original image except for noise. This is a basis for resampling, interpolation, coding and compression.
- Quality measures. There is a strong need for a rational procedure for judging analysis and description results in relation to the specified task.

3 IMAGE SEGMENTATION

It is evident from our experience that the requirements on the quality of the segmentation are very demanding (Gülch, 1991, 1992, Stokes, 1992, 1994). Even if combinations of conventional segmentation techniques are used, the result is in our experience not directly suited for parsing. Many methods are of the ad hoc type leaving the user with no information about the quality of the result. Nor do they consider the goal which is to be achieved. Segmentation methods existing today usually accept any kind of boundaries between neighbouring segments. A fundamental requirement which has developed in this project is that boundaries produced by a segmentation should be limited to that subset which is recognizable by the parsers, this in order to make the interpretation process decisive. This means the segmentation should be object related by explicitly using knowledge about the objects in the segmentation step.

3.1 Grouping of Single Features

One way to derive higher level structures from single features is to combine them according to a given object model. In our experience this approach is most suitable either to assist the measurement by the human operator or for tasks with parametric object models, where no complete object description is required. We see great difficulties in using them when having generic object models only.

Edge Focusing and Adjustment of Straight Line Segments. Edge focusing combines good performance in detecting edges with good performance in localizing edges using a coarse to fine approach, requiring only one threshold, that is dependent on the size of objects to be expected in a scene (Gülch, 1991). Advanced adjustment of straight line segments (Förstner, 1991) is applied to the edge elements and their attributes. The process is steered by providing global thresholds for minimal line length and maximal deviation of new line elements from the already adjusted line. This method gives high quality line segments that assist the interactive digital line photogrammetry and that are used for the attempt to solve line correspondence given specific object models.

Combinations of Segmentation Techniques and Choice of Best Model. Several segmentation techniques have been combined for the classification of junction models (Roesler, 1990). The focus of attention on potential junctions is steered by analyzing a combination of different feature extraction methods. With the Minimum Description Length (MDL) criterion a classification into no-model, a corner, a T-junction, a line crossing or a straight line is performed. Edgels not belonging to the model are regarded as outliers. Up to 50% of outliers could be handled. The approximation problem, the choice of models, the huge search space and the problems of combining different segmentation methods with correlated input data are, however, overwhelming. This approach was given up in favour of the area based segmentation described below.

3.2 Object Related Image Segmentation

A first attempt on area based segmentation using established region growing applied to aerial scenes gave many short line segments and too much irrelevant information with no relation to the objects looked for (Gülch, 1992). A further observation was a quite poor boundary representation in terms of geometric quality. If, instead, a representation is chosen where the relation to object types expected in the image is strong, parsing of the representation can be made simpler. An approach on object related segmentation, where the segmentation is adapted to a set of parsers for description of buildings is now being implemented (Wiman, 1995). This procedure produces a region map plus internal descriptions of first order segment trends and textures. It is intended to give segment boundaries consisting of straight lines and smooth curves. The simplest description of the regions and their boundaries is regarded to be the best one and is searched for.

An initial segmentation using region growing segmentation, extended with first degree trends for the segments, is performed. This is to reduce the search space, which would be prohibitively complex while working on a pixel basis. One user defined global threshold for the homogeneity is required. The image should be oversegmented at this stage, as the following steps only merge regions. In contrast to other MDL segmentation approaches, the resulting pixel chain of the boundary between nodes is vectorized by introducing break points where the boundary changes direction. The task is to make the initial vectors longer by removing break points and then to merge regions by removing entire strips and finally to adjust the remaining break points to an optimal position. A MDL measure is used to describe the description length of regions and the boundaries between nodes. Break points are removed in a top-down hierarchical way using strip-trees. A further, not yet implemented step, is the moving of points in a neighbourhood to find an optimal position. After finalizing one iteration of the MDL segmentation the result is fed into a new iteration step, thus triggering new actions and new simplifications.

The result of the initial region growing on an image patch (left part of fig. 2) is shown in figure 2 (middle). The vectorized boundaries are superimposed on the image. The object related image segmentation procedure was applied four times and the right part of fig. 2 gives the resulting vectors and break points superimposed on the image. The major object lines are straightened and a huge amount of insignificant structures have been removed.

Fig. 2: Object related image segmentation.
Left: Original image (256x256). **Middle:** boundaries of the regions after initial segmentation (threshold 10 grey val.) superimposed on the image. **Right:** result of segmentation after four iterations with vectors superimposed on the image.

For a parser to be successful, it is however necessary to go further: objects like buildings have strong relations between boundary parts. Line segments are for instance parallel or orthogonal in pairs. Texture measures will have an assistance role in the segmentation. The same procedure which has been used for choosing appropriate line segments and break points, i.e. the MDL criterion, can be used for incorporating texture and for choosing suitable simple geometrical structures like squares, rectangles, parallelograms, circles, etc., when these are appropriate. Thresholds need yet to be found for balancing these requirements. These extensions are under development.

Impact on Resampling, Image Coding and Compression. In this project a theory for interpolation in digital images has been developed. Interpolation should start with an image segmentation in order to obtain regions with homogeneous properties. Interpolation is then performed using only data within the appropriate segment. A method for interpolation in regular two-dimensional data sets (Stokes, 1994) is now being implemented. Image coding and compression are important tasks due to the enormous quantities of data that occur in Digital Photogrammetry. Coding can be approached in the same way as image interpretation. Viewing an image as a set of segments with explicitly given boundaries and internal properties, the image can be recovered from these data except for a noise component. These data can therefore be considered as an image coding. If the recovered image should be of acceptable quality, the segmentation procedure must be of high quality. It must be capable of generating segment boundaries without degradation and also of taking care of textures of very different kinds.

4 Interpreting Image Segmentations

The major goal of interpretation of image segmentations is the determination of the geometric properties of the object and of the topological relations between parts of the object. We will present our experience with some semi-automated methods that require interaction for the measurement or are dependent on parametric object models. The main stream is an automated object description approach based on generic object models.

4.1 Interactive Object Reconstruction

Active Contour Models. Active contour models, or snakes are suited for semi-automatic methods to extract object contours where the user provides few seed points and the snake attracts to the contour by an energy minimization procedure. A compromise has to be found between the internal forces of the snake and the external forces, the image energies, which steer the behaviour of the snake. The original method has been extended by further image energies of type points and edges in a common framework (Gülch, 1991) and it allows now the introduction of breakpoints with the help of image energies.

Closed snakes have been used to extract roof parts in digital aerial images (Gülch, 1991, 1995a). Approximate values have been provided by a region growing method or by the human operator. Snakes have been applied to measure signalized ground control points in images for digital aerial triangulation or in absolute orientation (Gülch, 1994a). Approximate values are provided by a knowledge based region segmentation and a parametric model of the signal. The snake approach has performed as good as commonly used mask matching methods. Snakes have found to be suitable for the extraction of vegetation boundaries, to update forestry maps from satellite imagery (Gülch, 1995a).

The strategies to reduce the high number of iterations have to be elaborated further. For the extraction of man made objects additional knowledge on the shape of the objects has to incorporated. It has to be studied which type of knowledge is required and available for the different tasks and how it can be translated to geometric constraints in the concept. The snakes are thus regarded as a interactive tool for cases where the automated region segmentation gets into problems. The snakes are otherwise intended to support the interactive measurement in many cartographic feature extraction tasks.

3-D Line Matching by Digital Line Photogrammetry. 3-D digital line photogrammetry is an extension to the classical stereo point photogrammetry. Now homologous lines instead of points are searched for. No point to point correspondence is required and the parameters of 3-D lines are calculated by using information from several images with known orientation. It requires the development of geometric adjustment techniques, where the precise coordinates and line parameters are simultaneously calculated using all given relationships and measurements. Matching is performed by least squares adjustment in object space which allows to introduce 3-D constraints directly in the matching process. Zielinski (1993) uses a four parameter representation of a 3-D line. Constraints on a single 3-D line, like horizontal or vertical or on combination of lines, like parallel or intersection are introduced into the adjustment process. The method has been implemented and tested for simultaneous processing of multiple images and features of type straight line segments. It has been proved to be superior to point photogrammetry. The problem of detecting homologous lines among the extracted lines in the images is, however, unsolved and they have to be identified by the user.

An interactive multistage comparator (fig. 3) based on digital line photogrammetry is currently under development (Gülch, 1995a,b). The lines in the images are extracted automatically. They don't deliver a complete description, but they are very precise, which is sufficient to release the human operator from tedious measurements. The 3-D constraints are interactively given so far, but they could be provided by higher level interpretation. The input to the adjustment is either the 2-D line segments or the originally extracted edgels, both obtained by using the parser output. After the identification of all lines and corners and their mutual geometric relations, digital line photogrammetry is used to calculate the final coordinates and line parameters and give quality estimates.

Fig. 3 Multi-stage comparator based on digital line photogrammetry.
Extracted lines (black overlay) in six images of two strips by edge detection and straight line approximation, The extracted lines don't give a complete description, but they are of high quality and automatically extracted using two thresholds common to all images. The lines can be depicted by human operator and fed into the 3-D line adjustment.

4.2 Semi-automatic Reconstruction with Specific Object Models

Given a specific object model, line correspondences in multiple images can be derived under certain restrictions by clustering procedures, stochastic relaxation and MDL methods (Axelsson, 1994). The extraction of lines in the images is based on the line adjustment procedure described above. The orientation of the cameras is assumed to be known. Object depending restrictions are applied to derive some few main directions of the object, which are corresponding to vanishing points in the image. MDL methods are used to weight different type of information and to find a solution. Lines on the object not following the main directions will be difficult to incorporate in the developed optimization procedure. A further problem is the limited number of relations to be handled so far. This approach is due to the problems encountered intended not to give a final description of an object, but a coarse 3-D model that could help the higher level interpretation.

4.3 Concept for Automated Parsing with Generic Object Models

A concept for an automated description of buildings in multiple aerial images has been developed and is described in detail by (Gülch, 1992, Stokes, 1992). The orientation of the images and the approximate object location are known. The output of an object related image segmentation (cf. above) in each image is interpreted in object space using a series of moderately successful parsers. Those parsers are based on a generic model for buildings and they assign 3-D labels to the segmented object descriptors. Corresponding object descriptors in different images are compared with each other. This part is terminated when a consistency check indicates an acceptable object description. A final precise adjustment of lines in 3-D object space as described above, using the constraints derived from the parsing step yields object coordinates and their quality estimates.

Generic Object Models. For the parsing a generic model for objects is required. This model describes a building in terms of roofs, walls, doors, windows and garden regions, represented as simple geometrical areas and relations between these. The number of the parameters and the parameters are unknown and have to be determined and estimated.

Single Image Interpretation. The principle chosen here for parsing recognizes the fact that parsers, when they are successful, are so only within limited areas. A general parser would not only be bad, it would also be slow. A set of simple parsers, each one effective for a limited set of object types is used. If the result of a parse is unsatisfactory, the next parser is tried until the consistency check reports an approval. The parsing can be performed either using a line segment or a region segment parser. Sets of straight line segments are suitable for describing closed boundaries in a building, some of the lines being vertical and horizontal in object space. The line parsing procedure starts by searching for a vertical line connected to several horizontal line segments and introducing these into object space at an arbitrary datum. Vertical and horizontal line segments connected to already introduced line segments are then successively introduced. Logically weaker rules are applied when no more line segments can be introduced. As soon as a line segment has been introduced into object space, it can be used as a connection for new ones. In the case of a building lying on or close to the nadir point the vertical lines are generally invisible or very short in the segmentation output. In this case a region parser is tried that is based on a small set of aspects, describing composite parts of the roof rather than describing the complete roof in a single aspect. The parsers are applied iteratively, until no further lines can be labelled.

Multi Image Interpretation. By assuming the orientation parameters known and an approximate location in 3-D given, the correspondence problem is regarded marginal as whole structures with 3-D labels are merged from different views. However there might remain contradictions that have to be resolved. From the interpretations in multiple images a consistent object description is derived.

Further work with this set of parsers and the consistency checks has been postponed until a satisfactory object oriented segmentation procedure has been developed.

5 Quality Evaluation Of Information Extraction

The result of the object reconstruction will be a consistent object description plus the precision of coordinates of corner points and lines from digital line photogrammetry. As soon as such methods for description of objects become operational, demands will occur on which properties digital images must fulfil in order that operations be successful. Existing methods for estimating the quality of photographic images are not directly suitable for digital images. They require strong interaction with a user. The quality measures for digital images should rather describe the ability of specified operations to extract certain information in images. A project on variation in subjective image quality judgements in aerial photographs by human operators due to different user goals is described in (Boberg, 1992). The experience obtained is immediately usable when studying related problems for digital images where, however, the final goals are to achieve rational goal dependent measures. Strongly related to quality estimation is the question of how well an operation on a digital image has performed. The answer to this question again depends on the goal with the operation. Also here the kind of information expected to be present in the image should be considered. In order to be adopted for practical applications, it is advantageous if the algorithms involved provide internal measures of quality that can assist human operator validation. Work is going on to realize such traffic light programmes, which signalize the user success, failure or problems.

6 Conclusions

An overall strategy for image analysis, including object oriented segmentation, parsing, measurement and quality judgement has been developed. Prior knowledge on object properties should be used extensively both for the segmentation and the parsing step. Comparative studies on various feature extraction methods have been performed. The requirements on improved segmentation outputs are yet not fulfilled, but the object related segmentation outputs are very promising to the extent that results have been obtained. Generic models for buildings and corresponding parsers are developed, but need to be implemented and extended to other types of objects expected to be present in scenes. Digital line photogrammetry is the natural approach to the measurement procedure once interpretation has been performed. It is currently used as the core in the development of a multistage comparator with automatically measured image lines and interactive selection of correspondence and constraints in 3-D. The concept of snakes has been improved. They are a suitable tool for interactive mapping or editing of outputs from automated processes.

This project is based on information fusion from multiple images and object space adjustment, wherever such data, knowledge, procedures etc. are available and suitable for the objects and tasks involved.

ACKNOWLEDGEMENTS

The project on 'Automatic interpretation in digital images' has been financed by STU TFR and BFR. The projects on 'Development of a method for resampling satellite scenes' and 'Resampling and interpretation of satellite scenes using segmentation' have been financed by the Swedish National Space Board and the Swedish Space Corporation. The work on line photogrammetry and image quality has been financed by KTH.

REFERENCES

Axelsson, P. (1994) *Line correspondence in aerial images using stochastic relaxation and minimum description length criterion,* Proceedings ISPRS Com. III Symposium, Munich, Germany, September, IAP Vol. 30, Part 3/1.

Boberg, A. (1992) *Subjective vs. objective image quality*, Proceedings XVII ISPRS Congress, Washington D.C., USA, August, IAP Vol. 29, Part B1.

Förstner, W. (1991) *Statistische Verfahren für die automatische Bildanalyse und ihre Bewertung bei der Objekterkennung und -vermessung*, Habilitationsschrift, DGK, Reihe C, Heft Nr. 370, München, 1991.

Gülch, E. (1991) *Automatic Extraction of Geometric Features from Digital Imagery*, Ebner/Fritsch/Heipke(Eds), Digital Photogrammetric Systems, Wichmann, Karlsruhe.

Gülch, E. (1992) *A Knowledge Based Approach to Reconstruct Buildings in Digital Aerial Imagery*, Proceedings XVII ISPRS Congress, Washington D.C., USA, August 1992, IAP Vol. 29.

Gülch, E. (1994a) *Using feature extraction to prepare the automated measurement of control points in digital aerial triangulation*, Proceedings ISPRS Comm. III Symposium, Munich, September 1994, IAP Vol. 30-3/1.

Gülch, E. (1994b) *Erzeugung digitaler Geländemodelle durch automatische Bildzuordnung*, Dissertation. DGK-Reihe C, Heft 418, Munich.

Gülch, E. (1995a) *Cartographic Features from Digital Images*, 2nd Course on Digital Photogrammetry, Bonn, Germany, February 6-10.

Gülch, E. (1995b) *Line Photogrammetry - A Tool for Precise Localization of 3D Points and Lines in Auotmated Object Reconstruction*, SPIE'95 Conference on 'Integrating Photogrammetric Techniques with Scene Analysis and Machine Vision II'. Orlando, Florida, April.

Roesler C. (1990) *Comparing Line Models Using The MDL Method,* Report, Department of Geodesy and Photogrammetry, KTH, Stockholm.

Stokes, J. (1992) *Parsing Segmented Images*, Proceedings XVII ISPRS Congress, Washington D.C., USA, IAP Vol. 29, Part B3.

Stokes, J. (1994) *Resampling satellite scenes*, Proceedings ISPRS Comm. III Symposium, Munich, September, IAP Vol. 30-3/2.

Wiman, H. (1995) *Object Related Image Segmentation using the Minimum Description Length Principle*. Accepted for publication in 9SCIA'95, Uppsala.

Zielinski, H. (1993) *Object Reconstruction with Digital Line Photogrammetry,* Dissertation, Department of Geodesy and Photogrammetry, KTH, Stockholm.

Automatic Extraction of Man-Made Objects
from Aerial and Space Images
Monte Verità, © 1995 Birkhäuser Verlag Basel

Linear Feature Extraction with Dynamic Programming and Globally Enforced Least Squares Matching

Armin Gruen   Peggy Agouris   Haihong Li

Institute of Geodesy & Photogrammetry
Swiss Federal Institute of Technology
ETH - Hoenggerberg, CH-8093 Zurich, Switzerland
Ph.: +41-1-633 3038, Fax: +41-1-633 1101
e-mail: {agruen *or* peggy *or* li}@p.igp.ethz.ch

Abstract

In this paper we present a method for *semi-automatic extraction of road networks from SPOT imagery* using wavelet-transformed images and cost functions expressing local gray value variations and global continuity constraints. For larger scale images and for various object types, we propose a global approach for *semi-automatic object outline detection* which uses local gray value variations to precisely identify edge locations. In this new procedure of globally enforced least squares template matching, the mathematical foundation is provided by least squares matching, while global continuity is enforced through the introduction of object-type-dependent shape constraints.

1 Introduction

In general, object extraction from images consists of two main operations:

- ❑ *identification* of an object within an image, which involves image interpretation and object classification, and
- ❑ *tracking* the object by precisely determining its outline.

It is well known that there exist no *universal* edge detectors which could be applied to a digital image function to both identify and track object outlines with sufficient success. Instead, there is a trade-off between qualitative accuracy, associated with identification, and geometric accuracy, associated with tracking. According to these measures, one can still classify existing operators into two broad categories [Fischler et al., 1981]:

- ❑ *type I operators,* offering high accuracy in properly identifying classes of objects without particularly dealing with precise outline determination, and
- ❑ *type II operators,* which do not aim at accurate identification, but offer high precision in detecting outlines when adequate approximations of the object location are available.

In an effort to optimize both measures, operators from these two classes can be combined in complex strategies for object extraction [Suetens et al., 1992].

According to our semi-automatic object extraction philosophy, the identification task of a type I operator is performed manually on a single image, while a special automatic digital module performs the tracking task of a type II operator. Monoplotting, i.e. the use of a single image, is the mode of choice as it provides operational ease, but general 3-D approaches, based on the simultaneous use of two or more images, are also available. More specifically, a human operator is used to identify an object from an on-screen display of a digital image, and to provide a rough approximation of its outline. Typically, this approximation consists of loosely identifying on-screen outline nodes (e.g. corners for houses, breakpoints for curvilinear objects etc.). Subsequently, these points are used as approximations for automatic precise edge positioning. By repeating this process, objects in an image can be identified and precisely positioned. The degree of automation can vary according to the extent of the required human operator contribution (e.g. how many nodes have to be provided for successful object extraction and how close to the actual outline breakpoints they should be). Judging from experience, and considering the established photogrammetric production processes, such use of a human operator within a broader object extraction strategy should be considered optimal, as humans perform the identification task flawlessly and almost effortlessly. Thus, their intervention optimizes accuracies without imposing time burden, while the task of precise object outline positioning and tracking, shown by experience to be the most time-consuming and error-prone part of data capture, is performed automatically in a fast and objective manner.

In the next sections we present the following methods:

- road extraction from SPOT or small scale aerial imagery, using wavelet transformation and dynamic programming, and
- globally enforced least squares template matching for object outline extraction.

2 Road Extraction Using Wavelet-Transformed Images

Our semi-automatic module for road network extraction from SPOT and small scale aerial images combines a wavelet decomposition for road sharpening with a linear feature extraction algorithm based on dynamic programming (Fig. 1). Images are preprocessed with *Wallis filter* for enhancement and to facilitate subsequent object extraction processes by locally forcing the gray value mean and contrast (dynamic range) to fit certain target values [Baltsavias, 1991]. Image preprocessing through Wallis filtering is particularly important for panchromatic SPOT images which have a big contrast globally but most grey values stretch locally over a small range.

After preprocessing, a *wavelet transformation* is applied to the image to emphasize features of interest, while suppressing other details. A wavelet $\Psi(\omega)$ is uniquely defined in frequency domain by specifying its associated filter function $H(\omega)$ [Grossmann & Morlet, 1984]. In spatial domain, the dilation by s of its corresponding wavelet $\psi(x)$

$$\psi_s(x) = s\psi(sx) \tag{1}$$

is used for the wavelet transformation of the image by applying it along each image coordinate direction [Mallat, 1989]. The wavelet $\psi(x)$ can be interpreted as the impulse response of a band-pass filter. Therefore, a wavelet transformation is a convolution with

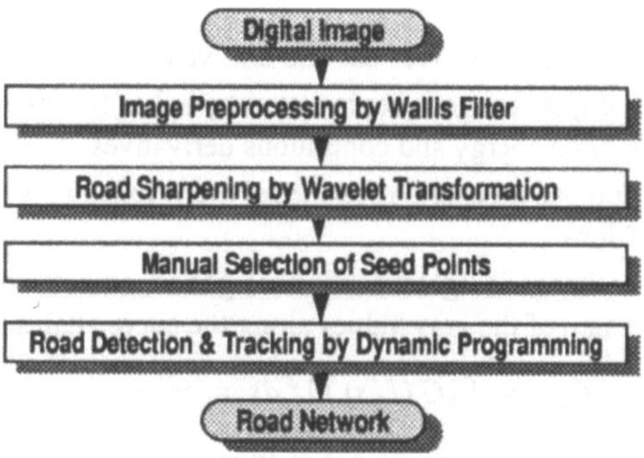

Fig. 1. Semi-automatic road extraction

a dilated band-pass filter. The scale parameter s defines the size of the features to be enhanced by this transformation. When scale decreases, finer details are emphasized. Various types of $\psi(x)$ (or correspondingly $\Psi(\omega)$) can be used, according to the characteristics of the sought features. In our experiments, road networks were enhanced using

$$H(\omega) = \left[\cos\left(\frac{\omega}{2}\right)\right]^3 \quad \text{and} \quad \Psi(\omega) = \omega^2\left[\sin\left(\frac{\omega}{4}\right)/\left(\frac{\omega}{4}\right)\right]^5 . \qquad (2)$$

The wavelet transformation of an image generates a new image, in which road networks are more prominent (compare Figures 2 *left* and 2 *right*). These networks are then precisely identified through *dynamic programming* [Ballard & Brown, 1982]. Its application to road extraction involves the definition of a function which embodies the notion of "best road segment", and the solution of an optimization problem to extract the actual road. By evaluating this merit function for possible road paths, one selects the sequence of eight-connected pixels which maximizes it. The disadvantages of this approach can be synopsized as being time consuming due to pixel-by-pixel tracking, having difficulties in bridging gaps, and being sensitive to noise and appropriate only for low curvature features.

To overcome these disadvantages, a road can be extracted as a set of n segments defined by the seed points $P^{\circ}_0, P^{\circ}_1, \ldots, P^{\circ}_n$ which are manually provided on-screen. The merit function E is then expressed as the summation of smaller terms

$$E\left(P_0, P_1, \ldots, P_n\right) = E_0\left(P_0, P_1\right) + E_1\left(P_1, P_2\right) + \ldots + E_{n-1}\left(P_{n-1}, P_n\right) . \qquad (3)$$

The original seed positions can be altered during the solution. A preset value δ defines the maximum allowable change in the position of seed points ($\left|P_i - P^{\circ}_i\right| = \left|V_i\right| \le \delta$) and it is used to eliminate tracking errors associated with misleading local maxima. Each term $E_i\left(P_i, P_{i+1}\right)$ expresses how well an image path connecting points P_i and P_{i+1} satisfies the merit function.

In our approach to road extraction, and drawing from experience, we make the following assumptions for a curve ζ corresponding to a road outline on a digital image:

- Curve ζ can be represented by a vector function $f(s)$, which maps the arc-length s to points (x, y) in the image;
- Curve ζ has continuous derivatives and the unit vector $n(s)$ is normal to the curve.
- The preprocessed digital image is represented by a 2-D function $G(x, y)$, which is measurable and has finite energy and continuous derivatives.

Based on these assumptions, a generic road model merit function is formulated through the following observations:

- A road pixel in the image is lighter than its neighbors on both road sides. This suggests that a squared sum of the gray values along the curve attains a maximum. i.e.

$$E_{p1} = \int [G(f(s))]^2 ds = max \quad . \tag{4}$$

- Gray values along a road usually do not change much within a short distance. This stems from the knowledge that road materials do not vary much and their spectral properties remain similar within a short distance. Thus,

$$E_{p2} = \sum \int_{\Delta s_i} \left[G(f(s)) - G_m\left(\Delta s_i\right)\right]^2 ds = min \tag{5}$$

where Δs_i is a short segment of the curve ζ, and $G_m\left(\Delta s_i\right)$ is the average value of G over Δs_i:

$$G_m\left(\Delta s_i\right) = \int_{\Delta s_i} G(f(s))\, ds / \left|\Delta s_i\right| \quad . \tag{6}$$

- A road is a light linear feature. In fact, this is a generalization of the two previous properties, but formulated in a more global manner. This property can be considered through the appropriate evaluation of the formula

$$E_{p_3} = \int w(d(s))\, [G(f(s) + d(s)\, n(s))]^2 ds = max \tag{7}$$

where $d(s)$ is the distance between curve ζ and the linear feature near it, and $w(d(s))$ is a Gaussian weight function decreasing as the distance $d(s)$ increases.

- In terms of basic geometric behavior, a road is usually smooth and does not have small wiggles. In fact, most roads consist of straight lines connected with smooth curves, normally circular arcs. This property can be mathematically represented as

$$E_g = \int |f''(s)|^2 ds = min \tag{8}$$

which implies that the curve ζ can be represented as a cubic spline.

- The local curvature of a road, and thus its local change, have an upper bound. This property, derived from the typical smooth traffic flow requirements, is expressed as

$$C_g = |f''(s)| < T_1 \tag{9}$$

with T_1 a given threshold.

❑ The width of a road does not change significantly. This is kept for practical reasons. We don't give a mathematical representation for this property here because it will not be used explicitly in our implementation, but is involved to define a linear feature in the third of the above properties.

The six properties given above, three in radiometric and three in geometric terms, form the generic road model merit function involved in our road extraction scheme. More specifically, the radiometric properties can be discretized as

$$E = \sum E\left(p_i, p_{i+1}\right) \tag{10}$$

and for the geometric ones we have

$$E_g = \sum_i [2 - 2\cos\left(\alpha_i - \alpha_{i+1}\right)]/|\Delta s_i| \tag{11}$$

$$C_g = |\alpha_i - \alpha_{i+1}| < T_1 \tag{12}$$

where α_i is the direction of vector between points p_{i-1} and p_i, and $|\Delta s_i|$ is the distance between them, i.e.

$$\alpha_i = atan\left(\frac{y_i - y_{i-1}}{x_i - x_{i-1}}\right) \quad \text{and} \quad |\Delta s_i| = \sqrt{\left(x_i - x_{i-1}\right)^2 + \left(y_i - y_{i-1}\right)^2} \; . \tag{13}$$

The generic road model can be formulated through the following merit function

$$E=\sum\left(1 + \cos\left(\alpha_i - \alpha_{i+1}\right)\right) [E_{p1}\left(p_i, p_{i+1}\right) - \beta E_{p2}\left(p_i, p_{i+1}\right) + \gamma E_{p3}\left(p_i, p_{i+1}\right)]/|\Delta s_i| \tag{14}$$

and an inequality constraint

$$C_i = |\alpha_i - \alpha_{i+1}| < T_1 \tag{15}$$

where β and γ are two positive constants and T_1 is a threshold for direction change between two adjacent vectors. This multistage optimization problem is solved by dynamic programming [Gruen & Li, 1994]. Its advantages lie on its efficiency in handling long curves, robustness in the presence of noise, and ability to bridge gaps.

The previously described semi-automatic road extraction strategy from wavelet-transformed SPOT imagery has been successfully implemented on a digital photogrammetric station. Fig. 2 and 3 show a typical example of obtained results. Fig. 2 (*left*) depicts a portion of a SPOT panchromatic image of Moudon, Switzerland. The seed points provided by the operator along the road to be extracted are marked on it by black discs. The white rectangles denote potential problematic areas, i.e. intersections with other roads which might cause an erroneous change of direction, or ambiguities due to passage through an urban area. Fig. 2 (*right*) shows the wavelet transformed image. The transformation has made the existing road network appear more prominently, thus facilitating the subsequent automatic extraction of the road of interest, the result of which is shown in Fig. 3. It can be easily observed that no problems were finally encountered in the extraction despite the presence of the aforementioned problematic areas. Additional experiments with other types of potential problems (e.g. gaps, low texture) verified the efficiency and effectiveness of the technique [Gruen & Li, 1994].

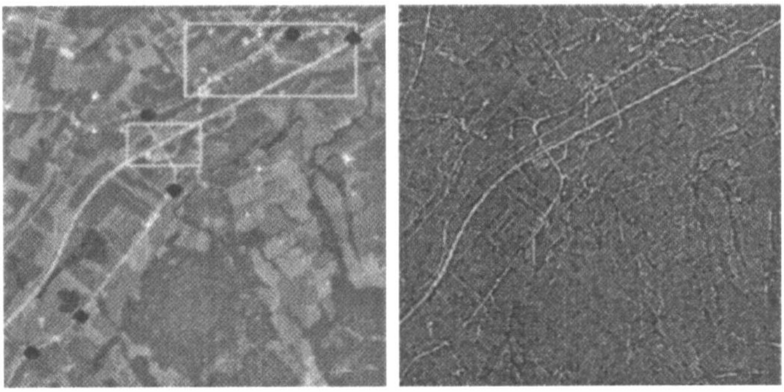

Fig. 2. Portion of a SPOT panchromatic image, with marked seed points and potential problematic areas (*left*), and its corresponding wavelet transformation (*right*)

Fig. 3. The semi-automatically extracted road

3 Least-Squares-Based Semi-Automatic Object Outline Detection

3.1 Least Squares Template Matching

This modified matching method, developed in our Institute and used for linear feature detection and tracking, is based on *least squares matching* [Gruen, 1985]. A real or synthetic linear feature pattern (e.g. edge, ribbon, corner) is introduced as the reference template and is subsequently matched with image patches containing actual feature segments. Assuming $f(x,y)$ to be the feature template and $g(x,y)$ to be the actual image patch, observation equations are formed between conjugate pixels as

$$f(x,y) - g(x,y) = e(x,y) \quad . \tag{16}$$

By relating template and image patch through an affine transformation

$$x_i = a_{11} + a_{12}x + a_{21}y \qquad \text{and} \qquad y_i = b_{11} + b_{12}x + b_{21}y \tag{17}$$

and linearizing with respect to the affine transformation parameters, the observation equations for all involved pixels can be written in matrix form as

$$-e = Ax - l \quad ; \quad P \tag{18}$$

where l is the observation vector containing gray value differences of conjugate pixels, x is the vector of unknowns consisting of the affine transformation parameters, A is the associated design matrix including derivatives of the observation equations with respect to the unknowns and P is the corresponding weight matrix. The matching solution is then obtained by minimizing the squared sum of gray value differences as

$$x = (A^T P A)^{-1} A^T P l \tag{19}$$

and a new position of the image window is determined as the template conjugate through the updated affine transformation parameters.

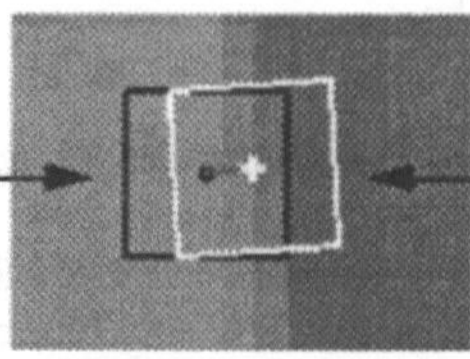

Fig. 4. Visualization of least squares template matching

However, due to the particular gray value distribution of the feature patches, a full set of affine transformation parameters cannot be obtained [Gruen & Stallmann, 1993]. Therefore, in case of an edge feature, only two shift (Δx, Δy) and one rotation (φ) parameter are employed to relate template and image patch (Figure 4). Since the edges to be measured are essentially uni-directional, a linear template edge would perpetually slide along the edge during matching. This effect is compensated by restricting the shift vector of the patch perpendicular to the local edge direction as

$$\Delta x \sin\theta + \Delta y \cos\theta = 0 \tag{20}$$

where θ is the angle formed between the edge normal and the y direction. In matrix form, this condition is expressed as a weighted constraint

$$-e_p = B_p x + t_p \quad ; \quad P_p \tag{21}$$

The least squares solution for the joint system of equations (18) and (21) is

$$\hat{x} = \left(A_I^T P A_I + B_p^T P_p B_p\right)^{-1} \left(A_I^T P l - B_p^T P_p t_p\right) = N^{-1} u \ . \tag{22}$$

Matrix A_I is analogous to matrix A of equation (18) but refers to the reduced set of unknown transformation parameters which is solved for by equation (22). Iterations are required due to the non-linearity of the mathematical model. From the final updated transformation parameters, an edge point is precisely located in the image as the conjugate of the prespecified template edge position.

This procedure has been extended into an edge tracking technique to automatically extract edge segments [Gruen & Stallmann, 1993]. The user gives an approximate position for the first edge point and then the algorithm precisely positions this point and tracks the edge. An approximate position for the next patch is determined using the previous matching result, the local edge direction and a user-defined incremental distance. Edge tracking stops either after the measurement of a prespecified number of points, or if matching fails when a conjugate window of the template can no longer be found in the image (e.g. the end of an edge segment is reached). Patch sizes typically range between *5x5* and *9x9* pixels. Smaller patch sizes result in greater sensitivity to noise, while larger sizes increase the probability of distracting interferences by features close to the current edge.

The great advantage of this method is that it offers very high positioning accuracies. In addition, its well established mathematical formulation allows statistical analysis of the results and realistic evaluation of its performance. At the same time though, it is a localized process and as such it is very sensitive to edge distortions, failing in the presence of edge gaps and outline breakpoints.

3.2 Globally Enforced Least Squares Template Matching

The limitations of least squares template matching, attributed to the use of highly localized radiometric information, can be overcome by its extension into a *globally enforced least squares template matching* strategy (Fig. 5).

An operator initially selects the feature class to which the object to be extracted belongs (e.g. house, road, land parcel etc.) and provides manually on-screen approximations of object outline breakpoints, which in essence define a polygonic approximation of the edge contour. By interpolating between the polygonic approximation nodes at user-specified equal intervals, numerous points are extracted, roughly outlining the current object of interest. These points are then used as approximations for the subsequent precise edge positioning through globally enforced least squares template matching. In case of an edge feature, the matching template is automatically generated as the 2-D ramp edge which best fits image profiles extracted perpendicular to the polygonic approximation at the interpolated positions. The matching solution, however, will no longer proceed independently for the various points of the same outline. Instead, the mathematical model presented in section 3.1 is extended to provide a simultaneous global matching solution, whereby globality refers to all matching candidates pertaining to a single object. Individual edge positions (denoted by the subscript i), detected by least squares template matching along a single object outline, will have to fulfill a certain geometric condition, describing the acceptable outline geometry. For example, it is typical for houses to have edges formed by straight linear segments, while smooth splines are suitable for roads. By defining such a geometric relationship, we are able to tie together the matching adjustment solution of independent points.

Linearity. Considering straight linear edges, and using the notation of the previous section, it is obvious that edge points along the same line should satisfy the condition

$$y_i - x_i tan\varphi_i = 0 \tag{23}$$

Through linearization equation (23) becomes

$$\left(y_i^o - x_i^o \tan\varphi_i^o\right) + \Delta y_i - \tan\varphi_i \Delta x_i - x_i \sec^2\varphi_i d\varphi_i = 0 \tag{24}$$

which in matrix form is expressed as

$$-e_l = B_l x + t_l \quad ; \quad P_l \; . \tag{25}$$

Vector x includes the transformation parameters of all matching templates referring to different locations along the same linear edge segment. The associated weights can vary

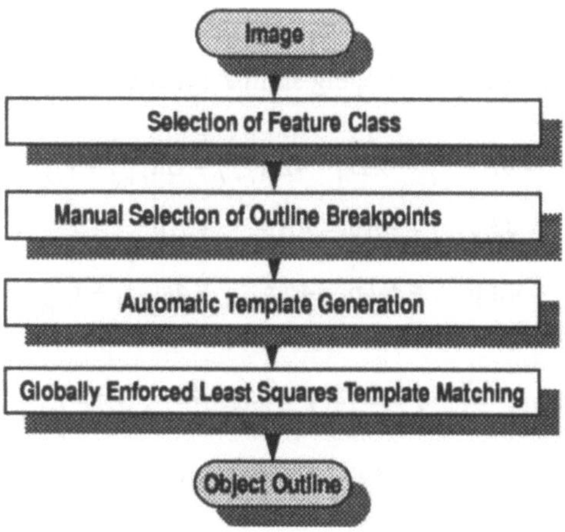

Fig. 5. Globally enforced least squares template matching strategy

from zero (i.e. constraint virtually removed) to infinity (i.e. constraint strictly enforced). By adding these constraints, the final solution is obtained in analogy to equation (22) as

$$\hat{x} = \left(A_l^T P A_l + B_p^T P_p B_p + B_l^T P_l B_l\right)^{-1}\left(A_l^T P l - B_p^T P_p t_p - B_l^T P_l t_l\right) \tag{26}$$

The above matrices include information from all involved templates, and so, matrices in equation (22) are actually submatrices of their counterparts in equation (26). Thus, the solutions of individual templates are tied together to preserve the linearity constraint.

Continuity of Derivatives. For curvilinear features (e.g. roads), spline-influenced principles are employed to ensure piecewise continuity and smoothness through the enforcement of derivative continuity. A discrete approximation of this constraint pertaining to the *continuity of first derivatives* can be given as

$$2\Delta x_i - \Delta x_{i-1} - \Delta x_{i+1} + t_{x_i} = 0 \tag{27}$$

$$2\Delta y_i - \Delta y_{i-1} - \Delta y_{i+1} + t_{y_i} = 0 \tag{28}$$

$$2d\varphi_i - d\varphi_{i-1} - d\varphi_{i+1} + t_{\varphi_i} = 0 \tag{29}$$

where

$$t_{x_i} = 2x_i^o - x_{i-1}^o - x_{i+1}^o \quad , \quad t_{y_i} = 2y_i^o - y_{i-1}^o - y_{i+1}^o \quad , \quad t_{\varphi_i} = 2\varphi_i^o - \varphi_{i-1}^o - \varphi_{i+1}^o \quad . \qquad (30)$$

Such equations relate the transformation parameters of each point i to the corresponding parameters of its previous (i-1) and its following (i+1) point along the same outline, leading to a tridiagonal sparsity pattern in the normal equation matrix.

In addition to the continuity of first derivatives, the enforcement of *second derivative continuity* is used to better approximate the behavior of cubic splines and is achieved through

$$2(2x_i - x_{i-1} - x_{i+1}) - (2x_{i+1} - x_i - x_{i+2}) - (2x_{i-1} - x_{i-2} - x_i) = 0 \ . \qquad (31)$$

By formulating and properly developing similar equations for y and φ, we produce the condition equations

$$\Delta x_{i-2} - 4\Delta x_{i-1} + 6\Delta x_i - 4\Delta x_{i+1} + \Delta x_{i+2} + t_{xx_i} = 0 \qquad (32)$$

$$\Delta y_{i-2} - 4\Delta y_{i-1} + 6\Delta y_i - 4\Delta y_{i+1} + \Delta y_{i+2} + t_{yy_i} = 0 \qquad (33)$$

$$d\varphi_{i-2} - 4d\varphi_{i-1} + 6d\varphi_i - 4d\varphi_{i+1} + d\varphi_{i+2} + t_{\varphi\varphi_i} = 0 \qquad (34)$$

where

$$t_{xx_i} = x_{i-2} - 4x_{i-1} + 6x_i - 4x_{i+1} + x_{i+2} \qquad (35)$$

$$t_{yy_i} = y_{i-2} - 4y_{i-1} + 6y_i - 4y_{i+1} + y_{i+2} \qquad (36)$$

$$t_{\varphi\varphi_i} = \varphi_{i-2} - 4\varphi_{i-1} + 6\varphi_i - 4\varphi_{i+1} + \varphi_{i+2} \qquad (37)$$

These equations relate the transformation parameters of each point i to the corresponding parameters of its two previous (i-1, i-2) and two following (i+1, i+2) points.

The first and second derivative continuity conditions can be introduced as geometric constraints in the global adjustment in lieu of equation (25), grouped in matrix form as

$$-e_d = B_d x + t_d \quad ; \quad P_d \qquad (38)$$

and the global solution is given by

$$\hat{x} = \left(A_1^T P A_1 + B_p^T P_p B_p + B_d^T P_d B_d\right)^{-1}\left(A_1^T P l - B_p^T P_p t_p - B_d^T P_d t_d\right) \qquad (39)$$

The resulting normal matrix will be pentadiagonally sparse, as shown in Fig. 6.

Results from the implementation of the described novel approach are shown in Fig. 7, which includes portions of an orthophoto image depicting the area around the ETH campus on Hoenggerberg. The black dots denote the locations of the outline points detected through this procedure and in the left part of Fig. 7 define a road border segment, while in the right part belong to a house edge. The infused globality allows the technique to both retain its precision in detecting the object outline and at the same time preserve the primary shape character of the outline through the guidance of the object-type-dependent shape constraints, avoiding distractions possibly due to diversions (e.g. in Fig. 7 *right*, the white rectangle encloses a possible distraction source area).

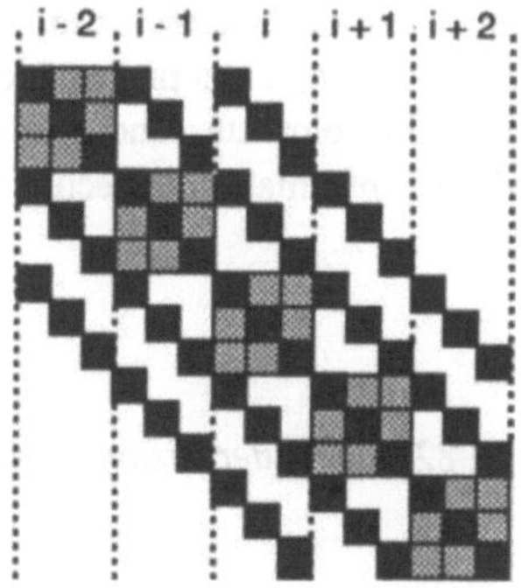

Fig. 6. Sparsity pattern of part of the normal matrix in globally enforced least squares template matching

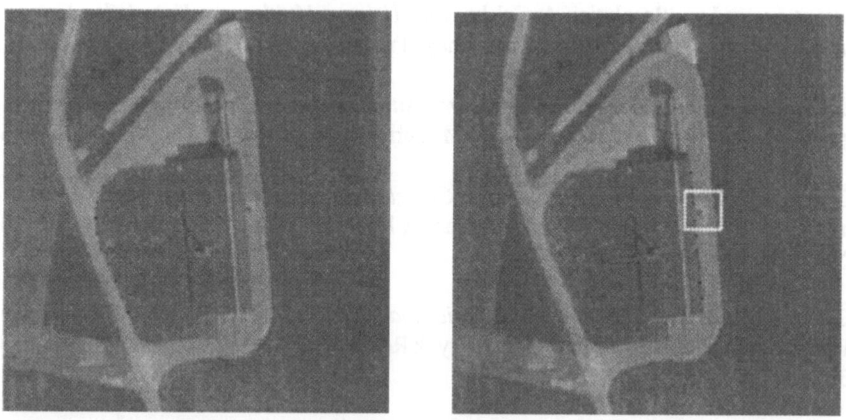

Fig. 7. Outline detection through globally enforced least squares template matching: a road border segment (curvilinear constraint) [*left*] and a house edge (linear constraint) [*right*]

4 Concluding Remarks

The method of dynamic programming is fully operational for road extraction from satellite and small scale aerial images with very satisfactory performance, while the method of globally enforced least squares template matching is still under development and at the early stages of testing. The potential advantage of the latter method lies on its better adaptability to a broad and diverse spectrum of image features through the possible choice of even fairly complex templates. Furthermore, through the use of a least squares estimation model, internal quality control is available.

In general, the semi-automatic object extraction algorithms presented in this paper share a common objective which is to relieve the human operator from cumbersome and error-prone precise positioning tasks, while at the same time employ human decisions for semantic reasoning, a task currently extremely difficult to automate with satisfactory results. They are object-dependent and complementing each other, promising a significant contribution towards the devision of a complete, functional and reliable automated qualitative and quantitative image-based information collection strategy.

References

Ballard D.H. & C.M. Brown (1982) *Computer Vision*, Prentice Hall, Englewood Cliffs, NJ.

Baltsavias E.P. (1991) *Multiphoto Geometrically Constrained Matching*, Report Nr. 49, Institute of Geodesy & Photogrammetry, ETH Zurich, Switzerland.

Fischler M.A., J.M. Tenenbaum & H.C. Wolf (1981) *Detection of Roads and Linear Structures in Low-Resolution Aerial Imagery Using a Multisource Knowledge Integration Technique*, Computer Graphics & Image Processing, Vol. 15, No. 3, pp. 201-223.

Grossmann A. & J. Morlet (1984) *Decomposition of Hardy Functions into Square Integrable Wavelets of Constant Shape*, SIAM Journal of Mathematics, Vol. 15, pp. 723-736.

Gruen A. (1985) *Adaptive Least Squares Correlation: A Powerful Image Matching Technique*, S. African Journal of Photogrammetry, Remote Sensing & Cartography, Vol. 14, No. 3, pp. 175-187.

Gruen A. & H. Li (1994) *Semi-Automatic Road Extraction by Dynamic Programming*, International Archives of Photogrammetry & Remote Sensing, Vol. 30, Part 3/1, pp. 324-332.

Gruen A. & D. Stallmann (1993) *High Accuracy Edge Matching with an Extension of the MPGC-Matching Algorithm*, in: "Applications of Geodesy to Engineering", IAG Symposia, Vol. 108, Springer Verlag, pp. 339-350.

Mallat S.G. (1989) *A Theory of Multiresolution Signal Decomposition: The Wavelet Representation*, IEEE-PAMI, Vol. 11, No. 7, pp. 674-693.

Suetens P., P. Fua & A.J. Hanson (1992) *Computational Strategies for Object Recognition*, ACM Computing Surveys, Vol. 24, No. 1, pp. 5-61.

Automatic Extraction of Man-Made Objects
from Aerial and Space Images
Monte Verità, © 1995 Birkhäuser Verlag Basel

Semi-Automatic Feature Extraction by Snakes

John C. Trinder

School of Geomatic Engineering
The University of New South Wales
Sydney NSW 2052
Ph: +612 385 4197, Fax: +612 313 7493
e-mail:j.trinder@unsw.edu.au

and

Haihong Li
Institute of Geodesy & Photogrammetry
Swiss Federal Institute of Technology
ETH - Hoenggerberg, CH-8093, Zurich, Switzerland
Ph: +41 1 633 3723, Fax: +41 1 633 1101
e-mail: li@p.igp.ethz.ch

Abstract

Implementation of a semi-automatic method of feature extraction, based on the active contour method or 'snakes', will be described and results obtained from the procedure demonstrated. The paper will define the method of snakes for 2 dimensional features on single images, and 3 dimensional features on two or more images. The results of tests on SPOT satellite images are described in terms of the accuracy of the extracted features and pull-in range, for a range of features, for both 2 dimensional and 3 dimensional geometry.

1. Introduction

This paper describes the principles and procedures for the implementation of a semi-automatic method for the extraction of linear features on remotely sensed images in 2D and 3D, based on active contour models or 'snakes'. Snakes are a method of interpolation by regular curves to represent linear features on images. The method requires the initial extraction of the locations of elements of linear features in a remotely sensed image, by appropriate image processing methods. In this paper, the procedures that have been implemented for the initial feature extraction are the Canny operator for single edges, and morphological tools for narrow linear features of 1 or 2 pixels in width. The procedure developed is semi-automatic, and hence is assisted by an operator to locate a selection of points along and near, but not necessarily exactly on the feature. The iterative computation then locates the feature as closely as the details in the image will allow. The paper will include: an explanation of the theory of snakes; the method of implementation of the snakes by B-splines in 2 dimensions (2D) and 3 dimensions (3D); tests of these procedures and conclusions.

2 Curve Fitting Algorithm by Snakes

2.1 Principles of Snakes

The implementation of feature delineation by snakes for satellite images in this paper, using interactive methods to determine approximate locations, is based on a framework of energy minimisation (Kass et al 1988). Fua and Leclerc (1990) have stated that the snakes method has two advantages: geometric constraints are directly used to guide the search for the edge; and the edge information is integrated along the whole length of the curve. That is, the method determines the structure of the feature. Fua and Leclerc have developed variations to the method of Kass et al.

Snakes derive their name from the fact that they are splines reacting with an image, which 'slither' around in position during the iterations to determine the feature. The curves used to model linear features in this paper are B-splines. The advantages of the splines are that they are smooth piecewise polynomials which maintain continuity between neighbouring sections or domains. In addition, they have the characteristic that they limit the influence of errors in the points used to locate the curve of order 'n', to 'n+1' domains of the curve. The definition of the B-splines is based on the knowledge of both a set of polynomial functions, which define the shape of the function, and a given set of control points or a control polygon, defined by a number of fixed locations along the curve. The B-spline function adopted is a 3rd order function, while the control polygon is determined from the image details representing the feature. The spline is subject to internal forces, referred to as internal energy, brought about by the inherent piecewise smoothness of the B-spline, while the external forces which relate to the image information, force the snake towards the feature.

The key concept of the algorithm is to minimize the total energy which relates to the snake's position. Total energy is the sum of the two kinds of energies : internal and external.

The total energy of snakes is expressed by a parametric representation $v(s) = (x(s), y(s))$, as,

$$\begin{aligned} E &= \int_{s_0}^{s_1} E(v(s))ds \\ &= \int_{s_0}^{s_1} [E_g(v(s)) + E_p(v(s)) + E_c(v(s), v_0(s))]ds \end{aligned} \tag{1}$$

where the internal or geometric energy Eg is derived from the geometric constraints of the object model. The photometric or external energy Ep represents the forces derived from the image which constrain the snake to the feature of interest. The control energy Ec defines the boundary conditions, which constrain the difference between the contour $v(s)$ and the initial curve $v_o(s)$

Normally, the geometric energy Eg is based on the first derivative (v_s) and second derivatives (v_{ss}) of the curve, which constrain the snake to be a smooth curve. It is defined by,

$$E_g = \alpha |v_s(s)|^2 + \beta |v_{ss}(s)|^2 \tag{2}$$

where α and β are constants which control the influence of the geometric energy against the photometric energy.

The photometric energy Ep, derived from the image, depends on the type of features to be extracted. For a narrow linear feature, it can simply be the square of intensity values of the image, multiplied with a positive or negative constant for lighter or darker features respectively. ie,

$$E_{pl} = w|I(x,y)|^2 \tag{3}$$

For a step edge, it can be calculated as,

$$E_{pe} = -|\nabla I(x,y)|^2 \tag{4}$$

thus attracting the snake to image points with high gradient values. In this implementation, the features in the image are defined by morphological tools for narrow features and the Canny operator for step functions. The 'energy image' is defined by a Chamfer image, derived from the feature image, in which the pixel values relate to their closeness to any surrounding edge.

Letting,

$$E_{ext} = E_p + E_c \tag{5}$$

(1) becomes,

$$\begin{aligned} E &= \int_{s_0}^{s_1} [\alpha|v_s(s)|^2 + \beta|v_{ss}(s)|^2 + E_{ext}(v(s))]ds \\ &= \int_{s_0}^{s_1} F(s, v, v_s, v_{ss})ds \end{aligned} \tag{6}$$

To minimise this energy the Euler-Lagrange necessary condition is derived as,

$$F_v - \frac{\partial}{\partial s}F_{v_s} + \frac{\partial^2}{\partial s^2}F_{v_{ss}} = 0 \tag{7}$$

Substituting the terms in the above equation, we obtain a pair of independent equations for x and y respectively,

$$-\alpha x_{ss} + \beta x_{ssss} + \frac{\partial}{\partial x}E_{ext} = 0 \tag{8a}$$

$$-\alpha y_{ss} + \beta y_{ssss} + \frac{\partial}{\partial y}E_{ext} = 0 \tag{8b}$$

where x_{ssss} and y_{ssss} are *4th* derivatives.

In the actual implementation, the curves are described as polygons with a set of vertices, defined by,

$$v(s_i) = (x(s_i), y(s_i)) \qquad i = 1, 2, \ \ldots, \ n \tag{9}$$

Equation system (8) for the snakes must be solved for these vertices.

Letting,

$$X = [x(s_i)] \tag{10a}$$

$$Y = [y(s_i)] \tag{10b}$$

(8a) and (8b) can be written in matrix form as,

$$AX + L_x(X, Y) = 0 \tag{11a}$$

$$AY + L_y(X, Y) = 0 \tag{11b}$$

where

$$L_x(X,Y) = \left[\frac{\partial}{\partial x} E_{ext} \Big|_{x_i, y_i} \right]$$

$$L_y(X,Y) = \left[\frac{\partial}{\partial y} E_{ext} \Big|_{x_i, y_i} \right]$$

$$A = -\alpha D_2 + \beta D_4$$

and D_i is the *ith* difference operator.

2. 2 Approximation by a cubic spline

Suppose a snake can be approximated by a cubic spline. That is,

$$x(s) = \sum_{i=1}^{n} N_i(s) X_i \tag{15a}$$

$$y(s) = \sum_{i=1}^{n} N_i(s) Y_i \tag{15b}$$

where X_i and Y_i are the parameters of the cubic spline in x and y directions respectively, and $N_i(s)$ is the normalised *4th* order B-spline between knots s_i and s_{i+4}.

For equally spaced simple knots, we have,

$$N_i(s)=\begin{cases} s^3/6h^3 & 0\le s-s_i\le h \\ (1+3s+3s^2-3s^3)/6h^3 & h\le s-s_i\le 2h \\ (4-6s^2+3s^3)/6h^3 & 2h\le s-s_i\le 3h \\ (1-3s+3s^2-s^3)/6h^3 & 3h\le s-s_i\le 4h \end{cases} \tag{16}$$

where h is the spacing between the knots.

In the actual implementation, the snake is located at discrete values so that the integration in (1) will be replaced with a summation. Letting,

$$N=[N_1(s)\ N_2(s)\ \dots\ N_n(s)] \tag{17a}$$

$$N_s=[N_1'(s)\ N_2'(s)\ \dots\ N_n'(s)] \tag{17b}$$

$$N_{ss}=[N_1''(s)\ N_2''(s)\ \dots\ N_n''(s)] \tag{17c}$$

$$X=[X_1\ X_2\ \dots\ X_n]^T \tag{17d}$$

$$Y=[Y_1\ Y_2\ \dots\ Y_n]^T \tag{17e}$$

$$E_p=[f(x,y)]^T[f(x,y)] \tag{17f}$$

the total energy can be written in matrix form as,

$$\begin{aligned} E &= \alpha X^T N_s^T N_s X + \beta X^T N_{ss}^T N_{ss} X \\ &\quad + \alpha Y^T N_s^T N_s Y + \beta Y^T N_{ss}^T N_{ss} Y \\ &\quad + [f(x,y)]^T [f(x,y)] + E_c \end{aligned} \tag{18}$$

E_c is derived from the initial positions of the feature by the operator.

Further development of these formulae will result in the equations used for the iterative solution of the snakes.

3. 3-D Feature Extraction By Snakes

If a feature is extracted from more than one image, its coordinates in 3-D space can be computed.

Suppose a snake in 3-D space can be approximated by a cubic B-spline. That is,

$$X(s)=\sum_{i=1}^{n}N_i(s)X_i$$
$$Y(s)=\sum_{i=1}^{n}N_i(s)Y_i \tag{19}$$
$$Z(s)=\sum_{i=1}^{n}N_i(s)Z_i$$

From the camera models, projection equations for each image can be written as,

$$\begin{aligned}x_i(s)&=P_{xi}(X(s),Y(s),Z(s))\\ y_i(s)&=P_{yi}(X(s),Y(s),Z(s))\end{aligned} \tag{20}$$

where $x_i(s)$ and $y_i(s)$ are the coordinates of the snake on the *ith* image.

The external energy of the snake can now be simply defined as a sum on each image,

$$\begin{aligned}E_{ext}&=E_{ext}(x_1(s),y_1(s))+,\ldots,+E_{ext}(x_m(s),y_m(s))\\ &=\left[F_1\ F_2\ \ldots\ F_m\right]\left[F_1\ F_2\ \ldots\ F_m\right]^T=F^TF\end{aligned} \tag{21}$$

where F is an expression for energy for 3D images, as expressed for 2D images, in equations (17f) and (18) and m is the number of images. These equations are developed for the 3D case as for the 2D case shown earlier in this paper.

The seed points of the feature are extracted by an operator on one of the images, say *ith* image. Then the approximate location of the snake in 3-D space can be computed from equation (20) with an interactive algorithm.

The Taylor's development of equation (20) can be written as,

$$\begin{aligned}&\frac{\partial P_{xi}}{\partial X}\Delta X+\frac{\partial P_{xi}}{\partial Y}\Delta Y+\frac{\partial P_{xi}}{\partial Z}\Delta Z+P_{xi}(X_0,Y_0,Z_0)-x_i=0\\ &\frac{\partial P_{yi}}{\partial X}\Delta X+\frac{\partial P_{yi}}{\partial Y}\Delta Y+\frac{\partial P_{yi}}{\partial Z}\Delta Z+P_{yi}(X_0,Y_0,Z_0)-y_i=0\end{aligned} \tag{22}$$

Assuming a DEM is available,

$$Z=S(X,Y) \tag{23}$$

or in the Taylor's development,

$$\frac{\partial S}{\partial X}\Delta X+\frac{\partial S}{\partial Y}\Delta Y+\Delta Z+S(X_0,Y_0)-Z_0=0 \tag{24}$$

Combining equations (22) and (24), a system with three equations for three unknowns is obtained. Solving this system the approximation of the snake is derived, which can be optimised in an interactive solution.

4. Implementation

The processing chain to derive the energy image from the original image is as follows :

- Preprocessing by image stretching
- Edge detection by:
  - Canny operator for single edges (Canny 1986).
  - Morphological tools for narrow features. To detect bright small elements, erosion followed by a dilation is applied; then the original image is subtracted from this result. For dark small elements, dilation is applied first, then erosion.
- Derivation of the chamfer image in which the pixel values relate to their closeness to any surrounding edge. For example, if an edge pixel has the grey value strength P_e, then for a non-edge pixel located x pixels away from it, its grey value $P_{non\text{-}edge}$ (in the chamfer image) should become P_e minus x. And since there are many edge pixels in an image, the recorded value $P_{non\text{-}edge}$ is the maximum value which can be provided from all the different edge pixels.
- Compute the snake by equations described earlier in this paper.

The results of tests on the processes will be described in the next section.

5. Experiment Results

5.1 Satellite Images

Visual tests of the 2D feature extraction process, for a single image, revealed that considerable success was achieved with the procedure. While the quality of the extracted features should be dependent on the initial feature extraction procedure, the snake interpolation generally works well when there are no other features close to the feature being extracted. If there are other nearby features in the image data, the snake may become distorted. This can be partially overcome if the operator locates the feature being extracted as accurately as possible in the region of the other features, but the snake may still be distorted. For wide features such as roads, the snake may jump from one side of the road to the other. For this case, a special tool should be developed for the parallel features. Tests on the 3D extraction of features, based on two overlapping images, revealed similar results to 2D extraction, with the additional advantage that the 3D coordinates in the ground coordinate system were also determined. Visual comparisons have also been carried out between the method of snakes, and the method of feature extraction based on dynamic programming (Gruen and Li 1994), revealing similar results.

In order to obtain numerical results of the quality of the feature extraction, a number of tests were carried out on two overlapping SPOT images over the Sydney region. The first pair of images, with a B/H of 1.0, was recorded on SPOT1 in 1986. They suffer from poor dynamic range and a considerable level of noise. The second set, with a B/H of 0.6, was recorded 5 months apart in 1994 from SPOT2. One of the images has high dynamic range, while the other suffers from poor dynamic range because of a considerable level of cloud cover. Small regions of cloud also appear on this image. Orthophoto maps in the region with a scale of 1:4,000 and 1m and 2m contours, were used as ground truth information for the 3D extraction. Accuracies of planimetric positions and elevations are of the order of 2m or better on the orthophotos.

The tests undertaken were:

(i) A study was made of the effect of the magnitude of the error in the initial position of the feature defined by the operator, on the accuracy of the computed position of the feature. This could be defined as the 'pull-in range' of the computation. For this test the position of a feature was initially defined manually on the screen as accurately as possible, and the snake computed. The manual location and snake computation were compared to check the quality of the extraction. This location was then considered as the 'reference' position for that feature for the purposes of comparisons with other tests of the feature location. The snake was then processed a number of times for the user input of the location of the feature ranging up the 10 pixels away from the 'reference' position of the feature. The results of the relative accuracy of the computed snake, compared with the 'reference' position of the snake were derived from the RMS of the deviations between the two features. These results are presented in Figures 1 and 2, for 2D and 3D extraction respectively, where 'Input RMS' refers to the RMS of the deviations of the input position of a feature by the operator, from the 'reference' position of the feature. 'Output RMS' refers to the RMS of deviations of the subsequently computed position of the snake, from the 'reference' position.

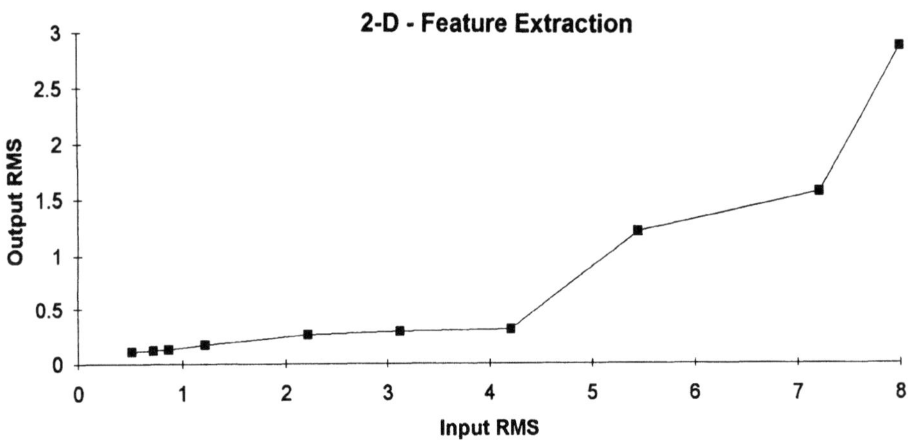

Figure 1

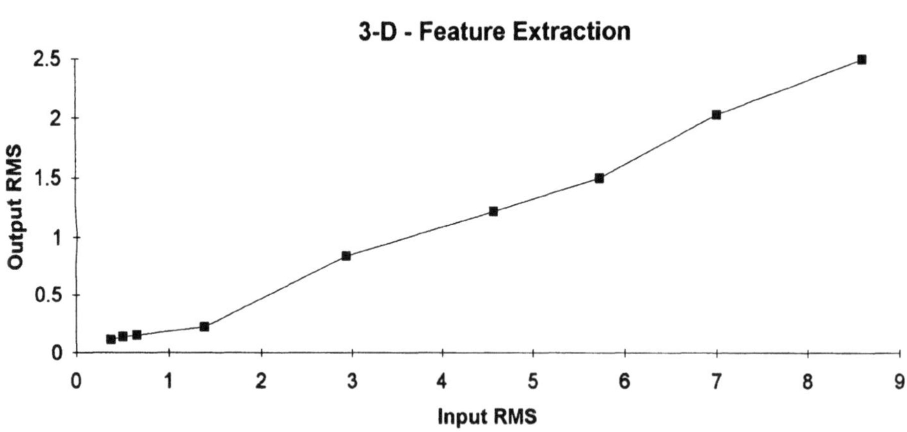

Figure 2

(ii) The absolute accuracy of the 3D feature extraction method was obtained by comparing the E, N and H coordinates of the 3D snake with the map coordinates of the same features. The results of these tests are shown in Table 1, based on tests of the indicated number of linear features on the images, and the number of points on the features. Tests on the elevations were only made on linear features which are approximately in the N-S direction.

Table 1
Absolute accuracy of extracted 3D features

	No of linear features	**Total no. of points on the features**	**Absolute Accuracy**
E and N	12	120	3m
H	9 (N-S only)	90	6m

Figures 1 and 2 reveal that features derived interactively by an operator within approximately 5 pixels of their correct positions, will generally be located by the snake computation with a relative accuracy of better than 0.5 pixel, or 5 metres on the ground for SPOT images. If the error in the location of the feature is larger than 7 pixels, the relative accuracy of the feature location will deteriorate to greater than 1 pixel.

A comparison of the results of tests on the 2D extraction with tests on 3D extraction revealed that the pull-in range for 3D extraction tended to deteriorate more quickly than for the 2D extraction. The relative accuracy for 3D extraction was 1 pixel when the input position of the feature was of the order of 5 pixels from its correct position. One of the advantages of using the overlapping images is that it may overcome the problems of missing data in one or both images. This possibility has not been adequately tested at this stage. Table 1 reveals that the absolute accuracy of the extracted 3D features is of the order of 3 m in E and N coordinates and 6m for elevations for the SPOT2 pair of images with a B/H of 0.6, on the flat terrain in the sections of the image tested.

5.2 Close Range Images

This test determined how well the snakes fit the object model, using a CCD camera image with some well defined object edges. The geometric model of the extracted curves in the image of a cylinder object was an ellipse, while the edges which were successfully extracted by the snake, were defined by a cubic spline. A set of points was sampled from the cubic spline and an ellipse fitted to these points by least squares, the residual errors and RMSE computed. This assessment can be considered as the relative accuracy of the snake. To assess the absolute accuracy, the results of the snake were compared with that obtained by a sub-pixel accuracy edge detection method (Huang and Trinder 1994). The differences between the snake and the ellipse, and an estimate of RMSE were computed. This test revealed relative accuracies of edge location of 0.2 to 0.3 pixel, and absolute accuracies of 0.5 to 0.6 pixel. Tests on straight lines on the object revealed relative accuracies of 0.2 pixel and absolute accuracies of 0.3 to 0.5 pixel.

6. Conclusions

(i) The tests revealed that considerable success has been achieved with the method of feature extraction using snakes, particularly for single features which are separated from other features in the image. It combines the skills of both the operator and the computer for efficient and accurate extraction of features. However, the method is dependent on the initial feature extraction procedure. If there are other features in the image data nearby, the snake may become distorted.

(ii) The pull-in range of the method can be in excess of 10 pixels for well defined features, but this range should normally be considered as approximately 5 pixels. Under these circumstances, the relative accuracy of the features extraction is of the order of 0.5 pixel for 2D data, and 1 pixel for 3D data. This accuracy has also been confirmed by tests on close range video images.

(iii) 3D feature extraction is slightly less efficient than 2D extraction, in terms of pull-in range. However, 3D extraction enables the determination of 3D coordinates of the features. The absolute accuracies of this process have been found to be of the order of 3m in E and N and 6m in H.

Acknowledgments

A number of people in addition to the authors of the paper have contributed to this research. Mr Alexis Vuillemin initially developed the procedures for the implementation of the snakes, which were subsequently further developed by Mr Haihong Li. Mr Yongru Huang, a post graduate student at UNSW, developed the code for the 3D snakes. Mr Brian Donnelly provided significant assistance in the provision of the orientations of the SPOT images and DEMs. Finally, it is acknowledged that the initial investigations of this research were funded by the Electronics Research Laboratory Defence Science and Technology Organisation (DSTO) in Salisbury, South Australia under the direction of Dr Vittala K. Shettigara, while the latter part of the research was supported by a grant from the Australian Research Council.

References

Canny J. (1986) 'A Computational Approach to Edge Detection', IEEE Trans Pattern Analysis and Machine Intelligence PAMI-8 pp

Fua P., and Leclerc Y. G. (1990) 'Model Driven Edge Detection', Machine Vision Applications (1990) 3:45-56

Fua P. and Hanson A.J.(1991) 'An Optimization Framework for Feature Extraction', Machine Vision Applications (1991) 4:59-87.

Gruen A. and Li H. (1994) 'Semi-Automatic Road Extraction by Dynamic Programming', International Archives of Photogrammetry and Remote Sensing, Vol. 30-3 pp 324-332.

Huang Y. and Trinder J. (1994) 'A Feature-Based Approach To Reconstruction Of 3-D Objects From Digital Images", International Archives of Photogrammetry and Remote Sensing, Vol. 30-3 pp 391-398.

Kass M., Witkin A., and Terzopoulos D., (1988) 'Snakes: Active Contour Models', Int. Journ. of Computer Vision, Vol 1, pp 321-331.

Rougon N. (1991) 'Kinetics of Interface Evolution with Applications to Active Contour Models', SPIE Vol. 1610 Curve and Surfaces in Computer Vision and Graphics II.

Rougon N. and Prêteux (1991) 'Deformable markers: mathematical morphology for Active Contour Models', SPIE's 1991 International Symposium on Optical Applied Science and Engineering Image. Algebra and Morphological Image Processing II, San Diego, California, Vol. 1568, pp 78-89.

Sperling G. (1970) 'Binocular Vision: A Physical and Neural Theory' Am J. Psychology, Vol 83 pp 461-534.

Williams D.J. and Mubarak Shah (1992), 'A Fast Algorithm for Active Contours and Curvature Estimation', Image Understanding, Vol. 55, No. 1, January, pp 14-26.

Automatic Extraction of Man-Made Objects
from Aerial and Space Images
Monte Verità, © 1995 Birkhäuser Verlag Basel

From Ziplock Snakes to Velcro^{TM} Surfaces

W. Neuenschwander*, P. Fua†, G. Székely*, and O. Kübler*

* Communication Technology Laboratory
Swiss Federal Institute of Technology
ETH
CH-8092 Zurich, Switzerland
Ph.: +41-1-632 5302, Fax: +41-1-632 1199
e-mail: waneu@vision.ee.ethz.ch

† SRI International
Artificial Intelligence Center
333 Ravenswood Avenue
Menlo Park, CA 94025, USA
Ph.: (415)-859 4480, Fax: (415)-859 3735
e-mail: fua@ai.sri.com

Abstract

The use of energy minimizing deformable models in various applications has become very popular. The issue of initializing such models, however, has not received much attention although the model's performance depends critically on its initial state. We aim at obtaining good convergence and segmentation properties from a minimum of a priori information.

We present a new approach to segmentation of 2– and 3–Dimensional shapes that initializes and then optimizes a deformable model given only the data and a very small number of 2–D or 3–D seed points respectively. This is a valuable capability for medical, robotic and cartographic applications where such seed points can be naturally supplied. In effect, the 2–D "snake" and the 3–D surface model are clamped onto the object boundary in manner reminiscent of a ziplock or velcro being closed.

We develop the method's mathematic framework and show results using 2–D cartographic data. Preliminary results in 3–D using volumetric medical data are shown as well.

1 Introduction

In recent years, deformable models have emerged as a very powerful tool for semiautomated object delineation and surface modeling as well as for 2– and 3–Dimensional image segmentation in applications as diverse as medical imaging, graphics, robotics or terrain-modeling.

The 2–D models, known as "snakes" have been originated by Terzopoulos, Kass, and Witkin (Terzopoulos *et al.*, 1987; Kass *et al.*, 1988) and have since given rise to a large body of literature (Fua and Leclerc, 1990; Cohen, 1991) among many others, that explores theoretical and implementation issues as well as new applications.

In most of these papers, however, it is assumed that the initial position of the snake is relatively close to the desired solution. In effect, initialization here amounted to a painstaking almost complete, manual delineation of the desired outline. While this is a reasonable assumption for applications such as motion tracking (Bascle and

Deriche, 1993) where the effort is limited to the first image of a whole sequence, it is ineffective for delineating complex objects from scratch.

Our efforts are aimed at alleviating the often repetitive task practitioners face when segmenting images. In particular, we aim at eliminating the need to outline the desired structure very precisely. In our implementation, the user needs only to supply a few discrete points through which the contour must pass; the system then propagates the information along the contour starting from these points. As a result, the snake is progressively clamped onto an image contour so that it smoothly connects those points and has the right orientation at their locations. This behavior is analogous to the closing of a ziplock, hence the name of our snakes. As illustrated in Figure 1, considerably fewer control points are needed than for conventional implementations which we refer to as *traditional snakes.*

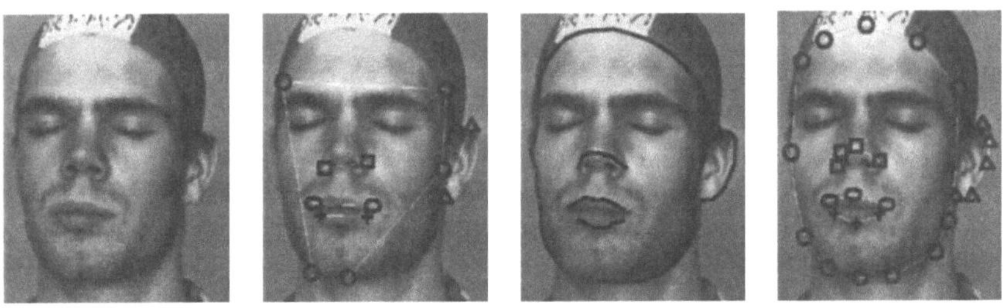

Figure 1: Outlining facial features. (a) A face image with low contrast contours. (b) Five sets of initial points, each denoted by a different symbol. Four of them are pairs of endpoints while the fifth is shown as a set of circles. (c) The contours delineated by Ziplock snakes. (d) The initial delineations that must be supplied to achieve the same result using traditional snakes.

In the same spirit, we present a method that allows a user to initialize and then optimize a 3–D surface model by supplying only a very small number of 3–D seed points and corresponding surface normals. This is a valuable capability because there are many applications, such as *Segmentation of 3–D shapes from volumetric data, Incremental construction of a world model by a mobile robot, Construction of composite models for high-resolution cartographic modeling* in which imposing initial conditions in this manner is both easy and natural.

Our technique relies on classical elastic models that are represented as triangulated meshes and deform themselves to minimize an objective function (Terzopoulos *et al.*, 1987). This method extends our 2–D ziplock approach (Neuenschwander *et al.*, 1994) to the 3–D domain. We show that a small set of 3–D seed points provides sufficient boundary conditions to solve the differential equation that governs the model's behavior in closed form, assuming that the data component of the objective function remains constant. As a result, it becomes possible to instantiate the model using these points alone by initially ignoring the data term and then propagating image information along the surface by progressively "turning it on". The triangulated mesh behaves like a piece of Velcro that is progressively being clamped onto the surface of interest, hence the name of our deformable surfaces.

2 Ziplock Snakes

The original snakes (Kass *et al.*, 1988) are modeled as time-dependent 2–D curves defined parametrically as $\vec{v}(s,t) = (x(s,t), y(s,t))_{0\leq s\leq 1}$, where s is proportional to the arc length, t the current time, and x and y the curve's image coordinates. The snake deforms itself as time progresses so as to minimize an image potential $E_I(\vec{v})$, with $E_I(\vec{v}) = -\int_0^1 P(\vec{v}(s,t))\,ds$ where $P(\vec{v}(s,t))$ is a function of the image. One typical choice is to take $P(\vec{v}(s,t))$ to be equal to the magnitude of the image gradient, that is $P(\vec{v}(s,t)) = |\nabla I(\vec{v}(s,t))|$, where I is either the image itself or the image convolved by a Gaussian kernel.

Whatever the choice of P, $E_I(\vec{v})$ is typically not a convex functional. To overcome this problem, Terzopoulos (Terzopoulos *et al.*, 1987) has proposed to introduce a regularization term $E_D(\vec{v})$ that is convex and to minimize a total energy term $E(\vec{v})$ that is the sum of $E_I(\vec{v})$ and $E_D(\vec{v})$. Using the elastic rod model, $E_D(\vec{v})$ is taken to be

$$E_D(\vec{v}) = \frac{1}{2}\int_0^1 \alpha(s)\left|\frac{\partial\vec{v}(s,t)}{\partial s}\right|^2 + \beta(s)\left|\frac{\partial^2\vec{v}(s,t)}{\partial s^2}\right|^2 ds\,, \tag{1}$$

where $\alpha(s)$ and $\beta(s)$ are arbitrary functions that regulate the curve's tension and rigidity. In the implementation described below $\alpha(s)$ and $\beta(s)$ are taken to be constant and supplied by the user. We have shown previously (Fua and Leclerc, 1990) that constant α and β can be chosen in a fairly image-independent way.

From variational calculus it is well known that if $\vec{v}(s)$ minimizes $E = E_D + E_I$ and is sufficiently regular, that is at least $C^4(0,1)$, then it must be a solution of the set of two coupled Euler differential equations

$$-\alpha\frac{\partial^2 v(s,t)}{\partial s^2} + \beta\frac{\partial^4 v(s,t)}{\partial s^4} = -\nabla P(v(s,t)) \tag{2}$$

where $v(s,t)$ stands for either $x(s,t)$ or $y(s,t)$. Note that, in order for this equation to have a unique solution, one must specify boundary conditions such as the values and derivatives of $v(s,t)$ for $s=0$ and $s=1$.

Discretizing Equation 2 using finite differences yields the linear system

$$K \cdot V = F_V \tag{3}$$

where V stands for the vector of either x or y coordinates, K is the stiffness matrix and F_V are the derived image forces. K is *not* invertible and these equations cannot be solved directly.

2.1 Initialization and Optimization using Boundary Conditions

To improve upon the snakes' convergence properties, we now introduce our ziplock optimization mechanism. The user is expected to specify endpoints in the vicinity of a clearly visible edge segment, which implies a well-defined edge direction.

To successfully optimize our snake, we start from an initial position that is approximately correct in the neighborhood of the endpoints. The easiest way to achieve this result is to solve the homogeneous equations that correspond to the system of Equation 2

$$-\alpha\frac{d^2v(s)}{ds^2}+\beta\frac{d^4v(s)}{ds^4}=0 \tag{4}$$

where v stands for either $x(s)$ or $y(s)$ and $0 \le s \le 1$ and has to fulfill the four boundary conditions: $v(0)=\hat{v}_0$, $v(1)=\hat{v}_1$, $v'(0)=\hat{v}'_0$, $v'(1)=\hat{v}'_1$. By construction, this solution has the specified tangent at the endpoints $\hat{v}_0, \hat{v}_1$ and is close to the right answer near these points.

While the tangent direction at the endpoints can be computed, its orientation cannot be determined therefore the interface provides the user with the possibility of flipping the orientation at both ends if necessary. By fixing the curve's endpoints and giving the curve's tangent at those points, the system of Equation 3 is reduced by 4 entries (see appendix of (Neuenschwander *et al.*, 1995)) to

$$K^* \cdot V^* = F^*_{V^*}, \tag{5}$$

where V^* stands either for the reduced X^*- or Y^*-vector, and K^* is a pentadiagonal stiffness matrix that *is* invertible. Of course, since $F^*_{V^*}$ depends on the snake's current position, the system is only semilinear and cannot, in general, be solved in closed form.

We start the optimization of the energy term by defining the initial snake as the solution of the homogeneous differential system of Equation 4. At this stage the snake "feels" absolutely no external potential forces. During the ongoing iterative optimization process the image potential is turned on progressively for all the snake vertices, starting from the fixed extremities. We distinguish between *passive* and *active* snake nodes, depending on whether the potential force field is turned on for that vertex or not.

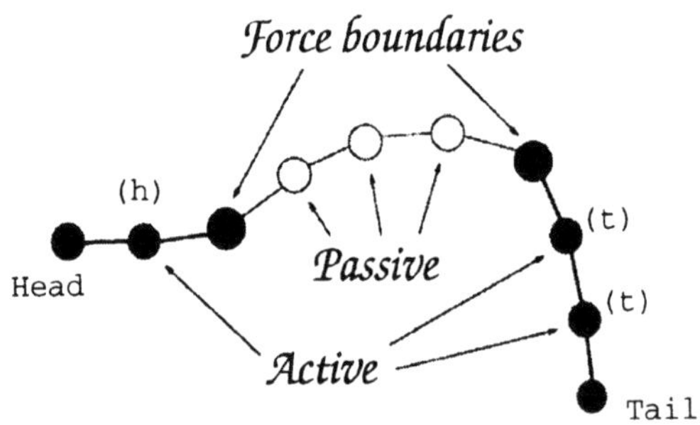

Figure 2: Schematic Ziplock Snake during optimization. A Ziplock Snake, fixed at `Head` and `Tail`, consists of two parts the *active* and the *passive* vertices. These areas are separated by moving *force boundaries*. The *active* part of the snake is divided up into two segments marked as `(h)` and `(t)` respectively.

As illustrated by Figure 2, we define the *force boundaries* as the location of the vertices farthest away from the endpoints that feel the image forces. These boundaries approach each other during the ongoing optimization process by moving forward.

The simplest way to optimize the snake would be to gradually move the force boundaries from the snake's head and tail towards its center and solve Equation 5 for each new position. However, because Equation 5 is semilinear, it cannot be solved in one single step. To enforce stability we introduce a viscosity term $\gamma(s,t)$ similar to the one used by traditional snakes and iteratively solve the Equation

$$(K^* + \gamma_t I) \cdot V_t^* = \gamma_t V_{t-1}^* + F_{V^*}^*\Big|_{V^* = V_{t-1}^*} \quad . \tag{6}$$

The initial coordinate vector $(X_0^*, Y_0^*)^T$ is the solution of Equation 4. The viscosity $\gamma(s,t)$ is initialized as $\gamma(s,0) \equiv 0$, $s \in [0,1]$ and recomputed each time the force boundaries move so that the initial displacement of each vertex is on the average of a given a-priori magnitude, typically 1 pixel, see (Fua and Leclerc, 1990). Each force boundary is moved individually but at most one vertex per iteration. A force boundary is moved only if the average motion of the *active* part of the snake, feeling the image forces and lying between the force boundary and the head (h) or tail (t), respectively, is below a fraction of a pixel. The two different parts are marked with (h) and (t) in Figure 2.

The potential of Ziplock Snakes has already been illustrated in Figure 1. In Figure 3 we use Ziplock Snakes to outline various features of a house in an aerial image. Using our snakes we can delineate rather complex object boundaries using a very small number of seed-points.

a) b)

Figure 3: Delineating man-made objects in aerial images. a) the two figures show complementary sets of pairs of seed-points used to initialize Ziplock Snakes. b) Outlined features by Ziplock Snakes. Courtesy of the Institute of Geodesy and Photogrammetry, ETH.

2.2 Extension to Ribbon Snakes

Following (Fua and Leclerc, 1990), we have implemented a tool for interactive road delineation, called *ribbon snake*. The model vector $\vec{v}$ is augmented by a third component, the varying width $w(s,t)$ of the road. The expression for the deformation energy (1) still holds for $\vec{v}(s,t) = (x(s,t), y(s,t), w(s,t))^T$, where the width is subject to "tension" and "rigidity" constraints as the two coordinate components. The ribbon forms the center of the road, while the assigned width defines two curves that are the actual deforming road boundaries. Note that the image information is taken into account along these two curves only. The sequence in Figure 4 depicts the initialization and the subsequent "Ziplock"-optimization of a ribbon snake model. Each ribbon is initialized by the two corresponding endpoints and the road direction, which is the average direction of the left and right road boundary.

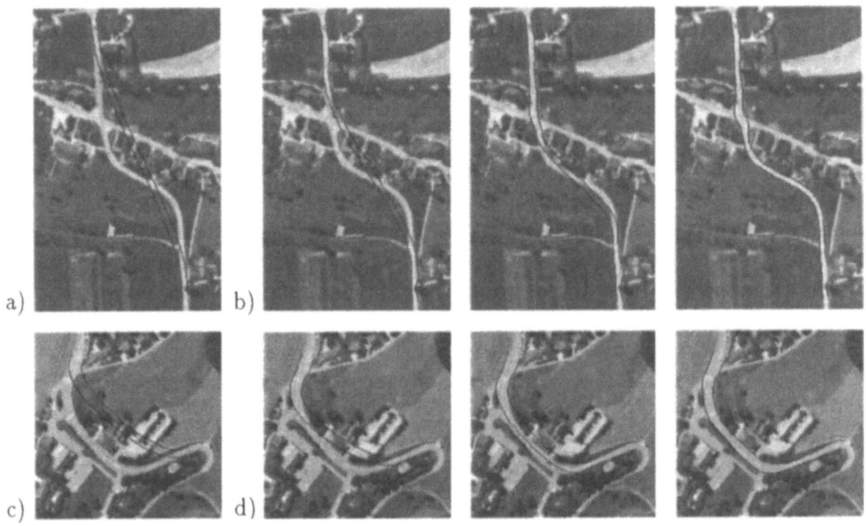

Figure 4: Road delineation in aerial images. a) & c) Initial ribbon snakes defined by the two fixed endpoints and the corresponding road direction. b) & d) The three images show intermediate and final results using the presented "Ziplock"-optimization mechanism. Note, that the ribbon snake can pass over other significant image features due to the fact that the image forces are not active between the force boundaries.Courtesy of the Institute of Geodesy and Photogrammetry, ETH and the Schweizer Landestopographie, Bern.

3 Deformable Velcro^{TM} Surfaces

Velcro surfaces as well as Snakes belong to the same framework of deformable models proposed by Terzopoulos (Terzopoulos *et al.*, 1987). The physical interpretation uses basics of elasticity theory and regards deformable surfaces as a weighted combination of membrane and thin plate. A generalized deformable surface model is defined as $\vec{v}(\boldsymbol{\omega},t) = (x(\boldsymbol{\omega},t), y(\boldsymbol{\omega},t), z(\boldsymbol{\omega},t))$ where $\boldsymbol{\omega} \in \Omega \subset \mathbb{R}^2$ is a suitable parametrization, t the current time, and x, y and z are the corresponding coordinate functions of the surface. The surface deforms itself so as to minimize its image potential energy $E_I(\vec{v}) = -\iint_\Omega P(\vec{v}(\boldsymbol{\omega}))d\boldsymbol{\omega}$ where P is the same potential function as derived in Section 2.

In the case of deformable surfaces, the convex regularization term $E_D(\vec{v})$ we use has the following form:

$$E_D(\vec{v}) = \iint_\Omega \tau(\boldsymbol{\omega}) \left[\left| \frac{\partial \vec{v}}{\partial \omega_1} \right|^2 + \left| \frac{\partial \vec{v}}{\partial \omega_2} \right|^2 \right] + \rho(\boldsymbol{\omega}) \left[\left| \frac{\partial^2 \vec{v}}{\partial \omega_1^2} \right|^2 + 2 \left| \frac{\partial^2 \vec{v}}{\partial \omega_1 \partial \omega_2} \right|^2 + \left| \frac{\partial^2 \vec{v}}{\partial \omega_2^2} \right|^2 \right] d\boldsymbol{\omega} \quad (7)$$

where $\rho(\boldsymbol{\omega}) = 1 - \tau(\boldsymbol{\omega})$ and $\boldsymbol{\omega} = (\omega_1, \omega_2)$. In the implementation described below, we take the surface tension parameter, τ, to be a constant between 0 and 1 supplied by the user. The Euler differential equation according to the optimization problem of the functional $E(\vec{v}) = E_I(\vec{v}) + E_D(\vec{v})$ can be written as

$$-\tau\,\Delta \boldsymbol{v} + (1-\tau)\Delta^2 \boldsymbol{v} = -\frac{\partial P}{\partial \boldsymbol{v}} \tag{8}$$

where $\boldsymbol{v}$ stands for either x, y or z. Note, these differential equations do not have a unique solution in the absence of boundary conditions, whose incorporation is crucial for a successful segmentation.

We use the same notations as introduced in Section 2 for the 3–D framework and refer to equations derived in 2–D. Note, however, that equivalence is restricted to the notation while these expressions differ in their detailed mathematical structure.

In the context of segmentation of anatomical organs in 3-D medical image data, we assume that the surface is topologically equivalent to a sphere. We tessellate the surface into triangular patches and therefore define the discrete model by a set of nodes. The user manually supplies a few "anchor-points" and supervises the tessellation by specifying the triangular facets and their refinement. Using finite differences, the governing Equation 8 becomes the linear system

$$K \cdot V = F_V \tag{9}$$

where now V stands for the vector of either x, y or z coordinates, K is the surface's stiffness matrix and F_V are the derived image forces. As it is the case for Snakes, the matrix K is *not* invertible and these equations cannot be solved directly.

3.1 Initialization and Optimization Using Boundary Conditions

First we address the acquisition of boundary conditions and their use to effectively initializing the deformable surface model. We then discuss the actual optimization.

Visual inspection of 3–D data sets simultaneously from the three cardinal directions is supported by appropriate user interfaces and allows to identify surface points interactively. The normal vectors are taken as the gradient's direction of the potential field at the selected points. This process yields a set $\mathcal{B}$ of boundary conditions defined by an arbitrary number m of 3–D anchor-points P_i ($4 \leq i \leq m$, non-coplanar) and the corresponding surface normals $\vec{n}_{P_i}$. Therefore $\mathcal{B}$ can be written as

$$\mathcal{B} = \{\vec{v}_1 = P_1, N_{\vec{v}_1} = \vec{n}_{P_1}\ ;\ \vec{v}_2 = P_2, N_{\vec{v}_2} = \vec{n}_{P_2}\ ; \ldots ; \vec{v}_m = P_m, N_{\vec{v}_m} = \vec{n}_{P_m}\} \tag{10}$$

To successfully optimize the surface, we must start from an initial shape that is approximately correct in the neighborhood of the selected anchor-points. This result can be achieved by solving the homogeneous equations that correspond to the system of Equation 8

$$-\tau\,\Delta \boldsymbol{v} + (1-\tau)\,\Delta^2 \boldsymbol{v} = 0 \tag{11}$$

where $\boldsymbol{v}$ stands for either x, y or z. The initial surface we compute is a solution of Equation 11 that satisfies the set $\mathcal{B}$ of boundary conditions defined above. By construction, this solution will pass through the anchor-points and have the specified normal vectors there, and be close to the final surface near these points.

The semilinear system of Equation 9 is reduced by incorporating boundary conditions (set $\mathcal{B}$) to

$$K^* \cdot V^* = F^*_{V^*}, \tag{12}$$

where now V^* stands either for the reduced X^*-, Y^*- or Z^*-surface vector, and K^* is the sparse stiffness matrix that *is* now invertible.

We start the optimization of the energy term by defining the initial surface as the solution of the homogeneous differential system of Equation 11. At this stage the surface "feels" absolutely no external potential forces. During the ongoing iterative optimization process the image potential is turned on progressively for all the surface nodes, starting from the seed points. We distinguish between *passive* and *active* surface nodes, depending whether the potential force field is turned on for that vertex or not. We define the expanding *force fronts* as the location of the nodes farthest away from their corresponding central seed point that feel the image forces. These fronts approach each other during the ongoing optimization process and propagate all over the surface by moving forward like concentric wave fronts. This propagation is illustrated in Figure 5.

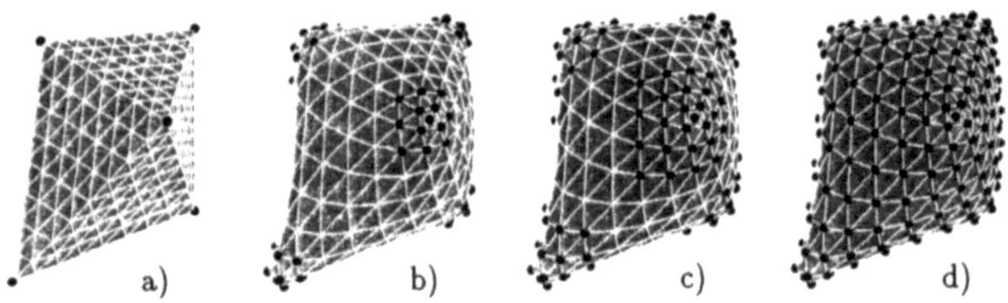

Figure 5: Propagation of the force boundaries. a) First polyhedral initialization performed by the user. b)-d) Solution of the Euler differential equation. At every seed point a force front expands on the surface. The black dots denote the various fix points while the light shaded vertices denote the active mesh nodes. Note, a case without image forces was used for this illustration.

As explained in Section 2, to enforce stability we introduce a viscosity term $\gamma(\boldsymbol{\omega}, t)$ and iteratively solve the equation

$$(K^* + \gamma_t I) \cdot V^*_t = \gamma_t V^*_{t-1} + F^*_{V^*}\Big|_{V^* = V^*_{t-1}} . \tag{13}$$

where now the initial coordinate vector $(X^*_0, Y^*_0, Z^*_0)^T$ is the solution of Equation 11. The viscosity $\gamma(\boldsymbol{\omega}, t)$ is initialized as $\gamma(\boldsymbol{\omega}, 0) \equiv 0$, $\boldsymbol{\omega} \in \Omega$ and recomputed each time the force boundaries move.

We illustrate the segmentation process on the example of the putamen, a relatively small nucleus of the deep gray matter of the brain. The original binarized dataset is shown in Figure 6a). First the user has to provide a reasonable number of anchor-points, which will be used to generate the initial approximation of the surface. At the same time they serve as seed points for the force front propagation of the surface evolution. We have built a special tool, which serves as a 3–D cursor in a gray-valued field, and is illustrated by Figure 6b). The tool simultaneously updates the three orthogonal planes that cross each other at the actual 3–D cursor position, which is shown as cross-hair on each of the planes.

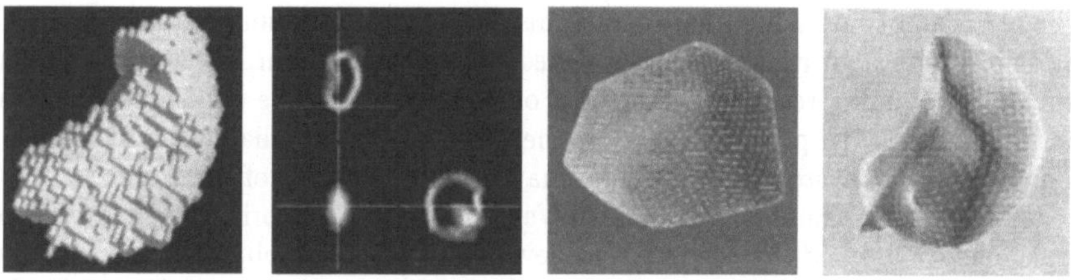

a) b) c) d)

Figure 6: a) The original dataset used as an example to illustrate the segmentation process. b) Tool for the selection of fix points in the dataset to be segmented. c) Initial surface resulting from the force-free solution of the homogeneous system of Euler-Lagrange differential equations, and its refined triangulation. d) The final result of the surface evolution, governed by the progressive turning on of the image potential forces. Note, that the resulting surface is smooth, but features which are selected by the user as anchor-points are preserved.

The Delaunay-tetrahedralization (Boissonnat, 1984) of the selected anchor-points is then computed and "sculpted" interactively by successive deletion of Delaunay tetrahedra from the convex hull, until all selected points lie on the surface. Finally the remaining tetrahedra are used to define an initial triangulation that is refined by a number of recursive subdivisions—4 in the example presented here resulting in 1794 vertices and 3584 facets—and used to compute the unperturbed initial surface, as shown in Figure 6c). This solution is then optimized according to the procedure proposed in this section. Figure 6d) shows the result of this surface evolution.

4 Conclusion and Future Work

We have proposed a snake-based approach to semiautomated delineation that allows a user to outline an open contour by specifying only very distant endpoints and allowing the computer to propagate edge information from the extremities toward the center. In other words, we have proposed a natural initialization procedure that is completely in line with both the practitioner's task and the mathematical problem so that the "expert user" (Kass *et al.*, 1988) of the original snake papers need not be that much of an expert anymore. Ziplock snakes have been ported into the RADIUS Common Development Environment (Mundy *et al.*, 1991) where they can be used to optimize or connect curves and ribbons in a semiautomated fashion while taking advantage of RCDE's extensive editing capabilities.

We have generalized the 2–D method to a 3–D procedure that allows initializing and optimizing a 3–D surface model given only a very small number of 3–D seed points and corresponding surface normals. This is valuable in semiautomated applications—such as medical ones—where seed points can be supplied manually by the user with reasonable ease. The user can influence the level of detail in the final representation by choosing the most appropriate force field, adjusting the tension parameter, and by providing more or less seed points for any given part of the surface.

The capability presented here should also be valuable for fully automated applications. Vision algorithms—such as the so-called "shape from X" methods—often provide high-quality results in some parts of a scene but may be unreliable in others. Our method has the potential to allow the use of the most reliable surface patches as anchors and the propagation of information to other parts of the surface.

Further development of deformable Velcro surfaces will primarily address the implementational and the user interface level. The governing objective is to develop the mathematical theory, the implementation, and the user interface of a tool for the segmentation of 3–D objects without requiring expert Computer-Vision knowledge for its use.

References

Bascle B. and R. Deriche (1993). Stereo Matching, Reconstruction und Refinement of 3D Curves Using Deformable Contours. In *International Conference on Computer Vision*, pp. 421–430, Berlin, Germany.

Boissonnat J.-D. (1984). Geometric structures for three-dimensional shape representation. *ACM Transactions on Graphics*, Vol. 3,pp. 266–286.

Cohen L.D. (1991). On Active Contour Models and Balloons. *Computer Vision, Graphics, and Image Processing: Image Understanding*, Vol. 53,No. 2,pp. 211–218.

Fua P. and Y.G. Leclerc (1990). Model Driven Edge Detection. *Machine Vision and Applications*, Vol. 3,pp. 45–56.

Kass M., A. Witkin, and D. Terzopoulos (1988). Snakes: Active Contour Models. *International Journal of Computer Vision*, Vol. 1,No. 4,pp. 321–331.

Mundy J.L., R. Welty, L. Quam, T. Strat, W. Bremmer, M. Horwedel, D. Hackett, and A. Hoogs (1991). The RADIUS Common Development Environment. In *Proc. of AIPR, Washington, DC.* (also *in Proc. of DARPA Image Understanding Workshop*, San Diego, California, 1992.).

Neuenschwander W., P. Fua, G. Székely, and O. Kübler (1994). Initializing Snakes. In *IEEE Computer Society Conference on Computer Vision and Pattern Recognition*, pp. 658–663.

Neuenschwander W., P. Fua, G. Székely, and O. Kübler (submitted, 1995). Ziplock Snakes. *International Journal of Computer Vision*, Vol. .

Terzopoulos D., J. Platt, A. Barr, and K. Fleischer (1987). Elastically deformable models. *Computer Graphics*, Vol. 21,No. 4,pp. 205–214.

Building Extraction

A Layered Abduction Model of Building Recognition

Toni Schenk

Department of Geodetic Science and Surveying
The Ohio State University
1958 Neil Ave., Columbus, OH 43210-1247
Tel. (614) 292-7126, Fax (614) 292-2957
e-mail: schenk@mps.ohio-state.edu

Abstract

This paper addresses the problem of recognizing buildings from large-scale aerial images on a rather conceptual level. After defining the problem and describing the assumptions and constraints, the object recognition problem is decomposed into different layers, beginning with the preprocessed images and progressing through intermediate levels of raw and segmented surfaces towards a geometric and semantic description of buildings. The interaction of the tasks on every layer is cast as a layered abduction model based on the hypothesis that visual perception is layered abduction. A brief motivation of this hypothesis is provided in the third section, together with a basic background of abduction. The remaining part of the paper is concerned with applying the general abduction model to the problem of building recognition.

1 Introduction

Great strides have been made in digital photogrammetry during the last few years. Several products, such as softcopy workstations, programs to automatically generate DEMs and digital orthophotos, are operational and increasingly being used by photogrammetrists. Despite of this remarkable progress digital photogrammetry is still in its beginning for only the easier problems have been solved. A great deal of effort in research and development has been directed towards early vision problems where the data driven character of processing imagery resembles much more closely traditional and therefore familiar techniques compared to the more symbolic type of information processing and abstract data representation that occurs in late vision.

Automatic extraction of man-made objects, the theme of the workshop, is an example of a hard problem, involving processes on all levels from early to late vision. Marr (1982) argues that hard problems, such as extracting and recognizing features, are complicated information processes that should be understood on the three levels of computational theory, representation and algorithm, and hardware implementation. It is interesting to observe that most attempts to, say, recognize buildings, are to be associated with the second or third level. Thus, one tries to find an answer in a bottom up fashion by exploring different algorithms and methods or by constraining the problem to special cases.

This paper is an attempt to contribute to the more theoretical aspect of recognizing man-made objects from aerial images. The next section summarizes our efforts in defining and structuring the problem. This is guided by Marr's vision paradigm (see Marr, 1982). In the following section we propose to analyze and solve the problem as a layered abduction model. This is entirely based on Josephson's hypothesis that perception is layered abduction (Josephson et al., 1994), according to which the information-processing mechanisms that occur in vision, hearing and understanding natural languages, are all variations of one basic computational mechanism.

2 Proposed Structure (Layers) of the Building Recognition Problem

Let us first define the building recognition problem (working definition).

given a set of n aerial images of the same urban scene, $N \geq 2$. The preprocessed images are of large scale, have known exterior orientation and are resampled to epipolar geometry.

problem recognize man-made objects (identify and locate) within certain three dimensional bounds.

It is assumed that the images are preprocessed, including radiometric adjustments, orientation and resampling to epipolar geometry. This is a very realistic assumption in photogrammetry since feature extraction is usually preceded by these steps (e.g. aerotriangulation).

The proposed approach to the building recognition problem is entirely based on the belief that it cannot be solved in one giant step. Rather, it must be decomposed into suitable modules, each with proper representations and interactions. Fig. 1 depicts the major layers (modules). This structure follows strongly Marr's vision paradigm which advocates to first reconstruct the surface before any interpretation or object recognition tasks follow. The surface, called 2.5 D sketch, emerges as a composite hypothesis of different cues (individual hypotheses), such as stereo, texture, shading, color, contour and motion. Stereo is performed by seeking corresponding features, including edges and blobs (feature-based matching). The construction of the raw surface is facilitated by the assumption of epipolar images with known orientation.

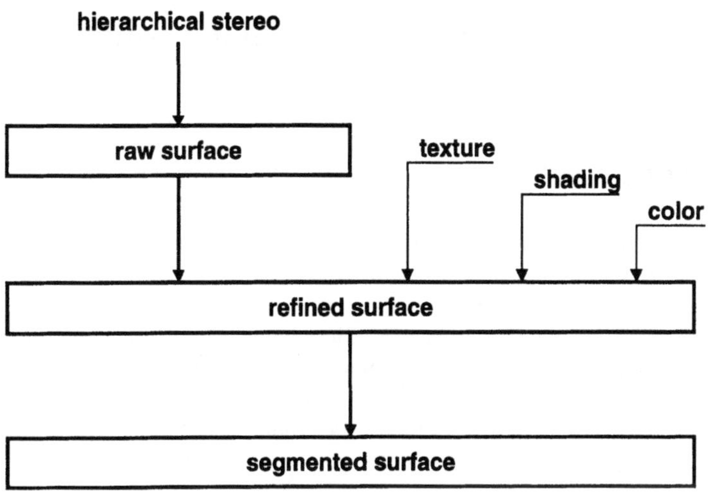

Fig. 1: Proposed layers of the object recognition problem.

The following is a concise description of our surface reconstruction system (see Schenk et al., 1991; Schenk et al., 1992). During the reconstruction process different kinds of surfaces are determined. They serve as input to the next higher level of surface representation.

current surface is obtained in the hierarchical approach of stereopsis. Stereopsis proceeds by matching increasingly higher resolution images. The matching is followed by surface densification with breakline detection and by a surface analysis. The current surface forms the corresponding level in the discrete scale space representation of the surface. Information made explicit on the current surface is used to guide subsequent iterations of the hierarchical approach.

raw surface is the scale space representation of the surface as obtained from stereo.

refined surface is the result of analyzing the scale space of the raw surface and of combining additional cues such as texture, shading and color. Thus, the refined surface corresponds conceptually to the 2.5 D sketch. Many surface properties, mainly breaklines, are made explicit.

segmented surface is a more explicit and symbolic description of the surface as compared to the refined surface from which it is derived by various segmentation processes. It consists of surface discontinuities, smooth surface patches and a separation of topographic surface, occluded areas and surfaces of features of certain dimensions. The segmented surface also includes compatibility checks with high level scene knowledge.

3 A Layered Abduction Model for Building Recognition

3.1 Background

Abduction can be described as finding the best explanation to a set of data, such as observations and facts. Although many researchers in philosophy and psychology associate perception with inference, only recently it has been occasionally proposed that perception relies on abduction (e.g. Charniak et al., 1985). Josephson takes a bold step further and hypothesizes that perception is abduction in layers (Josephson et al., 1994, Chapter 10). It appears that abductive inferences are widespread in scientific reasoning as well as in ordinary life. Indeed, we form what we take to be the best explanation of the light intensity pattern on the retina, or of an utterance of passage.

Information processing for vision as well as for natural language understanding is commonly believed to proceed in layers. A popular paradigm of vision suggests that information processing begins with retinal images by extracting features which are used in subsequent layers to reconstruct the visible surface before the cognitive tasks of recognition and interpretation are activated (see, Marr, 1982). The process is predominantly bottom up, from signal processing tasks to intermediate representations and finally to symbolic processes and abstract representations. However, top down processes are likely to occur too. For example, domain specific knowledge may be used for disambiguation.

Vision and language understanding can be thought of as a layered-interpretation task. Every layer receives the output from the previous layer and forms a composite hypothesis, that is, a coherent best explanation of the data. Whether or not the actual visual processing follows precisely an abduction model is not relevant for our problem. What matters is the fact that the process can be adequately described and computationally modeled.

3.2 Abduction Model

Abduction is a pattern of reasoning that can be characterized as finding the best explanation to a set of data. An abduction problem can be described as follows:

D_t	is a finite set of data to be explained, $D_1, D_2, \ldots, D_t$
D	is a set of data, $d_1, d_2, \ldots, d_n$
H_t	is a finite set of all individual hypotheses, $H_1, H_2, \ldots, H_t$
H	is a set of individual hypotheses, $h_1, h_2, \ldots, h_m$
h	is an individual hypothesis
e	is a map from subsets of H_t to subsets of D_t $\Rightarrow H$ explains D
p	is a map from subsets of H_t to a partially ordered set $\Rightarrow H$ has the plausibility $p(H)$

The criteria for the partially ordered set of hypotheses, $p(H)$, can be based on different measures, for example on probability, on fuzzy values or on a degree of fit.

H is said to be *complete* if $e(H) = D_t$. In that case, H explains all data. H is *parsimonious* if no other subset of hypotheses explains the data as well as H does. H is an *explanation* if it is complete and parsimonious.

As shown in detail by Bylander et al. (1991), it is intractable to determine whether there is more than one best explanation in ordered abduction problems. To be tractable, an abduction problem must have certain features and a reasoning method which takes advantage of them. A reasonable concession is to relax the optimal solution and accept a plausible solution instead. Such a solution may not explain all data and it may not be optimal.

3.3 Design Strategy

Based on the experience from building generic abduction mechanisms, primarily in the domains of diagnosis and natural language understanding, the design of an abduction system for building recognition takes into account those aspects of problem solving that appear to be generic to recognition tasks. The goal is to minimize a domain specific problem-solving strategy.

Computational models of information processing for vision follow an orderly progression of layers. Beginning with the preprocessed images as defined in the previous section, hypotheses are formed about low level features such as edges, and proceeding to higher level hypotheses about the different surfaces.

In accordance with Josephson et al. (1994), each locus of hypotheses formation is called an agora after the marketplace where the ancient Greeks gathered for dialog and debate. Here, hypotheses of certain types contend until, under good conditions, a consensus hypothesis emerges. For example the refined surface agora is the presumed location where surface hypotheses from stereo, texture, shading and color (or multispectral imagery) are contending. Each specific hypothesis accounts for certain specific data.

The agoras are organized according to the proposed layers of Fig. 1. The flow of information is bi-directional, with an emphasis of bottom up between earlier layers and an increasingly important top down flow in later layers. The main output of an agora is data to be interpreted by higher agoras. An example for a downward output is the expectation how the surface of an urban area should look like. This may then influence the evaluation of hypotheses at lower agoras.

3.4 Processing Strategy at Agoras

The information processing at each agora is decomposed into three types of activities.

evocation of hypotheses is triggered by the data and is typically bottom up. One set of data may give rise to several hypotheses. An example for an evocation stimulated top down is priming, that is an expectation from a higher level.

instantiation of hypotheses is the act of evaluating and scoring each hypothesis independently. At the same time a determination is made what data the hypothesis can account for. This is not necessarily identical to the data that evoked the hypothesis or that was used to score it. Hypotheses with low scores can be eliminated from further considerations, at least temporary.

composition occurs when instantiated hypotheses are evaluated with an emerging coherent best interpretation. Knowledge of interactions between hypotheses is necessary either to increase or to reduce the confidence of initially similarly weighted hypotheses. A proven strategy from earlier abduction machines begins the composition process by tentatively accepting the highest confidence conclusion that can be identified, and then propagating the consequences along known lines of hypothesis interaction. If this first attempt fails, the process is repeated with a lower confident conclusion.

4 Concluding Remarks

Recognizing buildings from large scale aerial images is commonly considered a hard problem, involving signal processing tasks and perception. It is unlikely that a one-step solution can be found. The proposed method in this paper is based on the assumption that the underlying information processing is modular (occurs in layers) and can be described and modeled as layered abduction. Abduction is inference to the best explanation. Whether or not the actual information processing of humans follows abduction is not relevant for our problem. However it appears that the building recognition problem can be usefully described and computationally modeled by abductive inference.

Current research draws heavily on previous experience with designing and building abduction machines in the domains of language understanding and diagnosis, and on experience gained with surface reconstruction. After completing the design, an abduction machine suitable for building recognition will be developed. It is hoped that these efforts will provide insights for developing a generic abduction model for object recognition. We expect that relationships between different objects (e.g. buildings and roads) offer additional knowledge that will be useful to disambiguate more complex cases.

References

Bylander T., D. Alleman, M.C. Tanner and J.R. Josephson (1991) The Computational Complexity of Abduction. *Artificial Intelligence*, vol. 49, 25–60.

Charniak E. and D. McDermott (1985) *Introduction to Artificial Intelligence*, Addison-Wesley, Reading, MA.

Josephson J.R., S.G. Josephson (1994) *Abductive Inference*, Cambridge University Press, New York, NY.

Marr D. (1982) *Vision*, Freeman and Company, New York.

Schenk T., C. Toth (1991) Reconstructing Visible Surfaces, *SPIE Proceedings Industrial Vision Metrology*, vol. 1526, 78–89.

Schenk T., C. Toth (1992) An Automatic System for DEM Data Collection. *ASPRS Proceedings, Spring Convention*, Baltimore, 215–226.

Wang Z. and T. Schenk (1992) 3D Urban Area Surface Analysis. *ISPRS Proceedings*, Washington, D.C., Vol. 29, part B3, 720–726.

Automatic Extraction of Man-Made Objects
from Aerial and Space Images
Monte Verità, © 1995 Birkhäuser Verlag Basel

Detection of Buildings from Monocular Images*

Chungan Lin, Andres Huertas and Ramakant Nevatia

Institute for Robotics and Intelligent Systems
University of Southern California
Los Angeles, California 90089-0273, U.S.A.
Ph.:(213) 740-6428, Fax: (213) 740-7877
e-mail: {chungan,huertas,nevatia}@iris.usc.edu

Abstract

We describe a system for detection and description of buildings in aerial scenes. This is a difficult task as the aerial images contain a variety of objects. Low-level segmentation processes give highly fragmented segments due to a number of reasons. We use a perceptual grouping approach to collect these fragments and discard those that come from other sources. We use shape properties of the buildings for this. We use shadows and walls to help form and verify the hypotheses generated by the grouping process. This latter step also provides 3-D descriptions of the buildings. Our system has been tested on a number of examples and is able to work with overhead or oblique views.

1 Introduction

The goal of this work is to detect and describe buildings from monocular views of arbitrary aerial scenes. This is a difficult but important task for many applications such as photo-interpretation, cartography and surveillance. Building detection is difficult for several reasons. The contrast between the roof of a building and surrounding structures such as curbs, parking lots, and walkways can be low. The contrast between the roofs of various wings, typically made of the same material, may be even lower. Low contrast alone is likely to cause low-level segmentation to be fragmented. In addition, small structures on the roof and objects, such as cars and trees adjacent to the building will cause further fragmentation and give rise to "noise" boundaries. Roofs may also have markings on them caused by dirt or variations in material. Shadows and other surface markings on the roof cause similar problems.

There are other characteristics of these images which may cause problems. Roofs have raised borders which sometimes cast shadows on the roof. This results in multiple close parallel edges along the roof boundaries and often these edges are broken and disjoint. At roof corners and at junctions of two roofs, multiple lines meet leading to a number of corners making it difficult to choose a corner for tracking. A roof cast a shadow along its side and often there are objects on the ground such as grass, trees, trucks, pavement, etc., which lead to changes in the contrast along the roof sides.

Consider the building in Fig. 1a (from a scene of Ft. Hood in Texas.) We use an overhead

* This research was supported, in part, by the Advanced Research Projects Agency of the United States Department of Defense and was monitored by the U.S. Army Topographic Engineering Center under Contract No. DACA76-93-C-0014.

a. b.

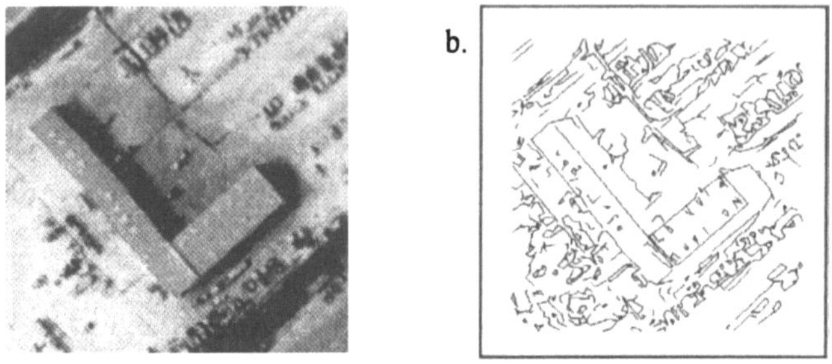

Fig. 1: (a) A building from Ft. Hood, Texas. (b) Line segments extracted from the image.

view as a running example, for simplicity. The building is easy for humans to see and describe, but it is difficult for computer vision systems. Fig. 1b shows the line segments detected in the image, using LINEAR, our linear feature extraction software (Nevatia, Babu, 1980, Canny, 1986). We are still able to see the roof structures of the buildings readily and easily, but the complexity of the task now becomes more apparent. The building boundary is fragmented, there are gaps and missing segments. There are also many extraneous boundaries caused by other structures in the scene.

There have been many previous attempts to solve this problem (Huertas, 1983, Huertas, Nevatia, 1988, Mohan, Nevatia, 1989, Huertas et al. 1993, Irvin, McKeown 1989, Liow, Pavlidis, 1990, Venkateswar, Chellappa, 1990). Building detection requires robust segmentation techniques and methods to infer the 3-D structure. These methods rely on edges or regions extracted from the image. Simple edge-based methods attempt to collect linked edge curves into the desired object boundaries, and succeed only for relatively simple scenes. Some edge-based methods have used some form of a contour tracing technique, see for example (Huertas, 1983, Huertas, Nevatia, 1988, Venkateswar, Chellappa, 1990). These are essentially local techniques that must make a decision of which path to trace at each local junction. Of course, all paths could be traced using backtracking but then the search space may become prohibitively large. Region based techniques construct closed curves that often do not correspond to the objects of interest.

Model based techniques can deal with fragmentation but require a-priori shape models. These systems have shown interesting performance but on limited examples. None of these systems can generate a description of the buildings at the level of shape descriptions of the different wings.

We have proposed, instead, to use a perceptual grouping approach. Cultural features such as buildings represent structures that are not random but have specific geometric properties. In this we restrict the shapes of buildings to be a single or a composition of rectangular parallelepipeds (thus allowing L, T and I shapes for example).

Previous systems have assumed that the viewpoint is more or less overhead. The system described here uses the viewpoint angles to deal with images acquired from an oblique viewpoint. The geometric constraints relevant to shape take into consideration, as a function of the viewpoint angles, the expected skewness of the rectangular surfaces that most buildings are expected to have. This property is used to organize the detected line segments into roof hypotheses. While the visible building sides (walls) can be hypothesized similarly, we only use them to help validate roof hypotheses. We believe that this approach leads to many fewer hypotheses than would be generated by a complete contour tracing scheme.

Our approach combines several of the techniques from previous work. Our perceptual grouping approach comes from the stereo analysis work described in (Mohan, Nevatia 1989). Here we use monocular cues and a very different hypotheses selection technique. Our shadow analysis method is an extension of the approach first described in (Huertas, 1983, Huertas, Nevatia, 1988, Huertas et al., 1993.)

Fig. 2 shows the components in our system. The system uses the line segments approximating the intensity boundaries to compute linear structures junctions among them. A hierarchy of features is constructed, leading to the formation of parallelogram hypotheses. These consist of instances of shapes that potentially correspond to building roofs and parts of building roofs. Next, promising hypotheses are selected and verified to correspond to roofs of buildings. Shadow and information, if available, is used to help form, select and verify hypotheses. Evidence of building walls in oblique views is also used to aid verification and to help derive estimates of the height of the structures, leading to a 3-D description of the scene.

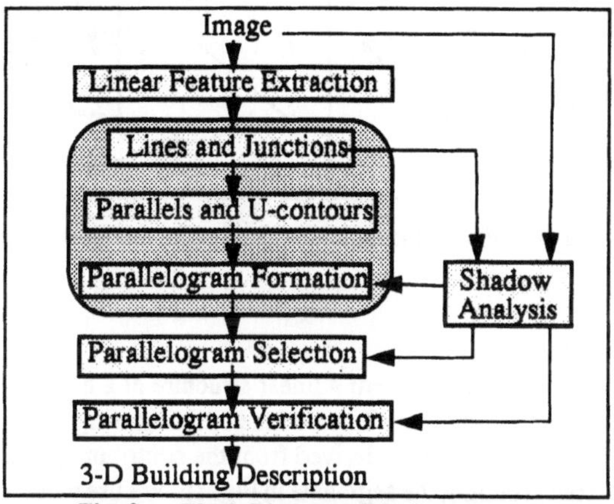

Fig. 2: Block Diagram of the System.

Our design philosophy has been to make only those decisions that can be made confidently at each level. Thus, we choose to generate as many hypotheses as seem feasible at the first level. Our selection process too is conservative and favors keeping hypotheses that may be viable. The verification process has the most global information and can make stronger decisions. Even here, if our system is to be embedded in a larger system, some of the decisions would be deferred to that system where more context is available for decision making.

The technique we describe in this paper, we believe, significantly extends the range of scenes that can be analyzed though many problems remain. Lastly, we show examples of scenes with many buildings, provided by the U.S. Government sponsored RADIUS program to demonstrate the effectiveness of our technique.

2 Generation of Hypotheses

The process of hypotheses formation is similar to the one described in (Mohan, Nevatia, 1989) with the appropriate extensions to oblique views and the use of strong shadow cues.

We construct a feature hierarchy which encodes the structural relationships specific to oblique views of rectangular shapes, presumably corresponding to the visible roof surfaces: Lines, skewed parallels, skewed U-contours, and skewed rectangles or parallelograms. Fig. 3 shows the viewing and sun angles involved. We expect that images from aerial scenes have a camera model and other data associated with them from which these angles can be derived. Next, we describe the hierarchy of features in our system:

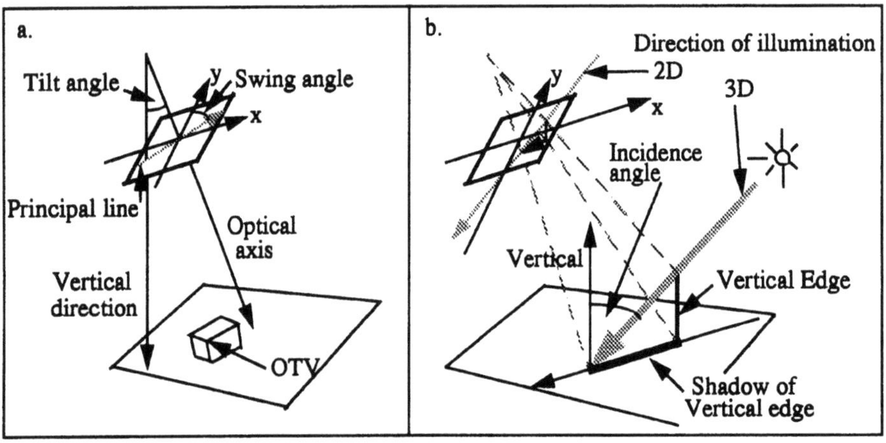

Fig. 3: (a) 3-D Viewpoint angles. (b) Sun angles and shadow geometry

Lines and Junctions

A group of close parallel lines represent a linear structure at a higher granularity level than the edges (see the common boundary between the building wings in Fig. 1a.) The resulting lines have a length and an orientation derived from the contributing elements. Fig. 4a shows the lines obtained from grouping the segments in Fig. 1b. We use these lines to detect L-junctions and T-junctions, also shown in Fig. 4a. For oblique views we also look for evidence of vertical edges in the immediate neighborhood of the L and T-junctions, thus allowing us to detect potential OTV's (Orthogonal Trihedral Vertices.)

Parallels and Skewed U-structures

Structures in urban scenes like buildings, roads and parking lots are often organized in regular grid-like patterns. These structures are composed of parallel sides. Thus, for each significant line-structure detected in the scene, there is not one but many lines parallel to it. For each line, we find lines that are parallel and satisfy a number of reasonable constraints. Note that the formation of a parallel structure also aids in the formation of new lines, as they suggest extension and contraction of the parallels to achieve full skewed overlap.

When the two lines in a parallel structure have their ends aligned as a function of the viewing angles, they strongly suggest the presence of a line with which the parallel structure would form a skewed U-structure. Even if the third line does not exist in the set of lines, we hypothesize it and generate the U-structure.

Skewed Rectangles or Parallelograms

Parallelogram structures are generated from the U-structures. The parallelograms formed in our example are shown in Fig. 4b. Thu number of parallelograms formed can be reduced by restricting their formation on the basis of size, as a function of image resolution, for example. Parallelograms are also generated from matching junctions along the direction of illumination (see Huertas et al., 1993.) We hypothesize the missing portions of a parallelogram having a corner with a matching shadow corner or evidence of an OTV.

a. b.

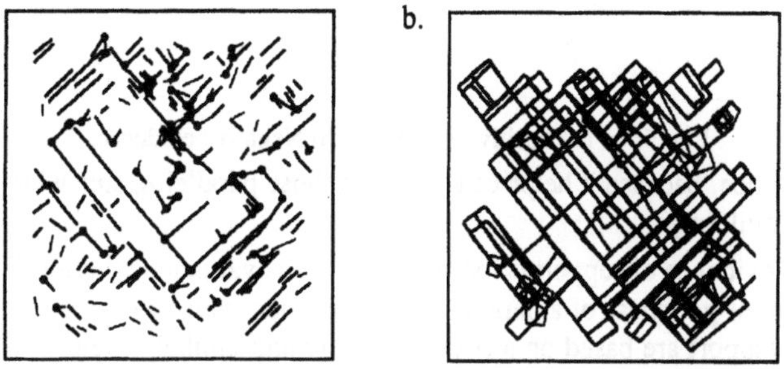

Fig. 4: (a) Linear Structures and Junctions. (b) Parallelogram generated

3 Selection of Hypotheses

After the formation of all reasonable parallelograms, a selection process is applied to choose those having strong evidence of support and minimum conflict among them. We use a criteria-based method which uses two kinds, *local* and *global* selection criteria. Local selection criteria determine whether or not a parallelogram is "good" on the basis of local supporting evidence. Good parallelograms are retained for global selection even if they overlap. Global selection criteria select the best consistent parallelograms from good parallelograms.

We apply local selection (or evaluation) criteria and global selection criteria differently. Local selection criteria work together to evaluate the goodness of a parallelogram, while global selection criteria work separately. Each global selection criterion acts like a filter. The set of retained parallelograms pass through all filters and the set of parallelograms coming out from the last filter will be the set of parallelograms selected by the selection process.

The local selection criteria are used to compute, for each parallelogram, a goodness value to be compared to a threshold value. Every evaluation criterion is weighted according to its importance. The goodness value is then measured by the sum of the weighted values calculated by the evaluation criteria. The problem of measuring the goodness of a parallelogram now becomes a problem of finding and formulating good evaluation criteria, and assigning appropriate weights. This allows us to remove parallelograms formed using weak evidence.

Whether a parallelogram is good or not depends on the evidence of support. We distinguish between positive evidence and negative evidence of support. The positive evidence of support includes the presence of edges, corners, parallels, OTV's and shadows (Fig. 5).

The negative evidence of support includes the presence of lines crossing any side of a parallelogram, existence of L-junctions or T-junctions in any side of a parallelogram, existence of overlapping gaps on opposite sides of a parallelogram, and displacement between four sides of a parallelogram and its corresponding edges support. Negative evidence is as important as

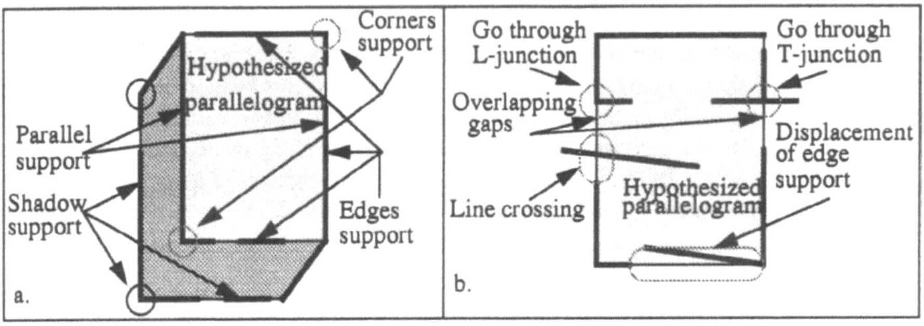

Fig. 5: (a) Positive evidence. (b) Negative evidence.

positive evidence, because it helps us to remove those parallelograms which are less likely to be part of buildings.

Each kind of evidence of support is formulated into an evaluation criterion. There are no formal definition of goodness of a parallelogram thus, our evaluation criteria formulated from evidence of support are based on analysis of likely and unlikely events. For example, four junctions are very unlikely to fall on the four corners of a parallelogram accidentally. Also, from the Gestalt Laws of Perceptual Grouping, the Law of Closure suggests that the existence of L-junctions or T-junctions on a side of a parallelogram will make a closure on part of the parallelogram and it means that the hypothesized parallelogram is not good. Some evidence of support are not always available such as the shadow evidence and the OTV evidence, but they are very important because it is very unlikely that some shadow features will appear around the hypothesized parallelogram by chance and the probability for three lines to form an OTV corner by chance is very small. We emphasize the importance of an evaluation criterion by assigning a higher weight to it.

Positive weights are assigned to those evaluation criteria formulated from positive evidence of support while negative weights are assigned to those evaluation criteria formulated from negative evidence of support. A weight should be assigned to each evaluation criterion according to the probability of existence of buildings under the condition of presence of the evidence of support from which the evaluation criterion is derived. However, we do not have the probabilistic analysis of goodness of a parallelogram, but the problem of optimal weights assignment for a given set of examples could be formulated into a search problem.

Good parallelograms surviving local selection may compete with each other. For example, some parallelograms could share the same edges or corners support and some parallelograms might overlap with each other. The goal of global selection criteria is to select a minimum set of parallelograms which best describe the composition of the scene. There are four global selection criteria in our system. They deal with duplicated, mutually contained, fully contained, and overlapping parallelograms. They examine competing parallelograms and choose one if appropriate. The selection is based on relative properties of each parallelogram, the amount and kind of overlap, and whether they share support or not. Note that a parallelogram fully contained in another is not necessarily removed. If a parallelogram does not overlap with any other parallelogram then it is not in competition, and is kept.

The flexibility of our scheme allows for straightforward incorporation evaluation of goodness and addition of filters expressing additional global relationships. The parallelograms selected in our Ft. Hood example after both the local and global selection criteria have been applied are shown in Fig. 6a.

4 Verification of Hypotheses

The purpose of verification is to validate the selected hypotheses to correspond to buildings. Our validation step includes **shadow verification process and wall verification process**. A hypothesis could be validated by either shadow and/or wall evidence. Also, these evidence provide our system the 3-D information to create a 3-D model of the structures.

4.1 Shadow Verification Process

The purpose of shadow verification process is to establish the correspondences between shadow casting elements and shadows cast, and the use of these correspondences to verify a hypothesis. We assume that the ground surface in the immediate neighborhood of the structure is fairly flat and level. The shadow casting elements are given by the sides and junctions of the selected parallelogram hypotheses. The shadow boundaries are searched for among the lines and junctions extracted from the image.

There are a number of difficulties that prevent the accurate establishment of correspondences however. Building sides are usually surrounded by a variety of objects such as loading ramps and docks, grass areas and sidewalks, trees, plants and shrubs, vehicles, light and dark areas of various materials. Nearby structures may reflect light into the shadowed areas making the objects in it more visible, and so on. To deal with these problems we have adopted some geometric and projective constraints and special shadow features.

The potential shadow evidence is extracted from image elements and knowledge of the sun angles: Lines parallel to the projected sun rays in the image may represent potential shadow lines cast by vertical edges of 3-D structures, lines having their dark side on the side of the illumination source are potential shadow lines. Junctions among the lines above are potential shadow junctions, and neighborhood pixel statistics give relative brightness.

The shadow verification process finds all shadow evidence from the extracted potential shadow evidence within a given distance of a roof hypothesis. This distance can be set to be an arbitrary value, a range of values, or derived from building height information if available.

We evaluate the shadow evidence and give a score as a weighted sum of the evidence of shadow lines cast by roof, shadow lines cast by vertical lines, shadow junctions and the shadow region statistics. Equation (1) is used to compute a score for the shadow evidence of a hypothesis p given a building height H. h_i is one of the evaluation functions of different shadow evidence and u_i is the corresponding weight.

$$S(p, H) = \sum_i u_i \cdot h_i(p, H) \tag{1}$$

4.2 Wall Verification Process

Generally speaking, some walls of buildings will be visible in oblique view images. As obliqueness increases wall information becomes more useful and shadow information becomes more difficult to handle, if available at all. Walls, after all, are part of the structures and some strong assumptions can be made, such as being vertical.

Collecting and evaluating wall evidence proceeds similarly to the shadow processing described above. We search within an arbitrary window, or within minimum and maximum weight values, if these are known. The system can either do an exhausted search over the search range or look for some evidence first and do a smart search just over the range with higher probability of existence of wall evidence.

The purpose of wall process is to find all wall evidence at every possible building height for each and every roof hypothesis. Given a building height, we can compute the height of walls in 2-D and estimate the expected wall boundary for a hypothesis. All evidence around the wall boundary will be collected and a score will be evaluated for the wall evidence.

Search and Evaluation of Wall Evidence

Given a roof hypothesis and the viewing angles we determine which sides should be visible. The swing angle gives the vertical direction from which building sides are hypothesized. We delineate the wall boundary for a given building height and activate a search process to collect all evidence around the delineated wall boundary.

The evaluation process evaluates the wall evidence collected from the previous step. Basically the score is a weighted sum of the evidence of ground-boundary, vertical-boundary, and corners. Equation (2) is the evaluation function of the wall evidence of a hypothesis p given a building height H. k_i is one of the evaluation functions of different wall evidence and v_i is the corresponding weight.

$$W(p, H) = \sum_i v_i \cdot k_i(p, H) \tag{2}$$

4.3 Combination of Shadow and Wall Evidence

Next we determine at what building height can each hypothesis find the best combined shadow and wall evidence. For this, we formulate a function to evaluate a score for the shadow and wall evidence of a hypothesis (p) for a given building height (H). The corresponding shadow and wall scores are given from equations (1) and (2). We use the equation (3) from certainty theory to combine these two scores.

$$\begin{aligned} C(p, H) &= S(p, H) + W(p, H) - S(p, H) \times W(p, H) \\ \text{where} \quad & 0 \le S(p, H), W(p, H) \le 1 \end{aligned} \tag{3}$$

The possible building heights are limited by the range between the minimum building height and the maximum building height. For each hypothesis, the building height gives the highest combined score will be considered the most possible building height of the hypothesis and the score will be called the confidence value of the hypothesis. If the confidence value of a hypothesis is greater than a given threshold value, the hypothesis will then be verified. The parallelograms verified in our Ft. Hood example are shown in Fig. 6b.

5 3-D Description of the Scene

Verified roof hypotheses have a height associated with them, allowing us to derive a 3-D model. The verified roofs are used to generate an "elevation map" with the pixels encoding structure height. The map can be rendered and viewed from an arbitrary viewpoint. Other 3-D representations such as wire frame models, are easily derived also. A 3-D arbitrary view computed from the parallelograms verified in our Ft. Hood example is shown in Fig. 6c.

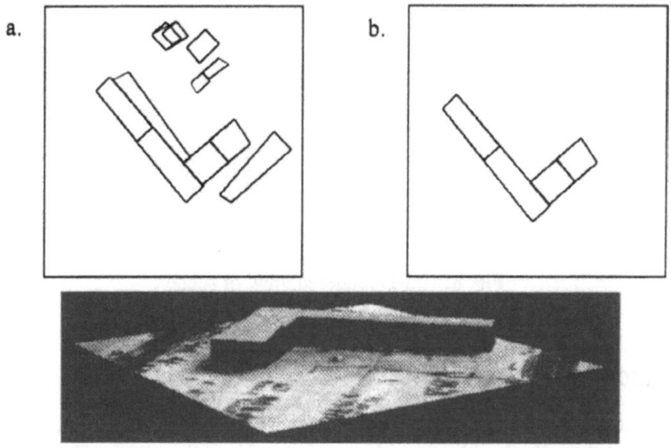

Fig. 6: (a) Selected roof hypotheses. b) Verified roof hypotheses. (c) 3-D rendered view.

Fig. 7: (a) Nadir view scene. (b) Oblique view scene.

6 Results

Our system has been tested on a number of examples provided by the RADIUS program with very good results. Due to lack of space, we show only the final result for two images of a modelboard scene. Fig. 7a shows the result for a nadir view and Fig. 7b the result for an oblique view. These 1320x1100 images with about 40 structures takes about one hour on a Sparc10/30 to process. In Fig. 9 we show an automatically generated 3-D view of the scene, from an arbitrary viewpoint.

Fig. 8: 3-D rendered view of scene

7 Conclusion and Future Work

We plan to continue testing our system with a variety of scenes, and to incorporate capabilities to detect and describe buildings with a variety of building roofs. Currently we assume that the detected and verified structures lay on the ground. Some structures however are located on top of other structures. That level of refinement of the description requires an additional step in our system and is one of the subjects of our current and future work.

References

Canny J. (1986), "*A Computational Approach to Edge Detection*", IEEE-PAMI, Vol. 8, No. 6, pp 679-698.

Chung C., Nevatia R (1992) *Recovering Building Structures from Stereo*, In Proceedings of the IEEE Workshop on Applications of Computer Vision, pp 64-73, Palm Springs, Calif., December 1992.

Huertas A. (1983), *Using Shadows in the Interpretation of Aerial Images*, University of Southern California USC-ISG Technical Report 104, October 1983.

Huertas A., Nevatia R. (1988), *Detecting Buildings in Aerial Images*, Computer Vision, Graphics and Image Processing, Vol. 41, No. 2, pp 131-152.

Huertas A., Lin C., Nevatia R. (1993), *Detection of Buildings from Monocular Views of Aerial Scenes using Perceptual Organization and Shadows*, in Proceedings of the 1993 ARPA Image Understanding Workshop, Washington D.C., pp 253-260, April 1993.

Irvin R., McKeown D. (1989) *Methods for exploiting the Relationship Between Buildings and their Shadows in Aerial Imagery*, IEEE-SMC, Vol. 19, No. 6, pp 1564-1575.

Liow Y., Pavlidis T. (1990) *Use of Shadows for Extracting Buildings in Aerial Images*, Computer Vision, Graphics and Image Processing, Vol. 49, pp 242-277.

Mohan R., Nevatia R. (1989) *Using Perceptual Organization to Extract 3-D Structures*, IEEE-PAMI, Vol. 11, No. 11, pp 1121-1139.

Nevatia R. Babu R. (1980) *Linear Feature Extraction and Description*, Computer Vision, Graphics and Image Processing, Vol. 13, pp 257-269.

Venkateswar V., Chellappa R. (1990) *A Framework for Interpretation of Aerial Images*, In Proceedings of the International Conference on Pattern Recognition, pp 204-206, Atlantic City, NJ.

Automatic Extraction of Man-Made Objects
from Aerial and Space Images
Monte Verità, © 1995 Birkhäuser Verlag Basel

High-resolution Stereo for the Detection of Buildings

M. Berthod* L. Gabet† G. Giraudon* J.L. Lotti*

Abstract

After a brief overview of the research going on in the Pastis group at Inria, which covers retrieving depth and recovering symbolic information from remotely sensed data, we concentrate in this paper on a particularly difficult task, *high-resolution stereo*, which deals with images with a resolution of one meter or less, on which there are smooth textured *natural* areas as well as discontinuities, or even occultations, of man-made structures. Two approaches are proposed for extracting reliable dense depth maps: one relies on the use of several aerial images, with increasing disparities, and the other on an adaptive window correlation scheme, which prevents the correlation window to extend over radiometric, and thus depth discontinuities.

1 Introduction

The extraction of cartographic information from satellite or aerial imagery is now a very active research field. There are two kinds of such information: planimetric (2-D) and altimetric (3-D). In the Pastis team, at Inria, we have worked for about ten years on both types. Planimetric information, such as roads, buildings, rivers, urban areas and more generally land use classification, can be extracted automatically, at least in part, from SPOT images [10] or from aerial images [4]. Using complementary sensors such as SPOT and ERS-1, along with previously available cartographic data, broadens of course the scope of informations that can be retrieved. We have devised a knowledge-based system, based on a *blackboard architecture*: MESSIE [1] which fuses, at a symbolic level, data from different sources (eg. SPOT, ERS-1 ...) to extract meainigful entities such as rivers, bridges .... Satellite or aerial data can also provide 3-D information. Stereo from SPOT data has been operationnal [13] for seven years. More recently, interferometry from SAR images has triggered a lot of efforts [7]. Higher resolution poses different sorts of problem, anyway. This paper presents the problem of the reconstruction of sub-metric resolution 3D images in urban areas using automated stereovision techniques and proposes two approaches which contribute to the solution of this problem.

*INRIA, BP 93, 06902 Sophia Antipolis Cedex (France)
†ISTAR, Espace Beethoven, Rte des Lucioles Sophia Antipolis Cedex (France)

2 State of the art

Typically, approaches in stereovision can be roughly divided into two categories : feature-based and aera-based (see for example [2]). The last one is based on photometry and usually uses algorithms based on correlation. It assumes that objects are textured and that the surface is continuous. It yields dense maps, but those are often erroneus near strong discontinuities. Okutomi and Kanade [11] propose a solution which is based on the computation of an adaptative correlation window the size of which is determined by noise modeling and information associated with the image gradients. This method allows refining an initial disparity map. In the next section, we present an adaptive window method which uses contour information constraint.

An other possibility (fusion-based methods) is to combine matches coming from photometry-based algorithms and from algorithms based on structured attributes. These works [5, 6] aim at combining the advantages of both approaches: a dense map and reliability near discontinuities. We may quote also model-based approches: [9] (surface model) and [3] (CAD model).

Finally, in [12], Okutomi et al presents a stereo matching method which uses several stereo pairs with increasing B/H. This work is theoretically well argued and based on a photometric method which attempts to minimize a cost function of the disparities on all the stereo pairs. The aim is to obtain a dense map which is both as precise as possible (large B/H) and as reliable as possible (small B/H). In section 4, a method which combines the last approach with fusion-based method is presented.

3 Contour Based Adaptive Window Algorithm

The goal of the Contour Based Pyramidal Resolution method *CBPR*, presented below, is to obtain an initial disparity map with good discontinuities localization, we extend our work on the *CBDL Contour Based Disparity Limits* algorithm, using a pyramidal model (with a reduction coefficient of 2).

In the following sections, we assume that images have been resampled along the epipolar geometry.

3.1 CBDL

CBDL uses the smallest limited contour window between left and right images to compute the disparity. This prevents occluded pixels to be taken into account in the correlation score (Fig. 1).

We compute the disparity using four excentred windows for each image pixel P (Fig. 1.5). An order constraint is then applied, before a cycle of left-right validating and completion stages gives a dense map. The contours used to limit the window are provided by a Deriche filter with manually thresholded hysteris.

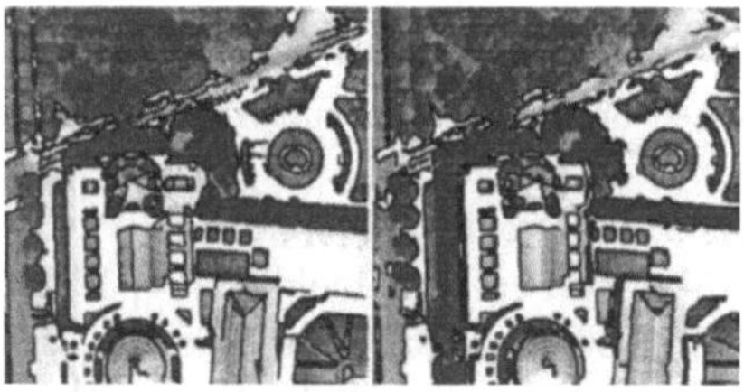

1)Left and right Contours with intensity, image of Grand-Palais.

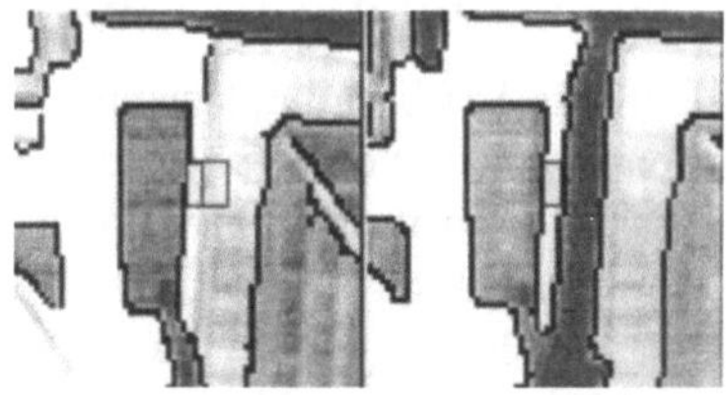

3)The contours of the right image constrain the window size in the left one.

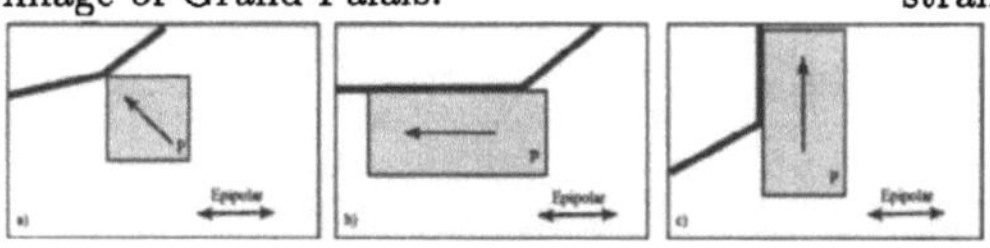

2.a) maximum square from P. 2.b) grow parallel to epipolar line if possible. 2.c) try perpendicular side if (2.b) is not possible.

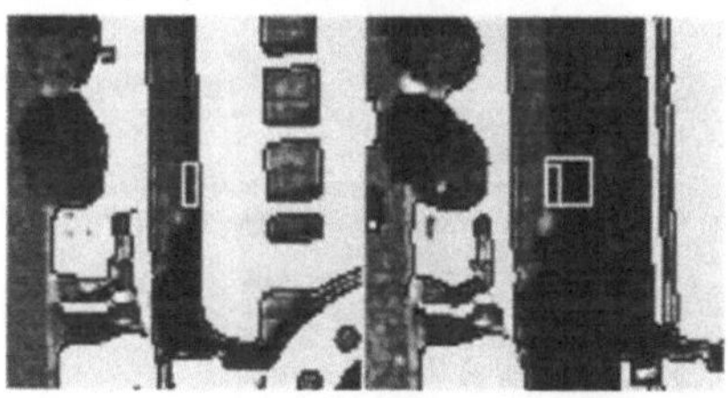

4)Relative motion of contours indicates an occluded area. So we need to take the smallest limited window.

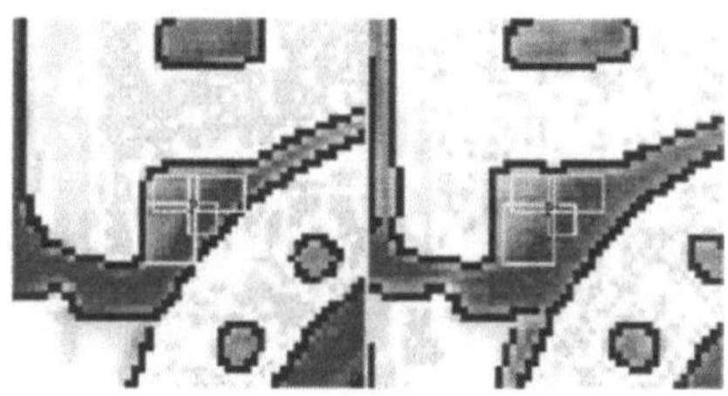

5)Four minimal windows associated to the pixel P for disparity d.

Figure 1: Window growing steps (Black line is contour, P the observed pixel in the corner.

3.2 CBPR

In *CBPR* [8], we use the disparity of *CBDL* method at each level. If none of the four excentred windows is validated, a centered one with a smaller size is tried. From estimations of the previous level of pyramid, we know that the disparity d is in a range $[d \pm Disp_{E}rr]$, where $Disp_{E}rr$ is fixed to 3 pixels. If this estimated disparity d is not confirmed in the pyramid space due to the lack of texture at this level, we propagate the last validated disparity estimation.

3.3 Technique to refine

At this point, we have a true localization of the disparity discontinuities on the dense initial map with pixel precision, a grey level correction between left and right images and anestimation of noise. So we may apply Kanade correction on it with non constant window size [11] to obtain sub-pixel value.

3.4 Experimental Results

The algorithm has been implemented and tested on images with natural and man-made non-polyhedric objects (eg. domes).
The classical disparity map based on correlation with fixed window size 7x7 is compared to a map obtained by *CBPR* on 512x512 the Pentagon building d=[-12,12] Figure 2. We show in Figure 3 a test on a 1000x200 image part of a Marseille survey d=[-3,30], with a DEM obtained by the multiple-baseline/multiple-algorithms method on this area. All the isolated black spots represent occulted or non matched pixels.

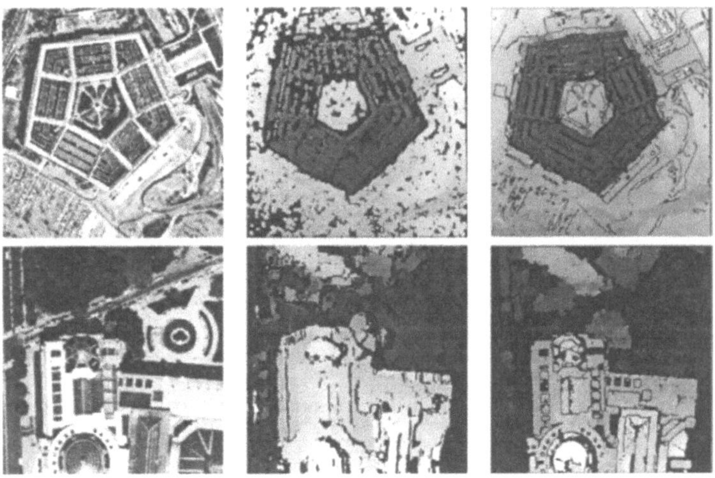

(a) (b) (c)

Figure 2: Experimental results: (a) Left images. (b) Classic cross correlation method with fixed window 7x7. (c) CBPR method.

We have also tried [8] the algorithm *CBDL* and *CBPR* on: SHARMAH area (400x400 d=[-12,7]), image of Death Valley (256x256 d=[-20,20]), Villefranche town area (400x400 d=[-12,12]), image of Paris center (600x600 d=[-3,42]), image of Grand Palais (400x400 d=[-3,53]), image of Fruits (512x512 d=[-25,30]) ...

Conclusion

We have a good localisation of disparity discontinuities with contour constraint windows *CBDL* on the correlation computation and *CBPR* method. Of course, a sub-pixel precision method can be applied on it. The results already obtained are encouraging, and we hope to confirm the interest for this technique of contour limitation by extracting CAD building structure from this map. [1]

Left image.

Right image.

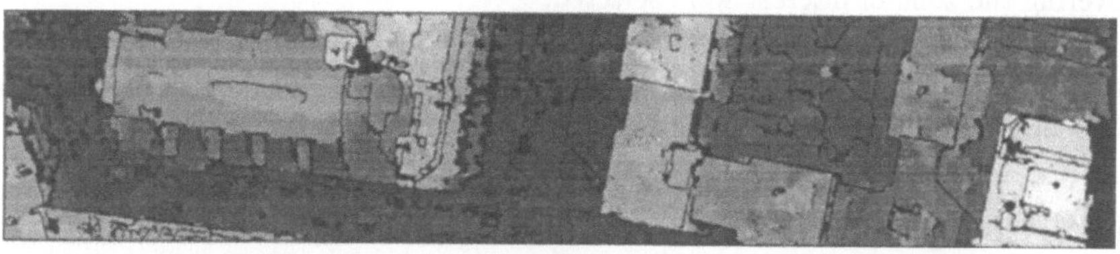

Left disparity automatic threshold with completion map CBPR.

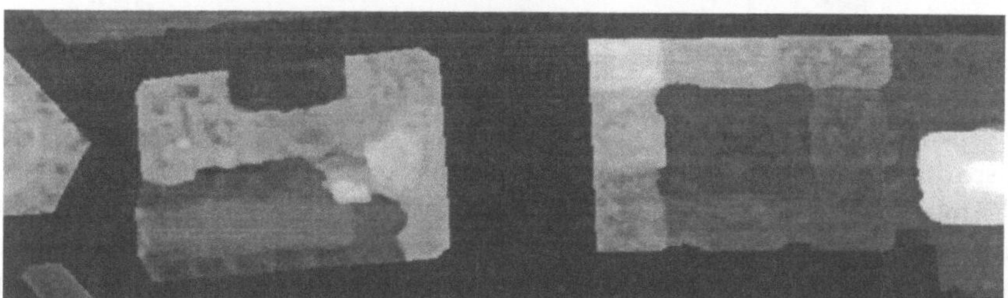

DEM from multiple-base/multiple-algorithms method.

Figure 3: Experimental Results on a part of Marseille Images.

4 A Multiple-baseline/Multiple algorithms method

The original method presented in this section is based on three principles. Firstly, it uses an image sequence to generate several stereo pairs with various B/H ratios

[1]Acknowledgements : the authors would like to acknowledge Professor G.Medioni, B.Hotz and the ISTAR Company for providing stereo pairs.

in order to obtain good precision and few errors due to hidden parts. Secondly, it generates a DEM where each pixel has the most probable and precise elevation by merging results obtained from different B/H stereo pairs and different matching algorithms based on correlation techniques. And thirdly, the DEM is segmented into two classes, terrain-DEM and buiding-DEM, each of which is refined by taking into account the constraints typical of each class. And finally, these two DEMs are combined to generate a final DEM which presents the best continuity for ground surface, and respects sharp building discontinuities.

4.1 Data

A 30 B&W aerial image track of Marseille (France) was acquired (scale 1:8000, 20 cm/pixel). The overlay between image n and $n+1$ is of 87% and, between image n and $n+3$, of 61%. This guarantees a good geometric precision.The 1 km x 1 km zone of interest, i.e., 5000x5000 pixels, was chosen for the diversity of its urban structures: the old town and the new town. Four successive representative images covering the zone of interest were selected.

The geometry of the images was modeled by aerotriangulation with ground control points on 1:5000 maps from characteristic points (sidewalks, bases of buildings, etc.) at ground level.

4.2 Using the method

The 7 main steps of the method are explained as below.

The computation of a DEM for six possible stereo pairs with different matching algorithms : Three matching algorithms were used. All three are based on intercorrelation with a fixed window size of 7x7 or 9x9. Each computation is independent from the others. The results are in the form of disparity maps. The disparities between the two images are given with respect to a reference image (here the right image). The reference image is different for all pairs so that the disparity maps cannot be compared to each other. So, to compare them, the disparity maps (x, y, disp) are projected in absolute cartographic coordinates (X, Y, Z). The disparity maps make the elementary DEM once they are projected.

The generation of a DEM "filtered" : 18 DEMs were computed from the 6 pairs and the 3 algorithms. The pairs with $87\%, 74\%$ and 61% covering respectively have a $0.14, 0.28$ and $0.42 B/H$, i.e., a matching error of one pixel generates an absolute error of $1.43m, 0.71m$ and $0.47m$ respectively. Comparing the DEMs aims at using all the difference characteristics between the DEMs in order to generate the most reliable (minimizing errors), the most precise and complete (minimizing the hidden zones) filtered DEM. The algorithm used in order to make this comparison is based on a majority vote method on all the DEMs, made pixel by pixel with a similarity threshold ($2m$). The most precise height is given priority when selecting the final height for the filtered DEM. The practical analysis of the results shows that most errors of individual DEMs do not appear in the filtered DEM (see fig. 4).

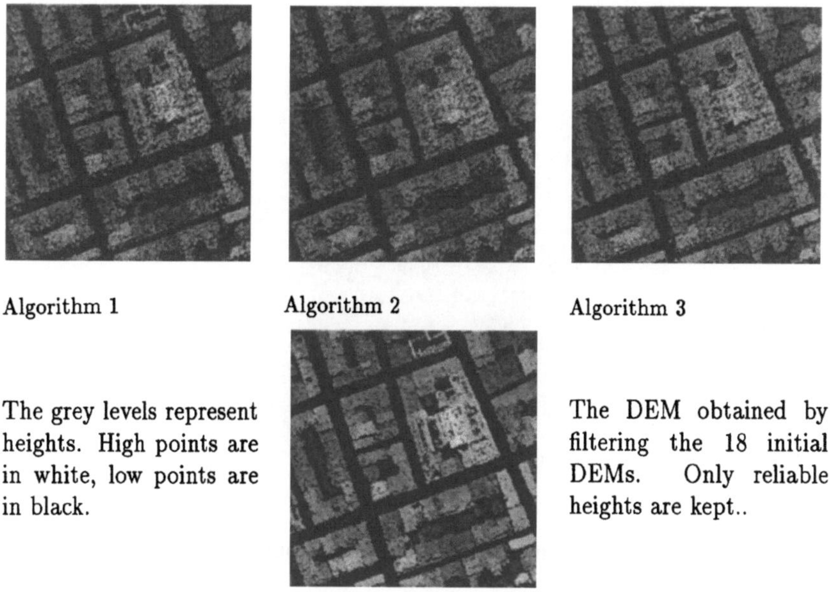

Algorithm 1 Algorithm 2 Algorithm 3

The grey levels represent heights. High points are in white, low points are in black.

The DEM obtained by filtering the 18 initial DEMs. Only reliable heights are kept..

Figure 4: For one pair (image 1 and image 2) comparison of the DEMs computed from the three algorithms and the filtered DEM. Zone size : 200m x 200m.

The segmentation of DEMs and the building selection : To eliminate remaining errors, the DEM is segmented into surfaces of homogeneous height to obtain two types of surfaces: ground and buildings. A surface is considered homogeneous if the local height differences are less than 4 m. The segmentation relies on the following criterion: a surface belongs to the ground group if its height is lower or equal to the neighbouring surfaces, if not, it belongs to the building group. A surface is said to be higher than its neighbouring if 90% of the contact points between the two surfaces are higher.

The approximation of building blocks with polygons : the analysis of the building-DEM shows that the height of the buildings is correctly recovered. On the other hand the outside edges of the buildings are poorly localized (from 2 to 5 pixel error. Most buildings or blocks of buildings are simple geometric shapes (squares or rectangles), and could be defined very precisely by convex contours. These convex shapes are created in two steps: creation of one convex polygon per region and post-processing by fusing the polygons which intersect (for block of buildings). This fusion is iterated as many times as necessary until no more polygons intersect. Finally, each polygon is assigned the average height of the building. However, concave-shaped buildings are very poorly defined with this method and must be outlined manually. This method is fairly efficient for isolating building-blocks. The polygons defining the buildings within a block are suppressed when fusing the polygons.

For evaluation, all building contours were manually extracted and compared

Figure 5: Building, Bear ground.

Figure 6: Building regions automatically obtained and manually corrected.

to the contours automatically computed. To compare the polygons, three criteria were used: the contour shape, the maximum, and the average shift in pixels between contours. The contour shapes are comparable for 70% of the buildings. The polygons with comparable shapes have maximum deviations of 5 pixels between the contours and the error average is of the order of 2 pixels.

Building processing

The polygons previously computed are used here to correct the building-DEM. Each building is constrained by the limits of the polygon. Missing altimetric information for buildings, due to non-correlated zones, is filled in by a median filter.

The validation of the building heights is very difficult. The height of the buildings was measured for about thirty points with a stereo viewer from the 60% covering pair, and compared to the corresponding value in the DEM. The results show a 1.5 m standard deviation in Z.

Ground processing

The ground-DEM obtained after selecting the homogeneous surfaces contains not only bare-ground type information, but small reliefs along the streets such as parked cars, bus shelters, trees, telephone booths, etc. are also integrated to the DEM. In order to generate a bare-ground DEM, all these small high frequencies have to be eliminated from the ground-DEM. The amplitude of these high frequencies is low, always less than $4m$. Filtering of the ground-DEM is based on the hypothesis that the ground surface is very regular when free of high frequencies, and is performed in three steps. The information-free zones (building sites ...) are interpolated by a regularisation technique. Then the DEM is convolved by a filter in order to suppress high frequencies. Finally, an interpolation by regularisation is made once more to fill in the gaps left by the filtering. With the help of 1:5000 maps, the height of about twenty spot heights was compared to the height of the corresponding points in the bare-ground DEM. The result gives a standard deviation of 40 cm for the height differences. The maximal errors are 1 m.

The final DEM : The final DEM is computed very simply : the building-DEM is overlaid on the ground-DEM in built-up areas.

4.3 Results

The results we presented for the whole zone in fig. 5 and fig. 6 show the feasability of generating urban data bases in one meter resolution from digital methods of automated stereovision.

References

[1] V. Clement, G. Giraudon, S. Houzelle, and F. Sandakly. Interpretation of remotely sensed images in a context of multisensor fusion using a multispecialist architecture. *IEEE Trans. on Geoscienceand Remote Sensing*, 31(4):779–791, 1993.

[2] U. Dhond and J. Aggarwal. Structure from stereo - a review. *IEEE Transactions on Systems, Man and Cybernetics*, 19(6):1489–1510, 1989.

[3] W. Forstner. Model based detection and location of houses as topographic control point in digital images. *International Archives of Photogrammetry and Remote Sensing*, 27:505–517, 1988.

[4] P. Garnesson, G. Giraudon, and P. Montessinos. An image analysis system, application for aerial imagery interpretation. In *International Conference on Pattern Recognition*, Atlantic City, jun 1990.

[5] M. Hannah. A system for digital stereo image matching. *Photogrammetric Eng. Remote Sensing*, 55(12):1765–1770, dec 1989.

[6] Y. Hsieh, D. McKeown, and F. Perlant. Performance evaluation of scene registration and stereo matching for cartographic feature extraction. *IEEE Transactions on Pattern Analysis and Machine Intelligence*, 14(2):214–238, feb 1992.

[7] D. Labrousse, S. Dupont, and M. Berthod. Synthetic aperture radar interferometry: a markovian approach for phase unwrapping. In *SPIE Conference Orlando 95*, apr 1995.

[8] J. Lotti and G. Giraudon. Adaptive window algorithm for aerial image stereo. In *12th International Conference on Pattern Recognition*, volume 1, pages 701–703, Jerusalem, oct 1994. IAPR, IEEE.

[9] H. Maitre and W. Luo. Using models to improve stereo reconstruction. *IEEE Transactions on Pattern Analysis and Machine Intelligence*, 14(2):269–277, feb 1992.

[10] N. Merlet and J. Zerubia. Classical mechanics and road detection in spot images. Research Report 1889, INRIA, mar 1993.

[11] M. Okutomi and T. Kanade. A locally adaptive window for signal matching. *International Journal of Computer Vision*, 7(2):143–162, 1992.

[12] M. Okutomi and T. Nakahara. A multiple-baseline stereo method. In *IEEE Image Understanding Workshop DARPA*, pages 409–427, 1992.

[13] L. Renouard. Création automatique de m.n.t. à partir de couples d'images spot. Technical report, Congrés, SPOT 1, Utilisation des Images, Bilan, Résultats. CNES, Paris, 1987.

Automatic Extraction of Man-Made Objects
from Aerial and Space Images
Monte Verità, © 1995 Birkhäuser Verlag Basel

3-D Reconstruction of Urban Scenes from Sequences of Images

Olivier Faugeras, Stéphane Laveau, Luc Robert
and
Gabriella Csurka, Cyril Zeller*
INRIA. 2004, route des Lucioles. B.P. 93. 06902 Sophia-Antipolis. FRANCE.
e-mail: Stephane.Laveau@sophia.inria.fr

Abstract

In this paper, we address the problem of the recovery of the Euclidean geometry of a scene from a sequence of images without any prior knowledge either about the parameters of the cameras, or about the motion of the camera(s). We do not require any knowledge of the absolute coordinates of some control points in the scene to achieve this goal. Using various computer vision tools, we establish correspondences between images and recover the epipolar geometry of the set of images, from which we show how to compute the complete set of perspective projection matrices for each camera position. These being known, we proceed to reconstruct the scene. This reconstruction is defined up to an unknown projective transformation (i.e. is parameterized with 15 arbitrary parameters). Next we show how to go from this reconstruction to a more constrained class of reconstructions, defined up to an unknown affine transformation (i.e. parameterized with 12 arbitrary parameters) by exploiting known geometric relations between features in the scene such as parallelism. Finally, we show how to go from this reconstruction to another class, defined up to an unknown similitude (i.e. parameterized with 7 arbitrary parameters). This means that in an Euclidean frame attached to the scene or to one of the cameras, the reconstruction depends only upon one parameter, the global scale. This parameter is easily fixed as soon as one absolute length measurement is known. We see this vision system as a building block, a vision server, of a CAD system that is used by a human to model a scene for such applications as simulation, virtual or augmented reality. We believe that such a system can save a lot of tedious work to the human observer as well as play a leading role in keeping the geometric data base accurate and coherent.

*Stéphane Laveau is supported by a grant under DRET contract No 91-815/DRET/EAR. This work was also partially funded by the EEC under Esprit project 6448, Viva and Esprit project 8878, Realise.

1 Introduction

The problem which is tackled in this paper and for which we propose a number of partial solutions is the following : we want to reconstruct a three-dimensional model of a static environment viewed by one or several cameras whose motions or relative positions are unknown and whose intrinsic parameters are also unknown and may vary.

The sequence of images that is used can be either a video sequence, or a film, or a number of snapshots taken from usually fairly distinct viewpoints. In the first two cases, which we will denote as the M(ovie)-situation, it is, as explained in section 2.1 possible to use the continuity in time of the images to help simplify the problem. In the third case, which we will denote by the S(napshot)-situation, this is not possible, and we must work a little harder, as also explained in the same section.

Solving this problem at relatively low cost is extremely important for such applications as image synthesis, simulation, virtual and augmented reality where 3-D models are required and are obtained today through lengthy manual interaction with the images. Our techniques build upon the knowledge which has been acquired in computer vision and photogrammetry in the last 20 years or so and can potentially reduce by a significant factor the amount of manual interaction which is currently necessary to get 3-D models of the world in the computer.

The Esprit project 8878, *Realise*, has set up as one of its goals the partial automation of the acquisition of computerized 3-D models of urban scenes from sequences of images taken from the ground or from an aircraft flying at a low altitude. As was mentioned before, no hypotheses are made upon the relative positions of the cameras, their intrinsic parameters, all of them are assumed unknown, or upon the presence in the environment to be modeled of control points with known coordinates in some fixed frame of reference. We nonetheless show in this article that the complete projective, affine, and Euclidean geometry (up to a global scale factor) of the scene can be accurately captured by a combination of techniques which have been developed over the years in computer vision and photogrammetry. We would like to emphasize the fact that these techniques encompass a wide range of traditionally distinct subjects such as feature detection (edges, corners, junctions) using non-parametric (image-based) and parametric (snake-like) models, tracking of image features in a sequence of images, geometric modeling of image correspondences at the projective, affine, and Euclidean levels with a clear distinction between those levels and the amount of information they require, robust estimation of algebraic instantiation of this geometry (i.e. the perspective projection matrices).

This is only one aspect of the *Realise* project, the geometric aspect; this project also aims at developing a deeper understanding of the photometry and colorimetry of real scenes in order, for example, to improve significantly the quality of the rendering of simulated scenes. This important topic is not covered in this article.

Before we describe in detail the techniques which lead from a set of images to a 3-D model of the scene they represent, we sketch in figure 1 the general flow of information in our method. We start with either a set of N snapshots (S-situation)

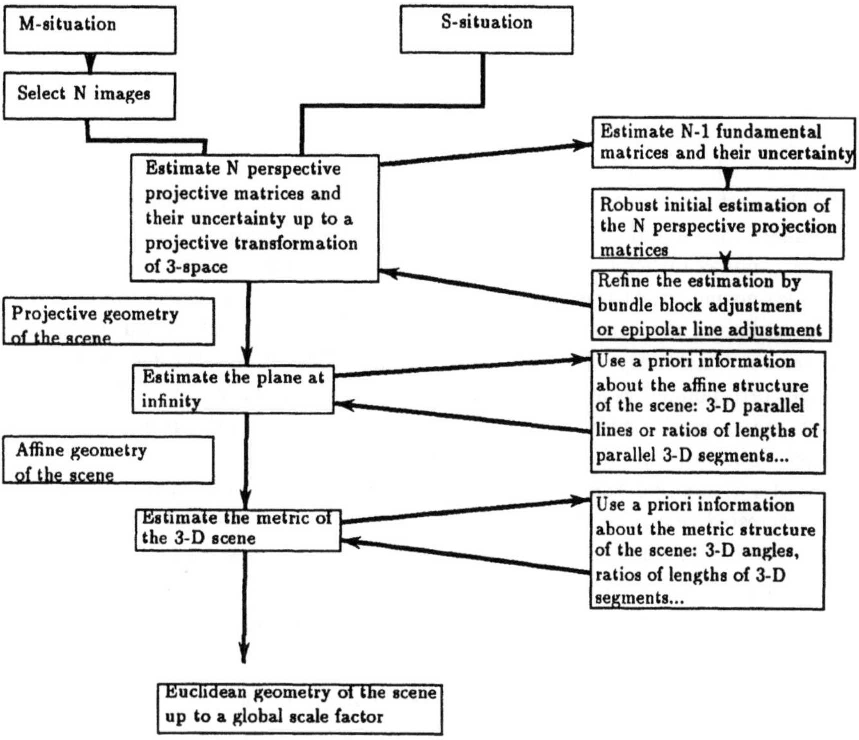

Figure 1: Flowchart of the method described in the paper: this is the backbone of the visual server which is being developed for the Esprit project *Realise*.

or with a set of video sequences from which we select N. In the second case, we can use a variety of feature trackers to establish a number of correspondences between the N images, as explained in section 2.1. Note that we do not require that all features be tracked in all sequences, a notoriously non-realistic hypothesis. Also, even though this makes the next process of estimating the perspective projection matrices easier, it is by no means mandatory.

The next step is to estimate the perspective projection matrices of each of the N images. We found that this was the best way of representing the geometry of a set of cameras rather than, for example, the fundamental matrices describing the epipolar geometry. At this stage the perspective projection matrices are defined up to an unknown projective transformation, i.e. they can be multiplied on the right by the same arbitrary 4×4 full rank matrix. This reflects the fact that if only image correspondences are available, only the projective geometry of the 3-D scene, i.e. those properties which are invariant under the action of the group of projective transformations, can be recovered. Since projective geometry is now widely used in computer vision and has been in computer graphics since the early days, we refer the interested reader to the corresponding literature [Semple and Kneebone, 1952, Mundy and Zisserman, 1992, Faugeras, 1993]. The estimation of the perspective projection matrices is done in three steps:

1. We estimate the $N-1$ fundamental matrices defining the epipolar geometry between consecutive images,
2. these are then used to obtain a first estimate of the perspective projection matrices,
3. which is then refined using possibly several methods, one of them, being the famous bundle adjustment [Brown, 1958],[Brown, 1971],[Gruen, 1978], [Gruen and Beyer, 1992].

Once this estimation has been completed, the 3-D scene can be reconstructed up to an arbitrary projective transformation. To be complete, let us mention that we track the uncertainty at all levels, starting from the pixel level, through the level of the fundamental matrices, the perspective projection matrices, and the reconstructed 3-D points. This is described in detail in section 2.

The next step, if required by the application, is to estimate the affine geometry of the scene, i.e. those properties of the scene which are invariant under the action of 3-D affine transformations. This can be achieved in several ways as described in [Faugeras, 1995] depending upon whether or not one can control the motion of the sensors. In the application described in this article we only use a priori information about the scene such as parallel lines or known ratios of collinear line segments. This is described in section 3.

The final step, again, if required by the application, is to estimate the Euclidean geometry of the scene, i.e. those properties of the scene which are invariant under the action of 3-D Euclidean transformations, i.e. similitudes. This can again be achieved in several ways as discussed also in [Faugeras, 1995]. In the application described in this article we use a priori information about the scene such as angles of lines, planes, or known ratios of line segments, to take a few examples. This is also described in section 3.

To summarize, by carefully distinguishing the various geometric levels at which the scene can be reconstructed, i.e. projective, affine, and Euclidean, we are able to determine the minimum amount of information necessary to access a given level from image measurements only. Note that, contrary to the case of the so-called "Self-Calibration" methods in photogrammetry [Gruen and Beyer, 1992], our method does not require the knowledge of the Euclidean coordinates of a small number of control points in the scene. Nonetheless, it does require, in the version presented in this article, some knowledge about the 3-D geometry of the scene such as parallel lines and angles. But we have shown in previous work [Maybank and Faugeras, 1992, Luong, 1992, Faugeras et al., 1992] that even this assumption is not necessary. We use it here because there is such a rich geometric information in images of urban scenes and because the system we are developing is partially interactive.

This last point is extremely important since it is the stepping stone of the philosophy of the design of the *Realise* project. Far from being after a complete automation of the modeling of the scene, the project aims at providing to a human agent working with a CAD system from a set of images of the environment he wants to model, the most advanced tools in computer vision to help him solve such problems as accurate (i.e. subpixel) detection of image features, matching of those features across views, estimation of the geometry of the set of views, computation of the 3-D coordinates of scene points, curves, and surfaces. This framework is depicted in figure 2.

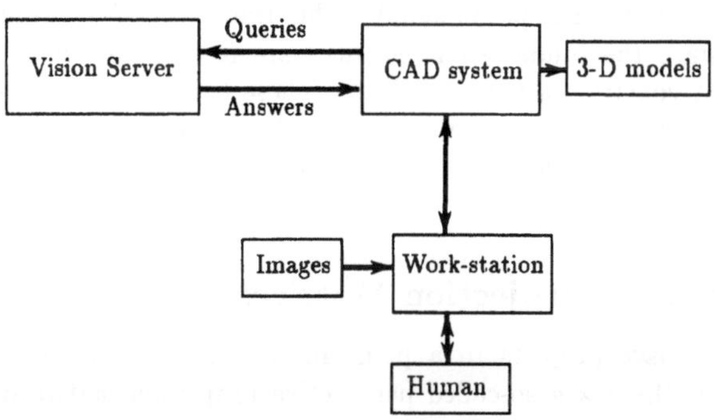

Figure 2: The principle of the interactive recovery of the 3-D geometry of the scene in the Esprit *Realise* Project.

We now briefly review the necessary background material that will be used in the rest of the article.

1.1 Epipolar Geometry

Recently, it has been discovered that the full calibration of the cameras (intrinsic and extrinsic parameters) is not needed to obtain a useful reconstruction of a scene viewed by a stereo system [Faugeras, 1992, Hartley et al., 1992]. This theory make use of epipolar geometry which can be retrieved easily from point correspondences in pair of images.

Since these first attempts at an uncalibrated stereovision, a lot of work has been done on the estimation of the epipolar geometry of two images [Luong et al., 1993a], [Luong et al., 1993b], [Luong and Faugeras, 1994c], [Luong and Faugeras, 1994b], [Luong and Faugeras, 1994a], [Hartley, 1992], [Hartley, 1993], [Olsen, 1992], [Boufama and Mohr, 1995]. Robust programs which work automatically are now publicly available. We will consider this problem as solved for the rest of this article; the interested reader is referred to the bibliography.

We will use the *fundamental* matrix representation of the epipolar geometry. In this representation, 2 points in correspondence in images 1 and 2 (expressed in homogeneous coordinates) $\mathbf{m}_1$ and $\mathbf{m}_2$ satisfy the following projective relation: $\mathbf{m}_2^T \mathbf{F}_{12} \mathbf{m}_1 = 0$. The fundamental matrix is defined up to a scale factor and satisfies $\mathbf{F}_{21}\mathbf{e}_{21} = \mathbf{F}_{12}\mathbf{e}_{12} = 0$, with $\mathbf{e}_{ij}$ being the epipole in image i generated by image j (or equivalently, the image in i of the optical center of camera j).

From this, we can conclude that the fundamental matrix depends upon at most seven parameters. It is shown elsewhere [Luong, 1992] that it depends exactly on seven parameters. Therefore, the set of all possible fundamental matrices between N cameras would depend on $7N(N-1)/2$ parameters if there were no constraints between them. However, in our simple perspective model, each camera depends on a fixed number of parameters (we use 6 for the pose and orientation and 5 for the intrinsic or internal parameters). This leads to $O(N)$ parameters for the cameras and since the fundamental matrices are represented by $O(N^2)$ parameters, there exist constraints between them. These constraints can be enumerated [Faugeras and Mourrain, 1995a, Faugeras and Mourrain, 1995b], but they are difficult to use and we prefer the more compact representation of the projective camera matrices.

1.2 Perspective Projection Matrices

A point $\mathbf{M}$ in space projects to a point $\mathbf{m}$ in the image if and only if $\mathbf{m} = \mathbf{PM}$ where $\mathbf{P}$ is the 3×4 so-called perspective projection matrix of the camera. All quantities are defined up to an unknown scale factor. It has been shown by [Luong and Viéville, 1994] that given a set of fundamental matrices satisfying the constraints, one can find corresponding projection matrices. The solution is unique up to an unknown projective transformation in space if the optical centers are not aligned.

The relation of the projection matrices to the fundamental matrices is extremely simple: If we write $\mathbf{P}_i$ as $[\mathbf{M}_i|\mathbf{t}_i]$, the epipoles $\mathbf{e}_{ij}$ satisfy Equation (1) by definition (as images of an optical center).

$$\mathbf{e}_{ij} = \mathbf{t}_i - \mathbf{M}_i \mathbf{M}_j^{-1} \mathbf{t}_j \tag{1}$$

For the fundamental matrices, an elimination scheme leads to

$$\mathbf{F}_{ij} = [\mathbf{e}_{ij}]_\times \mathbf{M}_j \mathbf{M}_i^{-1} \tag{2}$$

where $[\mathbf{e}_{ij}]_\times$ is the 3×3 matrix representing the cross-product with the vector $\mathbf{e}_{ij}$.

It is understood that these equations are projective and therefore are defined only up to an unknown scale factor. We assume that $\mathbf{M}_i$ and $\mathbf{M}_j$ are invertible. If they are not, we can always transform them by a proper choice of a projective transformation to a new frame such that they are invertible and satisfy our assumptions.

1.3 Reconstruction

From the consistent epipolar geometry, we can recover the 3-D scene up to an unknown projective transformation of space. This is not as far from a Euclidean reconstruction as it seems. The set of projection matrices for N cameras depends upon $11N - 7$ parameters in the Euclidean case, whereas only $11N - 15$ in the projective case.

These 8 additional free parameters are the *low* price to pay for not knowing the internal parameters of the cameras and their relative positions in space. We will later see how these unknown parameters can be recovered using very little information. As stated previously, the epipolar geometry of a set of cameras is best represented by a set of projection matrices that we will in the sequel assume to be defined up to a projective transformation in space.

2 Robust recovery of the geometry

In the M-situation, we select (presently manually but we plan to automate this process in the near future) a subset of the images in the sequence so that we end up in the S-situation. The important difference is that the intermediate images can be used, as explained below, to simplify the process of establishing correspondences between the views.

2.1 Obtaining correspondences between images

The algorithm used to compute the projection matrices needs correspondences and a few epipoles in order to work. We first obtain feature points using very simple corner detectors (we use [Harris and Stephens, 1988], but other possibilities are [Kitchen and Rosenfeld, 1982], [Noble, 1988], [Guiducci, 1988], [Förstner and Gülch, 1987]) and we refine their position using a model-based approach [Blaszka and Deriche, 1994].

In the S-situation, we then establish correspondences between the corners using grey-level correlation between neighboring regions of those feature points. For a given point in one image several candidate matches are in general possible in another image. In order to reduce the number of hypotheses, we make use of relaxation

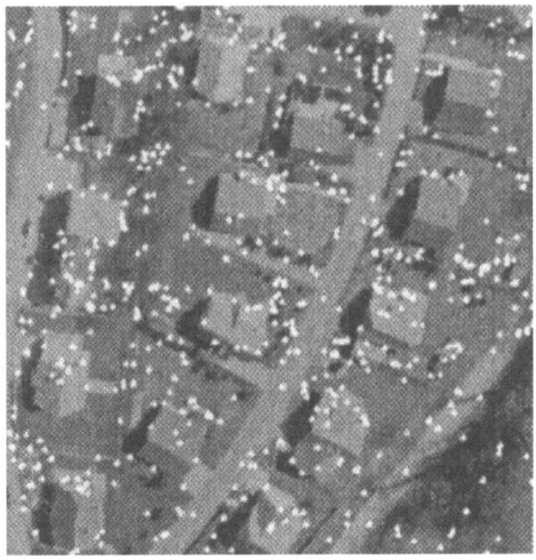

Figure 3: Points extracted from image 5888 which have been matched by the Image-Matching algorithm with at least one point in another image.

methods [Zhang et al., 1994]. Figure 3 shows a subset of the correspondences which have been automatically obtained between two of the four Ascona images.

In the M-situation we track the feature points in the sequence whenever possible (small motion between frames). If a given point can be tracked all the way between two of the selected view, a correspondence is established. Tracking of a point of interest is performed by predicting its position from one image to the next and searching in a small neighborhood of the predicted point for an actual point. An example of such a tracking is shown in Figure 4. We can of course also use the method of the previous paragraph.

Figure 4: Tracking of points of interest in the library sequence.

2.2 Estimating the fundamental matrices between pairs of images

At this stage, we have obtained a number of correspondences between some images. Correspondences between pairs of images are input to a program that reliably and robustly estimates the fundamental matrices between those pairs [Zhang et al., 1994]. In particular this program has the capability of rejecting some of the correspondences as outliers.

2.3 Estimating the uncertainty of the fundamental matrices

The uncertainty associated with points of interest (typically between 0.1 and 1 pixel) is propagated to the fundamental matrices. In order to compute an estimate of this uncertainty, we parameterize the fundamental matrix with the minimum number of parameters, namely 7, and compute the covariance matrix of the corresponding vector of size 7. There are several technical difficulties in doing this. First, the parameterization using 7 parameters is nonlinear, not unique, and has singularities. We therefore have to find the best one in the sense that it is the most remote from singularities. Second, the criterion which is minimized in order to estimate the fundamental matrix is also nonlinear and does not provide an analytical expression of the solution as a function of the point correspondences. We therefore have to use the implicit function theorem to actually compute the covariance matrix of the parameter vector of the fundamental matrix. The details of those computations can be found in [Csurka et al., 1995].

2.4 Recovering the geometry of the N cameras

Up to this stage in the processing, we have estimated the fundamental matrices of consecutive pairs of images as well as obtained a number of point correspondences between the views. Nonetheless, we usually still have false matches to account for. The set of false matches for the epipolar geometry between pairs of images is a strict subset of the false matches for the projective geometry: this simply means that the images of a point can satisfy the epipolar geometry between pairs of images and still be incorrect when considering the complete set of images. The geometry estimation algorithm has been designed to deal with this problem.

We have the choice of the projective basis in which we want to express the perspective projection matrices, knowing that the choice of a particular basis will not have any influence on the final results. We are going to use this property to compute our projection matrices. Using the theory developed in [Faugeras, 1992], from 5 points in correspondence in a pair of images and the epipoles, we can obtain the projection matrices, expressed in the projective basis defined by those points. This step is just a matter of writing equations of the type:

$$\mathbf{m}_i = \mathbf{P}_j.\mathbf{E}_i, i \in \{1,\ldots,5\}, j \in \{1,2\} \qquad (3)$$

where the $\mathbf{E}_i$ represent the canonic 3-D projective basis: $[0,0,0,1]^T$, $[0,0,1,0]^T$, $[0,1,0,0]^T$, $[1,0,0,0]^T$, $[1,1,1,1]^T$. The 20 scalar equations given by (3) need to the be completed by two equations exploiting the fact that the epipoles are known.

In order to obtain a set of projection matrices, we first choose 5 points in correspondence in the N images. These points are usually chosen in the scene. We then proceed pairwise. For each consecutive pair of images, we compute the projection matrices in the basis of the 5 chosen points. For this, we make use of the coordinates of the points in the images and of the coordinates of the epipoles that we determined previously. Of course, there can be conflicts: the projection matrix $\mathbf{P}_j$ computed from the pair $(j-1, j)$ can be different form the one computed with $(j, j+1)$[1]. We do not consider this as a major problem because there are usually not very different and because this initial estimate is just a starting point for a refinement procedure. Only one of the possible projection matrices is kept.

Of course, running through this process only once has very little chance to succeed because of the possible outliers. If one of the correspondences is erroneous, then the projection matrix and its neighbors will be useless[2]. To overcome this problem, we use robust methods.

2.4.1 Least Median of Squares

The Least Median of Squares (LMedS) is a classic method in outlier detection. A very good introduction can be found in [Rousseeuw and Leroy, 1987]. We need an quality measure of the set of projection matrices for each point. We define r_i as the distance between a given point and the reprojection of the reconstructed 3-D point with these projections.

$$r_i = \sum_j d(\mathbf{m}_{ij}, \mathbf{P}_j.\mathbf{M}_i) \tag{4}$$

$\mathbf{M}_i$ is obtained with the reconstruction algorithms mentioned in section 3. The LMedS method estimates the parameters by solving the non-linear minimization problem:

$$\min(\text{med}_i(r_i^2)) \tag{5}$$

That is, the estimator must yield the smallest value for the median of squared residuals computed for the complete set of points. Of course, it is not reasonable to generate all the possible subsets of 5 point correspondences. Rather, we use a Monte-Carlo technique to draw m random subsamples of $p = 5$ different point correspondences. For each subsample J, we estimate the set of projection matrices $\mathbf{P}_j^J$ by the methods previously described. For each set $\mathbf{P}_j^J$, we can determine the median of the squared residuals denoted $M^J = \text{med}_i(r_i^2)$, with respect to the whole

[1] This can only be due to the epipoles, because the coordinates of the 5 points remain unchanged.

[2] Note that all other matrices will be correct. This quality of localness is desirable and cannot be achieved with iterative (image after image) techniques

set of point correspondences. We retain the estimate of the $\mathbf{P}_j$ leading to the minimal M^J.

The question now is: *how do we determine m?* A subsample is considered good if it consists of p good correspondences across the N images. Assuming that the probability of a point correspondence across 2 images being an outlier is ϵ, the probability of a point correspondence across the N images being an outlier is $1-(1-\epsilon)^{N-1}$. The probability that at least one of the m subsamples is good is given by

$$P = 1 - (1 - (1 - \epsilon)^{(N-1)p})^m \tag{6}$$

In our implementation, we assume that $\epsilon = 15\%$ and require $P = 0.99$, thus $m = 6907$. Note that the algorithm can be sped up by means of parallel computation, because the processing for each subsample is done separately.

The five points of a subsample may be very close to each other. Such a situation should be avoided because the estimation of the 3-D structure from such a projective basis is highly unstable and the result is useless. It is a waste of time to evaluate such a subsample. Bucketing techniques were developed to ensure that such configurations are avoided. The images are evenly divided in buckets and we impose that the points be drawn from different buckets (i.e. different regions of the image). The previous formula determining m still holds under the assumption that the outliers are uniformly distributed over the image.

2.4.2 Block estimation

Over a long sequence, it is very difficult or even impossible to find correspondences for the same five points across the whole sequence of images. We therefore split our estimation process over different consecutive blocks of images, with the precaution that the intersection of two consecutive blocks of images contains at least two images. Knowing the projective bases used in each such pair of blocks, we can compute the projective transformation from one to the other and apply it to the matrices of the second block. This process glues the blocks together.

2.5 Refinement

Once a correct set a projection matrices has been computed, we can refine it by two possible methods. Of course, the outliers found at the previous step are marked as invalid and are not taken into account any further.

2.5.1 Epipolar line adjustment

From the set of projection matrices, we can compute a consistent set of fundamental matrices. The points that we have matched must satisfy the epipolar constraints. We then minimize the sum of the distances between the points and the epipolar lines generated by their correspondents by varying the projection matrices. However, this method is slow because we have to recompute the epipolar lines at each step. In other words, there is no possible decoupling of the minimization because of its high non-linearity.

2.5.2 Bundle adjustment

This classical method in photogrammetry [Brown, 1958],[Brown, 1971], [Gruen, 1978],[Slama, 1980],[Gruen and Beyer, 1992] is very well suited to our problem. With our initial estimation, the optical rays used in the method approximately intersect because the reprojections of the 3-D point are close to the initial points. This method has the advantage of being fast. The only modification that we have made is that instead of reconstructing the actual Euclidean 3-D points, we reconstruct the points in a projective basis. Although our problem is over-parameterized (we allow the projective basis in space to change), the minimization converges because of the nice properties of the Levenberg-Marquardt algorithm. The average distance between one point and the reprojection of its reconstruction is initially around 1 pixel and typically goes down to 0.3 pixels.

The epipolar line adjustment is slower than the bundle adjustment but performs best in the bad cases when the set of projection matrices is not very well initialized.

Figure 5 shows the epipolar geometry obtained with the calibration data on the aerial images provided by ETH, the epipolar geometry obtained by our algorithm without any prior knowledge and a close-up. We worked with images of reduced size (1000×1000 pixels). The computation took 6.3 minutes for 10000 bases tried. The average image error was less than 0.3 pixels on the 183 points used for the refinement.

3 Reconstruction

As mentioned in section 2, the projection matrices that we have determined allow us to compute three-dimensional structure from image correspondences, *up to an unknown projective transformation.* This fact had been first stated in [Koenderink and van Doorn, 1991] in the case of two affine cameras (in this case, the unknown transformation in space is affine). The case of two projective cameras has been then presented in [Faugeras, 1992, Hartley et al., 1992].

From a practical standpoint, this means that from image correspondences, we can compute three-dimensional points represented by their homogeneous coordinates (4-vectors). Knowing the coordinates of a point, we can back-project it onto any of the cameras for which a projection matrix has been computed. We can even project it onto an arbitrary virtual camera: this way we can produce new views of the scene. This process, usually called view transfer [Barrett et al., 1992, Ullman and Basri, 1991], has been used for image synthesis [Faugeras and Laveau, 1994], [Laveau and Faugeras, 1994].

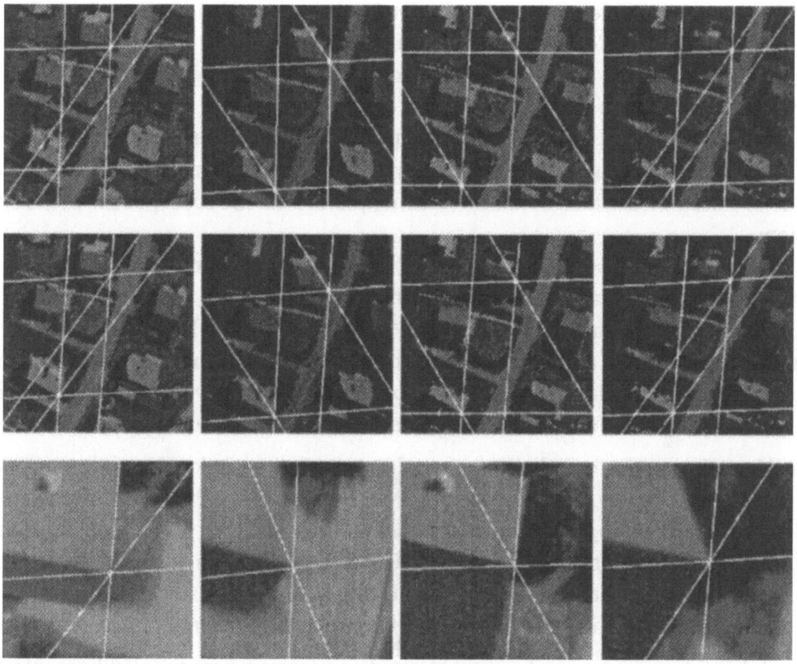

Figure 5: The top row shows the epipolar geometry obtained with the calibration data provided by ETH, the middle row shows the epipolar geometry obtained by the method described in the article without any prior knowledge and the bottom row demonstrates on four subimages the high quality of the estimated epipolar geometry.

However, for simple reasons, the projective reconstruction may not be sufficient: for instance within *Realise*, the generated building models are used for realistic rendering and virtual walk-through, requiring fast-rendering hardware which can only handle Euclidean descriptions.

Without any additional information, recovering Euclidean structure is impossible: all the geometric relations induced by point correspondences have been already used. We need to use additional information, either on the scene, or on the viewing system.

Approaches have been developed which deal with the former case [Maybank and Faugeras, 1992, Luong, 1992, Faugeras et al., 1992]: assuming that the intrinsic parameters of the camera do not vary across at least three views, they can be computed from a number of point correspondences in the three views. Though good results have been obtained, this approach tends to be very sensitive to noise.

In the latter case, we assume that some information is known about the scene:

- If we know the coordinates of at least five reconstructed points in general configuration (i.e., a projective basis) with respect to a Euclidean frame, we can compute the projective transformation which changes projective coordinates into Euclidean ones. This principle of using a few "anchor points" to derive Euclidean coordinates is also used in self-calibration [Gruen and Beyer, 1992]. It supposes that one has performed manual measurements on the real scene.
- A Euclidean frame can be characterized as a frame where parallel lines intersect at infinity, and where orthogonal lines are indeed orthogonal (the dot-product of their directions is zero). The first property characterizes affine structure, whereas the second one characterizes Euclidean structure up to an unknown scale. As shown below, using images of parallel lines we can recover affine structure; Using pairs of orthogonal direction, we then reach scaled Euclidean structure. Less restrictive than the above approach (here, no manual measurement is performed), this approach is perfectly suited to the requirements of *Realise*, because there are in general many images of parallel or orthogonal lines in views of buildings.

Let us now see in more detail the various stages of the process. In the first stage, we recover affine structure, in which parallelism is preserved. In the second one, we recover Euclidean structure in which orthogonality is preserved as well.

After the first stage described in the previous section, the world is modeled as a three-dimensional projective space $\mathcal{P}^3$. We use the standard embedding of a three-dimensional affine space $\mathcal{A}^3$ into $\mathcal{P}^3$ obtained by identifying $\mathcal{A}^3$ with $\mathcal{P}^3 \setminus \Pi_\infty$, where Π_∞ is a plane, called the plane at infinity and which can be thought as the set of directions of lines in $\mathcal{A}^3$. In particular, two parallel lines of $\mathcal{A}^3$, seen as lines of $\mathcal{P}^3$ intersect at a point of Π_∞ (called their point at infinity). Such a point is not as mysterious as it sounds since when viewed by a camera, the images of the two lines usually intersect at a point (called their vanishing point) which can be thought of as (in fact in some sense *is*) the image of the point at infinity of the two lines. Hence, in order to determine the plane at infinity, it is in principle sufficient to have in the scene three pairs of non coplanar parallel lines. Once the plane at infinity has been determined, an affine coordinate system can be chosen and affine coordinates of the points in the scene computed.

In the remainder of this article, we use the standard embedding of $\mathcal{A}^3$ into $\mathcal{P}^3$ which maps a point of affine coordinates $[x, y, z]^T$ onto its corresponding point of $\mathcal{P}^3 \setminus \Pi_\infty$ of homogeneous coordinates $[x, y, z, 1]^T$. This embedding simply means that the plane at infinity is the plane of equation $T = 0$ in the projective space with homogeneous coordinates X, Y, Z, T.

3.1 Parallel lines: affine structure

Two lines in space are parallel if and only if they intersect each other in the plane at infinity. A projective transformation preserves affine structure if and only if preserves parallelism, which means that it leaves the plane at infinity (the set of all points at infinity) globally invariant.

Thus, the problem of recovering affine structure is equivalent to finding a projective transformation H_a which maps the plane at infinity onto the plane represented by $[0, 0, 0, 1]^T$. This is a very simple operation provided that we can compute the coordinates of the plane at infinity in the initial projective frame. For this purpose, we first need to determine at least three non-aligned points on this plane, i.e. three non coplanar directions of lines. Since we observe images of lines, each of these points is computed as the vanishing points point of a set of parallel lines observed in the images. This is shown in Figure 6.

In this figure the three pairs of lines (L_1, L_2), (M_1, M_2), and (N_1, N_2) are respectively parallel and the images of their points at infinity V_L, V_M, V_N are the points of intersection v_l, v_m, v_m of the pairs of image lines (l_1, l_2), (m_1, m_2), (n_1, n_2), respectively.

From a practical standpoint, parallel directions are defined manually in the images. For each image, we compute a polygonal approximation of the edge chains extracted with a sub-pixel feature detector. Line segments representing parallel lines are selected in the different images. Each group of segments representing parallel lines in space allows computing one point at infinity. In Figure 7 we show the line segments in image 5888 with which points at infinity have been computed.

3.1.1 Computing points at infinity

Let us assume that we measure in the images the projections of parallel space lines $\langle D_i \rangle$. $\langle d_{ij} \rangle$, represented by the three-dimensional homogeneous vector $\mathbf{d}_{ij}$, is the image of $\langle D_i \rangle$ in view j. We want to compute their point of intersection $\mathbf{V} = \bigcap_i \langle D_i \rangle$ which we know to be in Π_∞. The image $\mathbf{v}_j$ of $\mathbf{V}$ in view j is the vanishing point of the image lines $\mathbf{d}_{ij}$. The problem that we need to solve is the following: given $\mathbf{d}_{ij}$, compute $\mathbf{V}$.

Since the image of $\mathbf{V}$ in camera j lies on line $\mathbf{d}_{ij}$, we have:

$$\forall i, j \qquad \mathbf{d}_{ij}^T \mathbf{P}_j \mathbf{V} = 0.$$

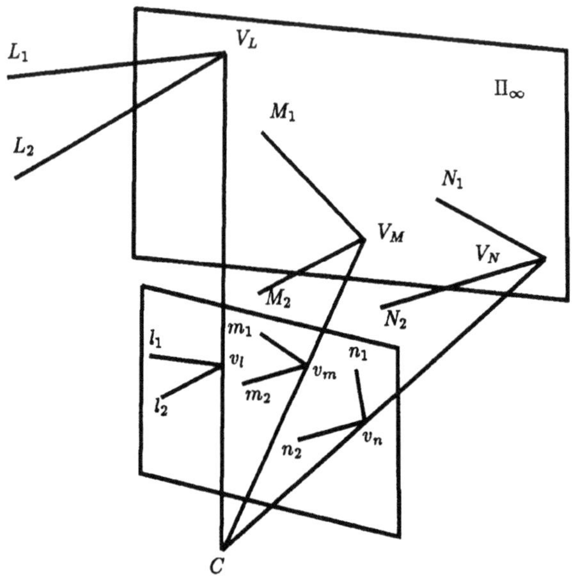

Figure 6: The plane at infinity is determined from the vanishing points of the images of three sets of non-coplanar parallel lines (L_1, L_2), (M_1, M_2), and (N_1, N_2).

Figure 7: Line segments of image 5888 used for the computation of Π_∞.

This is system of P linear, homogeneous equations in the four homogeneous coordinates of $\mathbf{V}$, where P is the total number of lines observed in the images. We solve it using SVD, and end up with the homogeneous vector $\mathbf{V}$.

Remark: Another method for computing $\mathbf{V}$ consists of first computing in each image the vanishing points v_j as solutions of the homogeneous system:

$$\forall i \qquad \mathbf{d}_{ij}^T \mathbf{v}_j = 0$$

Then, $\mathbf{V}$ is obtained using standard reconstruction of the computed vanishing points. This method proves to provides much worse results than the previous one. It is probably because vanishing points are computed independently in all the images, so they are not constrained to satisfy the epipolar constraint before they are reconstructed. In the previous method, they are constrained to be the image of one single point $\mathbf{V}$, so they necessarily satisfy the epipolar constraint.

3.1.2 Computing any point on the plane at infinity

The previous process can be applied to all the directions for which parallel lines are observed, yielding points at infinity $\mathbf{V}^k$. Provided that there are at least three non-aligned points at infinity, we can compute the plane at infinity Π_∞, represented by the four-dimensional homogeneous vector $\mathbf{\Pi}_\infty$, as the non-zero solution of the linear homogeneous system:

$$\forall k \qquad \mathbf{V}_k \mathbf{\Pi}_\infty = 0$$

Now that we once we know $\mathbf{\Pi}_\infty$, we can compute the point at infinity of any line as long as this line can be reconstructed in space, by computing the intersection of this line with the plane at infinity.

3.1.3 Deriving an affine reconstruction

To define the transformation which maps the plane at infinity onto $[0,0,0,1]^T$, we proceed as follows:

First, we compute a projective reconstruction of the scene, using standard multi-camera reconstruction based on SVD [Rothwell et al., 1995].

One reconstructed point of the scene, denoted by $\mathbf{C}$, is chosen as the origin: in the new frame, it has coordinates $[0,0,0,1]^T$. Then, three arbitrary independent directions are selected as coordinate axes, and their points at infinity $\mathbf{V}_X, \mathbf{V}_Y, \mathbf{V}_Z$ are computed using the method described above. They are respectively mapped onto $[1,0,0,0]^T, [0,1,0,0]^T, [0,0,1,0]^T$. To define a projective transformation in space, we need a fifth point mapping. We select another reconstructed point $\mathbf{S}$, which does not lie on any of the three planes defined by the origin and two of the three axes. In the new frame, this point is assigned arbitrary non-zero coordinates $[\alpha, \beta, \gamma, 1]$.

$\mathbf{H}_a$ is then computed as the projective transformation which maps the initial projective basis $\mathbf{V}_X, \mathbf{V}_Y, \mathbf{V}_Z, \mathbf{C}, \mathbf{S}$ onto the final one. Since $\mathbf{V}_X, \mathbf{V}_Y, \mathbf{V}_Z$ are all mapped onto points whose fourth component is zero, any point of the plane at infinity will also be mapped onto a point whose fourth component is zero, which is precisely what we needed for the reconstruction to be affine. Figure 8 shows an example of affine reconstruction. Two of the three directions chosen as coordinate axes form a very small angle (left). This implies a strong affine skew effect on the reconstructed scene (right).

Parameters α, β, γ have a very simple meaning: they represent scale factors along the three coordinate axes. For instance, if α is multiplied by a non-zero factor, then the reconstructed scene will be stretched by the same scale factor along the $\mathbf{V}_X$ axis. This is visible in Figure 9, which displays two affine reconstructions which differ by only one scale factor (along the direction of the top edge of the roof).

Figure 8: Line segments and directions used for defining the coordinate axes (left), top-view (middle) and side-view (right) of the affine reconstruction (see text).

3.2 Euclidean structure up to three scale factors

As we have seen in Figure 8, the choice of non-orthogonal reference directions may cause severe affine distortion of the reconstructed scene. A first step toward the recovery of Euclidean structure is to use three pairwise orthogonal directions.

In this case, the directions of edges parallel to the reference directions are preserved. In fact, the recovered structure is equivalent to Euclidean structure scaled with three scale factors along the three coordinate axes. As a consequence, two edges aligned with two orthogonal coordinate axes remain orthogonal in the final affine reconstruction, for any value of the scale parameters α, β, γ (e.g. previous section). This is for instance the case of the roof on which the two horizontal directions have been defined (left). The relative values of the scale factors used for the two displayed affine reconstructions (middle,right) are very different. This modifies drastically the aspect of the reconstructed roof (at the bottom-right in each view), but the principal directions remain orthogonal in both reconstructions.

Of course, angles between lines which are not aligned with the coordinate axes are not preserved. In particular, orthogonality is not preserved for such directions (cf the roof on the bottom-left). We will now see how this property can be used for recovering Euclidean structure up to one global scale factor.

3.3 Euclidean structure up to one scale factor

We now assume that some pairs of orthogonal lines are known a priori. The points at infinity of these lines, $\mathbf{V}_i$, are computed as described above. In a Euclidean frame, lines i, j are orthogonal if and only if $\mathbf{V}_i^T\mathbf{V}_j = 0$.

If we consider the orthogonal frame defined in the previous paragraph, finding Euclidean structure is equivalent to finding relative values of the scale parameters α, β, γ for which the dot-products $\mathbf{V}_i^T\mathbf{V}_j$ are zero for all pairs (i, j) of orthogonal lines.

If the three reference axes used for affine reconstruction are not orthogonal, three additional parameters ("skew" parameters) are introduced which account for the non-orthogonality of the reference affine frame. More precisely, instead of using the mapping defined in 3.1.3, we respectively map $\mathbf{V}_X, \mathbf{V}_Y, \mathbf{V}_Z$ onto $[1,0,0,0]^T$ (this has not changed), $[\lambda, 1, 0, 0]^T$, $[\mu, \nu, 1, 0]^T$.

We end up with the following criterion to be minimized over the scale parameters and the skew parameters: :

$$E(\alpha, \beta, \gamma, \lambda, \mu, \nu) = \sum_{i,j \text{ orthogonal}} (\mathbf{V}_i^T\mathbf{V}_j)^2$$

The global scale of the scene cannot be recovered. So, we search for the particular solution for which $\alpha = 1$. Minimizing $E(1, \beta, \gamma, \lambda, \mu, \nu)$ with the standard Levenberg-Marquardt iterative technique (the initial values of the five parameters are the ones used for affine reconstruction), we end up with a Euclidean reconstruction up to a global scale factor (see Figure 10).

This way we have computed a Euclidean reconstruction of the scene without any knowledge of the camera parameters, nor of the scene coordinates. Only information about point and line matches, parallel and orthogonal relations have been used.

4 Conclusion

We have described in this article the skeleton of a system based on computer vision that is going to be used to partially automate the 3-D modeling of urban scenes. The system can use any number of cameras and images of the scenes to be modeled and proceeds to estimate automatically the perspective projection matrices corresponding to all the images by matching image features such as corners, junctions, lines. The resulting matrices do not allow to recover a metric model of the scene since no metric information has been used so far, only a projective one which can be used for some applications. In order to go further, the system can use information provided by the user about the actual affine or Euclidean structure of the scene, such as parallel lines, ratios of lengths, and angles. This information allows the system

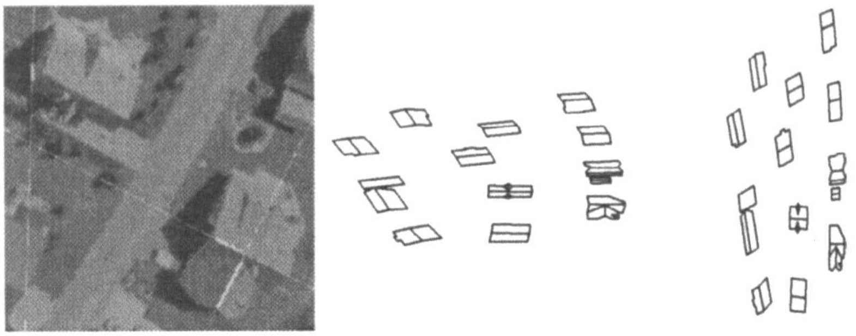

Figure 9: Line segments and directions used for defining the coordinate axes (left); two top-views (left) and (right) of the scene reconstructed with two different values of the scale-factor along one of the horizontal axes (see text).

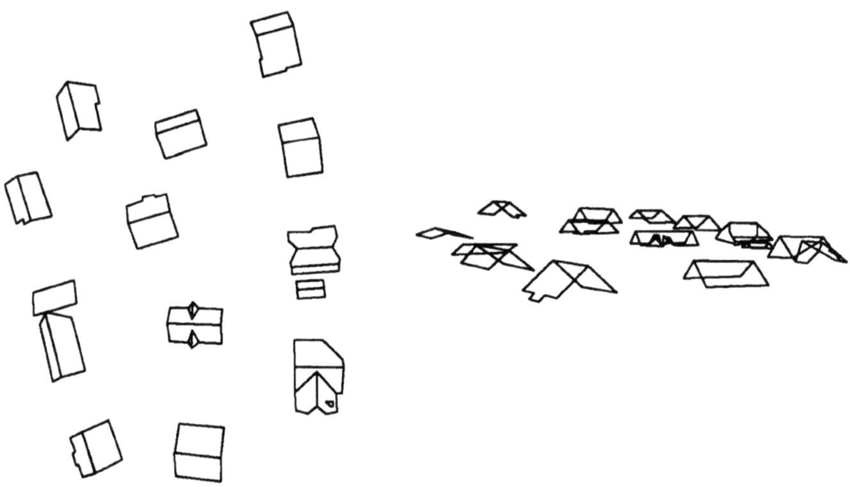

Figure 10: Top-view (left) and side-view (right) of the scene reconstructed after automatic adjustment of the scale factors in order to preserve orthogonality. The frame of reference is the same as in Figure 9.

to specialize its representation of the environment from a projective one to an affine one and finally a Euclidean one. The whole system uses sophisticated computer vision tools and has been developed as a flexible server, a vision server, that can be queried by a human user who is using a CAD system to develop a 3-D model of the scene.

One of the advantages of this system is that it does not require any prior knowledge about the cameras, which is handy in applications like video-based modeling for example. The user is then allowed to use his camera the way he likes, without any special set-up.

We think that this concept of an interactive approach in which the user can have at its fingertips the most sophisticated tools in computer vision and photogrammetry can save a lot of tedious work that is presently done by the human modeler.

Acknowledgement
The authors would like to acknowledge the fact that the work presented here would not have been possible without the contributions of Thierry Blaska, Rachid Deriche, Cyril Gauclin, Charlie Rothwell, and Zhengyou Zhang.

References

[Barrett et al., 1992] Barrett, E. B., Brill, M. H., Haag, N. N., and Payton, P. M. (1992). Invariant Linear Methods in Photogrammetry and Model-Matching. In Mundy, J. L. and Zimmerman, A., editors, *Geometric Invariance in Computer Vision*, chapter 14. MIT Press.

[Blaszka and Deriche, 1994] Blaszka, T. and Deriche, R. (1994). Recovering and characterizing image features using an efficient model based approach. Technical Report 2422, INRIA.

[Boufama and Mohr, 1995] Boufama, B. and Mohr, R. (1995). Epipole and fundamental matrix estimation using the virtual parallax property. In *Proceedings of the International Conference on Computer Vision*, Cambridge, Ma. IEEE Computer Society Press. To appear.

[Brown, 1958] Brown, D. C. (1958). A solution to the general problem of multiple station analytical stereotriangulation. Technical Report 43, RCA Data Reduction Technical Report, Patrick Air Force base, Florida.

[Brown, 1971] Brown, D. C. (1971). Close-Range Camera Calibration. *Photogrammetric Engineering*, 37(8):855–866.

[Csurka et al., 1995] Csurka, G., Zeller, C., and Zhang, Zhengyouand Faugeras, O. (1995). Characterizing the uncertainty of the fundamental matrix. Technical report, INRIA.

[Faugeras, 1992] Faugeras, O. (1992). What can be seen in three dimensions with an uncalibrated stereo rig. In Sandini, G., editor, *Proceedings of the 2nd Euro-*

pean Conference on Computer Vision, volume 588 of *Lecture Notes in Computer Science*, pages 563–578, Santa Margherita Ligure, Italy. Springer-Verlag.

[Faugeras, 1993] Faugeras, O. (1993). *Three-Dimensional Computer Vision: a Geometric Viewpoint.* The MIT Press.

[Faugeras, 1995] Faugeras, O. (1995). Stratification of 3-d vision: projective, affine, and metric representations. *Journal of the Optical Society of America A*, 12(3):465–484.

[Faugeras and Laveau, 1994] Faugeras, O. and Laveau, S. (1994). Representing three-dimensional data as a collection of images and fundamental matrices for image synthesis. In *Proceedings of the International Conference on Pattern Recognition*, pages 689–691, Jerusalem, Israel. Computer Society Press.

[Faugeras et al., 1992] Faugeras, O., Luong, T., and Maybank, S. (1992). Camera self-calibration: theory and experiments. In Sandini, G., editor, *Proc. Second European Conference on Computer Vision*, number 588 in Lecture Notes in Computer Science, pages 321–334, Santa-Margherita, Italy. Springer-Verlag.

[Faugeras and Mourrain, 1995a] Faugeras, O. and Mourrain, B. (1995a). On the geometry and algebra of the point and line correspondences between n images. In *Proceedings of the International Conference on Computer Vision*, Cambridge, Ma. IEEE Computer Society Press. To appear.

[Faugeras and Mourrain, 1995b] Faugeras, O. and Mourrain, B. (1995b). On the geometry and algebra of the point and line correspondences between n images. Technical report, INRIA.

[Förstner and Gülch, 1987] Förstner, M. A. and Gülch, E. (1987). A fast operator for detection and precise location of distinct points, corners and centers of circular features. In *Proceedings of the Intercommission Workshop of the International Society for Photogrammetry and Remote Sensing*, Interlaken, Switzerland.

[Gruen, 1978] Gruen, A. (1978). Accuracy, reliability and statistics in close-range photogrammetry. In *Proceedings of the Symposium of the ISP Commission V*, Stockholm.

[Gruen and Beyer, 1992] Gruen, A. and Beyer, H. A. (1992). System calibration through self-calibration. In *Proceedings of the Workshop on Calibration and Orientation of Cameras in Computer Vision*, Washington D.C.

[Guiducci, 1988] Guiducci, A. (1988). Corner characterization by differential geometry techniques. *Pattern Recognition Letters*, 8:311–318.

[Harris and Stephens, 1988] Harris, C. and Stephens, M. (1988). A combined corner and edge detector. In *Proceedings Alvey Conference*, pages 189–192.

[Hartley, 1993] Hartley, R. (1993). Euclidean reconstruction from uncalibrated views. In Mundy, J. and Zisserman, A., editors, *Applications of Invariance in Computer Vision*, volume 825 of *Lecture Notes in Computer Science*, pages 237–256, Berlin. Springer-Verlag.

[Hartley et al., 1992] Hartley, R., Gupta, R., and Chang, T. (1992). Stereo from uncalibrated cameras. In *Proc. International Conference on Computer Vision and Pattern Recognition*, pages 761–764, Champaign, Illinois. IEEE.

[Hartley, 1992] Hartley, R. I. (1992). Estimation of relative camera positions for uncalibrated cameras. In *European Conference on Computer Vision*, pages 579–587, Santa Margherita Ligure, Italy. Springer-Verlag.

[Kitchen and Rosenfeld, 1982] Kitchen, L. and Rosenfeld, A. (1982). Gray-level corner detection. *Pattern Recognition Letters*, pages 95–102.

[Koenderink and van Doorn, 1991] Koenderink, J. J. and van Doorn, A. J. (1991). Affine Structure from Motion. *Journal of the Optical Society of America*, A8:377–385.

[Laveau and Faugeras, 1994] Laveau, S. and Faugeras, O. (1994). 3-D scene representation as a collection of images andfundamental matrices. Technical Report 2205, INRIA.

[Luong, 1992] Luong, Q.-T. (1992). *Matrice Fondamentale et Calibration Visuelle sur l'Environnement-Vers une plus grande autonomie des systèmes robotiques.* PhD thesis, Université de Paris-Sud, Centre d'Orsay.

[Luong et al., 1993a] Luong, Q.-T., Deriche, R., Faugeras, O., and Papadopoulo, T. (1993a). On determining the fundamental matrix: Analysis of different methods and experimental results. Technical Report 1894, INRIA.

[Luong et al., 1993b] Luong, Q. T., Deriche, R., Faugeras, O. D., and Papadopoulo, T. (1993b). On determining the Fundamental matrix: analysis of different methods and experimental results. In *Israelian Conf. on Artificial Intelligence and Computer Vision*, Tel-Aviv, Israel. A longer version is INRIA Tech Report RR-1894.

[Luong and Faugeras, 1994a] Luong, Q.-T. and Faugeras, O. (1994a). Camera Calibration, Scene Motion and Structure recovery from point correspondences and Fundamental matrices. *The International Journal of Computer Vision.* Submitted.

[Luong and Faugeras, 1994b] Luong, Q.-T. and Faugeras, O. (1994b). The fundamental matrix: theory, algorithms, and stability analysis. *The International Journal of Computer Vision.* To appear.

[Luong and Viéville, 1994] Luong, Q.-T. and Viéville, T. (1994). Canonic representations for the geometries of multiple projective views. In Eklundh, J.-O., editor, *Proceedings of the 3rd European Conference on Computer Vision*, pages 589–599, Vol. I. Springer-Verlag, Lecture Notes in Computer Science 800-801.

[Luong and Faugeras, 1994c] Luong, T. and Faugeras, O. (1994c). A stability analysis of the fundamental matrix. In Eklundh, J.-O., editor, *Proceedings of the 3rd European Conference on Computer Vision*, pages 577–588, Stockholm, Sweden. Springer-Verlag.

[Maybank and Faugeras, 1992] Maybank, S. J. and Faugeras, O. D. (1992). A theory of self-calibration of a moving camera. *The International Journal of Computer Vision*, 8(2):123–152.

[Mundy and Zisserman, 1992] Mundy, J. L. and Zisserman, A., editors (1992). *Geometric Invariance in Computer Vision*. MIT Press.

[Noble, 1988] Noble, J. (1988). Finding corners. *Image and Vision Computing*, 6:121–128.

[Olsen, 1992] Olsen, S. (1992). Epipolar line estimation. In *Proc. Second European Conference on Computer Vision*, pages 307–311, Santa Margherita Ligure, Italy.

[Rothwell et al., 1995] Rothwell, C., Csurka, G., and Faugeras, O. (1995). A comparison of projective reconstruction methods for pairs of views. In *Proceedings of the International Conference on Computer Vision*, Cambridge, Ma. IEEE Computer Society Press. To appear.

[Rousseeuw and Leroy, 1987] Rousseeuw, P. and Leroy, A. (1987). *Robust Regression and Outlier Detection*. John Wiley & Sons, New York.

[Semple and Kneebone, 1952] Semple, J. and Kneebone, G. (1952). *Algebraic Projective Geometry*. Oxford: Clarendon Press. Reprinted 1979.

[Slama, 1980] Slama, C. C., editor (1980). *Manual of Photogrammetry*. American Society of Photogrammetry, fourth edition.

[Ullman and Basri, 1991] Ullman, S. and Basri, R. (1991). Recognition by Linear Combinations of Models. *IEEE Transactions on Pattern Analysis and Machine Intelligence*, 13(10):992–1006.

[Zhang et al., 1994] Zhang, Z., Deriche, R., Faugeras, O., and Luong, Q.-T. (1994). A robust technique for matching two uncalibrated images through the recovery of the unknown epipolar geometry. Research Report 2273, INRIA Sophia-Antipolis, France. submitted to *Artificial Intelligence Journal*.

Automatic Extraction of Man-Made Objects
from Aerial and Space Images
Monte Verità, © 1995 Birkhäuser Verlag Basel

Automatic Extraction of Buildings and Terrain from Aerial Images *

Robert T. Collins, Allen R. Hanson, Edward M. Riseman, and Howard Schultz

Department of Computer Science
Lederle Graduate Research Center
University of Massachusetts
Amherst, MA. USA 01003-4610

Abstract

A system has been developed to acquire, extend and refine 3D geometric site models from aerial imagery. The system hypothesizes potential building roofs in an image, automatically locates supporting geometric evidence in other images, and determines the precise shape and position of the new buildings via multi-image triangulation. Model-to-image registration techniques are applied to align new images with the site model, and model extension and refinement procedures are performed to acquire previously unseen buildings and improve the geometric accuracy of the existing 3D models. A correlation-based terrain recovery algorithm provides complementary information about the site, in the form of a digital elevation map.

1. Introduction

Acquisition of 3D geometric site models from aerial imagery is currently the subject of an intense research effort in the U.S., sparked in part by the ARPA/ORD RADIUS project (Gerson, 1992; Huertas, 1993; Collins, 1994; Roux, 1994). We have developed a set of image understanding modules to acquire, extend and refine 3D volumetric building models, and to provide a digital elevation map of the surrounding terrain. System features include model-directed processing, rigorous camera geometry, and fusion of information across multiple images for increased accuracy and reliability.

Site **model acquisition** involves processing a set of images to detect both man-made and natural features of interest, and to determine their 3D shape and placement in the scene. This paper focuses on algorithms for automatically extracting models of buildings (Section 2) and terrain (Section 3). The site models produced have obvious applications in areas such as surveying, surveillance and automated cartography. For example, acquired site models can be used for automated model-to-image registration of new images (Collins, 1993), allowing the model to be overlaid on the image to aid visual change detection and verification of expected scene features. Two other important site modeling tasks are **model extension,** updating the geometric site model by adding or removing features, and **model refinement**, iteratively refining the shape and placement of features as more views become available. Model extension and

---

* This work was funded by the RADIUS project under ARPA/Army TEC contract DACA76-92-C-0041 and ARPA/TACOM contract DAAE07-91-C-R035.

refinement are ongoing processes that are repeated whenever new images become available, each updated model becoming the current site model for the next iteration. Thus, over time, the site model is steadily improved to become more complete and more accurate.

2. Building Model Acquisition and Extension

This section focuses on algorithms for automatically extracting models of buildings in the site. To maintain a tractable goal for our research efforts, we have chosen initially to focus on a single generic class of buildings, namely flat-roofed, rectilinear structures. The simplest example of this class is a rectangular box-shape; however other examples include L-shapes, U-shapes, and indeed any arbitrary building shape such that pairs of adjacent roof edges are perpendicular and lie in a single plane.

2.1. Initial Model Acquisition

The building model acquisition process involves several subtasks: 1) line segment extraction, 2) building detection, 3) multi-image epipolar matching, 4) constrained, multi-image triangulation, and 5) projective intensity mapping. These algorithms will be presented by way of an experimental case study using images J1-J8 of the RADIUS model board 1 data set. Figure 1 shows a sample image from the data set. Each image contains approximately 1320 x 1035 pixels, with about 11 bits of gray level information per pixel. Unmodeled geometric and photometric distortions have been added to each image to simulate actual operating conditions. The scene is a 1:500 inch scale model of an industrial site. Ground truth measurements are available for roughly 110 points scattered throughout the model, which were used to determine the exterior orientation for each image. The residual resection error for each image is in the 2-3 pixel range, representing the level of unmodeled geometric distortion present in each image. This corresponds to a backprojection error of roughly 3-4.5 feet in (simulated) object space. This is a significant amount of error that presents a good test of system robustness.

Line Segment Extraction. To help bridge the huge representational gap between pixels and site models, feature extraction routines are applied to produce symbolic, geometric representations of potentially important image features. The algorithms for

Fig. 1: Sample image from the Radius model board 1 data set.

Fig. 2: Straight line segments produced by the Boldt algorithm.

acquiring building models rely on extracted straight line segments (Boldt, 1989). At the heart of the Boldt algorithm is a hierarchical grouping system inspired by the Gestalt laws of perceptual organization. Zero-crossings of the Laplacian of the intensity image provide an initial set of local intensity edges. Hierarchical grouping then proceeds iteratively; at each iteration edge pairs are linked and replaced by a single longer edge if their end points are close and their orientation and contrast values are similar. Filtering to keep line segments with a length of at least 10 pixels and a contrast of at least 15 gray levels produced roughly 2800 line segments per image. Figure 2 shows a representative set of lines extracted from the image shown in Figure 1.

Building Detection. The goal of automated building detection is to roughly delineate building boundaries that will later be verified in other images by epipolar feature matching and triangulated to create 3D geometric building models. The building detection algorithm is based on finding image polygons corresponding to the boundaries of flat, rectilinear rooftops in the scene (Jaynes, 1994). Briefly, possible roof corners are identified by line intersections. Perceptually compatible corner pairs are linked with surrounding line data, entered into a feature-relation graph, and weighted according to the amount of support they receive from the low-level image data. Potential building roof polygons appear as cycles in the graph; virtual corner features may be hypothesized to complete a cycle, if necessary. Rooftops are finally extracted by partitioning the feature-relation graph into a set of maximally weighted, independent cycles representing closed, high-confidence building roofs.

Figure 3 shows the results of building detection on image J3 of the model board 1 data set. The roof detector generated 40 polygonal rooftop hypotheses. Most of the hypothesized roofs are rectangular, but six are L-shaped. First, note that the overall performance is quite good for buildings entirely in view. Most of the major roof boundaries in the scene have been extracted, and in the central cluster of buildings (see area **A** in Fig. 3) the segmentation is nearly perfect.

Fig. 3: Results of building detection on image J3.

There were some false positives, i.e. polygons extracted that do not in fact delineate the boundaries of a roof. The most obvious example is the set of overlapping polygonal rooftops detected over the large building with many parallel roof vents (area **B**) Note that the correct outer outline of this building roof is detected, however. There are also some false negatives, which are buildings that should have been detected, but weren't. The most prevalent example of this is a set of buildings (area **C**) that are only partially in view at the edge of the image. Label **D** marks a false negative that is in full view. Two adjacent corners in the rooftop polygon were missed by the corner extraction algorithm. It should be stressed that even though a single image was used here for bottom-up hypotheses, buildings that are not extracted in one image will often be found easily in other images with different viewpoints and sun angles.

There are several cases that cannot be strictly classified as false positives or false negatives. Several split-level buildings appearing along the right edge of the image (area **E**) are outlined with single polygons rather than with one polygon per roof level. Some peaked roof buildings were also outlined, even though they do not conform to the generic assumptions underlying the system.

Multi-image Epipolar Matching. After detecting a potential rooftop in one image, corroborating geometric evidence is sought in other images (often taken from widely different viewpoints) via epipolar feature matching. The key problem in epipolar matching is disambiguation of multiple potential matches. One way to avoid ambiguity is to match higher-level structures that are more distinctive.

Rooftop polygons are matched by searching for each component line segment separately and then fusing the results. For each polygon segment from one image, an epipolar search area is formed in each of the other images, based on the known camera transformations and the assumption that the roof is flat. This quadrilateral search area is scanned for possible matching line segments, each potential match implying a different roof height in the scene. Results from each line search are combined in a 1-dimensional histogram, each match voting for a particular roof height, weighted by compatibility of the match in terms of expected line segment orientation and length. A single global histogram accumulates height votes from multiple images, and for multiple edges in a rooftop polygon. After all votes have been tallied, the histogram bucket containing the most votes yields an estimate of the roof height in the scene and a set of correspondences between rooftop edges and image line segments from multiple views.

Epipolar matching of a rooftop hypothesis is considered to have failed when, for any edge in the rooftop polygon, no line segment correspondences are found in any image. This criterion was chosen because the 3D line triangulation algorithm will fail to converge in this case. Based on this criterion, epipolar matching failed on eight rooftop polygons. Six were either peaked or multi-layer roofs that did not fit the generic flat-roofed building assumption, and the other two were building fragments with some sides shorter than the minimum length threshold on the line segment data. At this stage, six incorrect building hypotheses were removed by hand; detecting and removing such mistakes automatically is being actively investigated.

Multi-image Line Triangulation. Multi-image triangulation is performed to determine the precise size, shape, and position of a building in the local 3D site coordinate system. A nonlinear estimation algorithm has been developed for simultaneous multi-image, multi-line triangulation of 3D line structures. Object-space constraints are imposed for more reliable results. This algorithm is used for triangulating 3D rooftop polygons from the line segment correspondences determined by epipolar feature matching.

The parameters estimated for each rooftop edge are the Plücker coordinates of the algebraic 3D line coinciding with the edge - specific points of interest, like vertices of the rooftop polygon, are computed as the intersections of these infinite algebraic lines. Plücker coordinates are a way of embedding the 4-dimensional manifold of 3D lines into $\boldsymbol{R}^6$. Although the Plücker representation requires 6 parameters to be estimated for each line rather than 4, it simplifies the representation of geometric constraints between lines. For the generic flat-roofed rectilinear building class being considered here, we specify a set of constraints to ensure that pairs of adjacent lines in a traversal around the polygon are perpendicular, that all lines are coplanar, and that all lines are perpendicular to the Z-axis of the local site coordinate system. An iterative, nonlinear least-squares procedure determines the Plücker coordinates for all lines simultaneously such that all the object-level constraints are satisfied and an objective "fit" function is minimized that measures how well each projected algebraic line aligns with the 2D image segments that correspond to it.

After triangulation, each 3D rooftop polygon is extruded down to the ground to form a volumetric model. For the Model Board 1 site we represented the ground as a horizontal plane with Z-coordinate value determined from the ground truth measurements. More generally, we will soon be combining our symbolic building extraction routines with the digital terrain maps produced by the UMass Terrain Reconstruction System (Section 3). Outlines of the final set of triangulated rooftops are shown in Figure 4.

Fig. 4: The final verified and triangulated 3D rooftops.

To evaluate the 3D accuracy of the triangulated building polygons, 21 roof vertices were identified where ground truth measurements are known (numbered vertices in Figure 4). The average Euclidean distance between triangulated polygon vertices and their ground truth locations is 4.31 feet, which is reasonable given the level of geometric distortion present in the images. The average horizontal distance error is 3.76 feet, while the average vertical error is only 1.61 feet. This is understandable, since all observed rooftop lines are considered simultaneously when estimating the building height (vertical position), whereas the horizontal position of a rooftop vertex is primarily affected only by its two adjacent edges.

Projective Intensity Mapping. Backprojection of image intensities onto polygonal building model faces enhances their visual realism and provides a convenient storage mechanism for later symbolic extraction of detailed surface structure. Planar projective transformations provide a locally valid mathematical description of how surface structure from a planar building facet maps into an image. By inverting this transformation using known building position and camera transformations, intensity information from each image is backprojected to "paint" the walls and roof of the building model. Since multiple images are used, intensity information from all faces is available, even though they are not all visible from any single view. Multiple intensity maps for each polygonal building facet are combined using knowledge of the sun angle and camera viewpoint to remove visual artifacts caused by shadows and occlusion. The resulting intensity mapped site model can then be rendered to predict how the scene will appear from a new view (Figure 5).

Fig. 5: Intensity mapped model rendered from a new view.

2.2. Site Model Extension

The goal of site model extension is to find unmodeled buildings in new images and add them into the site model database. The main difference between model extension and model acquisition is that the camera pose for each image can be determined via model-to-image registration using the current partial site model, whereas for initial model acquisition the pose must be supplied in some other way. Our approach to model-to-image registration involves two components: 1) model matching to determine correspondences between model features and image features, and 2) pose determination to determine the precise geometric relationship between the image and the scene.

The goal of **model matching** is to find the correspondence between 3D features in a site model and 2D features that have been extracted from an image; in our case this involves determining correspondences between edges in a 3D building wireframe and 2D extracted line segments from the image. To find this correspondence, we are using a model matching algorithm described in (Beveridge, 1992). The result of model matching is a set of correspondences between model edges and image line segments and an estimate of the transformation that brings the projected model into the best geometric alignment with the underlying image data.

The second aspect of model-to-image registration is precise **pose determination.** We are using a robust pose estimation procedure (based on a least median squares minimization procedure) described in (Kumar, 1994). The final results of pose determination are a set of camera pose parameters and a covariance matrix that estimates the accuracy of the solution.

Model Extension Example. The model extension process involves registering a current geometric site model with an incoming image, and then focusing on unmodeled areas to recover new buildings that have been recently built, that were previously unseen, or that for some other reason are not present in the site model database. We illustrate this process using the partial site model constructed in Section 2.1, and image J8 from the Radius Model Board 1 dataset.

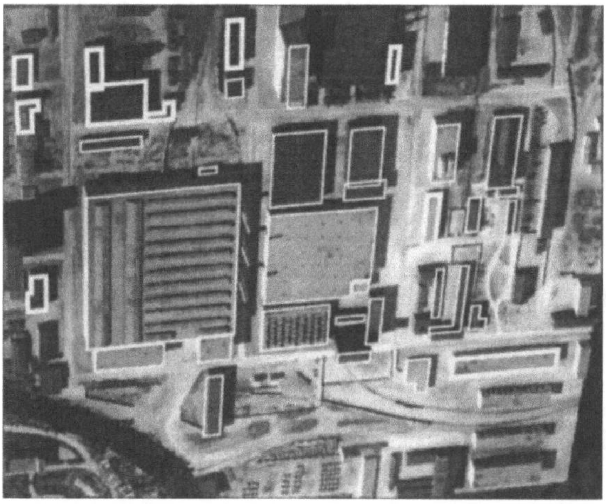

Fig. 6: Extended site model overlaid on J8 (Fig. 1).

Results of model-to-image registration of image J8 with the partial site model can be seen in Fig. 6, showing projected building rooftops from the site model (thin lines) overlaid on the image. Image areas containing buildings already in the site model were masked off, and the building rooftop detector was run on the unmodeled areas in the image, yielding 19 new rooftop hypotheses. The multi-image epipolar matching and constrained multi-image triangulation procedures from Section 2.1 were applied, again using images J1-J8, to verify these hypotheses and construct 3D volumetric building models. Only 10 hypotheses survived the verification and triangulation process. These were added to the site model database, to produce the extended model shown in Figure 6 (thick lines). The main reason for failure among building hypotheses that were not verified was that they represented buildings located at the periphery of the site, in an area which is not visible in very many of the eight views. If more images were used with greater site coverage, we expect that more of these buildings would be included in the site model.

3. Terrain Extraction

The geometric component of a site model consists not only of the building models and other cultural features but also an accurate model (digital elevation map or DEM) of the underlying terrain. The type of imagery of a site that can be expected in the RADIUS project can be characterized as being highly oblique, with widely separated views taken from (perhaps) different cameras at varying temporal intervals. In addition, the camera parameters (both extrinsic and intrinsic) may be unknown or only incompletely estimated. Even under the assumption of known camera parameters, these images present unique problems for correlation-based stereo reconstruction systems because of their oblique viewing geometry and the associated large base-to-height ratios[1].

When a disparity map is computed from widely separated images perspective distortion may result in a large number of false matches and poor reconstruction accuracy. For example, when the base-to-height ratio exceeds approximately 0.5, the performance of correlation-based matching algorithms begins to deteriorate, and when it becomes greater than 1, elevation errors caused by perspective distortion can become large. This

[1] For oblique geometries, the 'height' is the distance from the center of the camera baseline to a nominal point on the surface. In this case, the base to height ratio can vary considerably across the scene.

implies that the size of the correlation mask should be small to minimize the effects of perspective distortion. On the other hand, the size of the mask should be fairly large to provide increased robustness against random noise in the images. To develop algorithms that balance these competing factors we take advantage of the fact that pixels near the center of the correlation mask are less affected by perspective distortion.

Schultz (1994) has developed a correlation based stereo algorithm which incorporates several modifications to account for these effects. Briefly, these are:

(1) A weight is assigned to each element in the correlation mask that depends on its distance from the mask center. Gaussian weights are used with a variance that is either fixed or context dependent. By assigning position dependent weights, it is possible to place more emphasis on the central elements.

(2) The optimal match score is estimated from a series of match scores computed at subpixel disparity steps. By estimating the shape of the disparity function over a narrower disparity range, it is more likely that the match scores at the end of the intervals are statistically significant.

(3) An enhancement to standard multiresolution matching in which perspective distortion is iteratively removed from the images as the processing progresses from low to high resolution.

Fig. 7: ARPA/Martin Marietta UGV Demo C site reconstruction.

Fig. 8: ISPRS "FLAT" scene reconstruction.

At any resolution step, the disparity map is represented as the sum of an initial and incremental disparity map. The initial map (which is derived from the computed disparities at the previous level) is used to remove the perspective distortion at the current resolution level before the incremental disparity map is computed. This process of successive removal of perspective distortion makes it possible to match small features in images taken from widely varying viewpoints.

The resulting stereo analysis package has been tested on both synthetic and real images taken from widely disparate positions (that is, with high base to height ratios) and has been shown to be both accurate and robust in its reconstruction of elevation maps from two or more images (Schultz, 1994). Figure 7 shows a reconstruction of the ARPA Unmanned Ground Vehicle Demo C site at Martin-Marietta in Denver. The original images were taken looking straight down, with a base-to-height ratio of 0.63. Figure 8 shows a portion of the reconstruction of the FLAT Test Dataset 3 of the ISPRS Working Group III/3 data set.

4. Summary and Future Work

A set of IU algorithms for automated site model acquisition and extension have been presented. The algorithms currently assume a generic class of flat roofed, rectilinear buildings. To acquire a new site model, an automated building detector is run on one image to hypothesize potential building rooftops. Supporting evidence is located in other images via epipolar line segment matching, and the precise 3D shape and location of each building is determine by multi-image triangulation. Projective mapping of image intensity information onto these polyhedral building models results in a realistic site model that can be rendered using virtual "fly-through"" graphics. To perform model extension, the acquired site model is registered to a new image, and model acquisition procedures are focused on previously unmodeled areas. In an operational scenario, this process would be repeated as new images become available, gradually accumulating evidence over time to make the site model database more complete and more accurate.

Several avenues for system improvement are open. One high priority is to add capabilities for detecting and triangulating peaked roof buildings. Another significant improvement would be extending the epipolar matching and triangulation portions of the system to analyze why a particular building roof hypothesis failed to be verified. There are many cases where the rooftop detector has outlined split-level buildings with a single roof polygon. This currently causes the subsequent epipolar verification procedure to fail, since all lines in the polygon are assumed to be at the same height. However, a careful analysis of the height histogram in these cases reveals it to be bimodal, meaning that some lines have been found to be at one height, while some occur at another. Automatic detection of these situations, followed by splitting of the rooftop hypothesis into two separate hypotheses, one for each roof level, would result in an improvement in system performance.

In the near future we plan to combine our symbolic building extraction procedures with the correlation-based terrain extraction system described in Section 3. The two techniques clearly complement each other: the terrain extraction system will be used to determine a digital elevation map upon which the volumetric building models will sit, and the symbolic building extraction procedures will be used to identify building occlusion boundaries where correlation-based terrain recovery can be expected to behave poorly. A tighter coupling of the two systems, where an initial digital elevation map is used to focus attention on distinctive humps that may be buildings, or where correlation-based terrain extraction techniques are applied to building rooftop regions to identify fine surface structure like roof vents and air conditioner units, may also be investigated.

Acknowledgments

This paper would not have been possible without the creativity, dedication, and hard work of the RADIUS team of graduate students: Yong Qing Cheng, Chris Jaynes, Frank Stolle, and Xiaoguang Wang. We would also like to acknowledge the software and technical support of Robert Heller and Jonathan Lim, the video wizardry of Fred Weiss, and the administrative support of Janet Turnbull and Laurie Waskiewicz.

References

Beveridge J.R., E.Riseman, "Hybrid Weak-Perspective and Full-Perspective Matching," *Proc. Computer Vision and Pattern Recognition*, Champaign, IL, 1992, pp. 432-438.

Boldt M., R. Weiss, E. Riseman, "Token-Based Extraction of Straight Lines," *IEEE Transactions on Systems, Man and Cybernetics*, Vol. 19, No. 6, 1989, pp. 1581-1594.

Collins R., A. Hanson, E. Riseman, Y. Cheng, "Model Matching and Extension for Automated 3D Site Modeling," *Proceedings of the ARPA Image Understanding Workshop,* Washington, DC, April 1993, pp. 197-203.

Gerson D. "RADIUS : The Government Viewpoint," *Proceedings of the DARPA Image Understanding Workshop,* San Diego, CA, January 1992, pp. 173-175.

Huertas A., C. Lin, R. Nevatia "Detection of Buildings from Monocular Views of Aerial Scenes using Perceptual Grouping and Shadows," *Proceedings of the ARPA Image Understanding Workshop,* Washington, DC, April 1993, pp. 253-260.

Jaynes C., F. Stolle, R. Collins, "Task Driven Perceptual Organization for Extraction of Rooftop Polygons," *Proceedings ARPA Image Understanding Workshop,* Monterey, CA, November 1994, pp. 359-365.

Kumar R., A. Hanson, "Robust Methods for Estimating Pose and Sensitivity Analysis," *CVGIP: Image Understanding,* Vol. 60, No. 3, November 1994, pp. 313-342.

Roux M., D. McKeown, "Feature Matching for Building Extraction from Multiple Views," *Proceedings ARPA Image Understanding Workshop,* Monterey, CA, November 1994, pp. 331-349.

Schultz H. "Terrain Reconstruction from Oblique Views," *Proceedings ARPA Image Understanding Workshop,* Monterey, CA, November 1994, pp. 1001-1008.

Automatic Extraction of Man-Made Objects
from Aerial and Space Images
Monte Verità, © 1995 Birkhäuser Verlag Basel

Mid-Level Vision Processes for Automatic Building Extraction

Wolfgang Förstner
Institut für Photogrammetrie, Universität Bonn

Ascona Workshop on 'Automatic Extraction
of Man-Made Objects froma Aerial and Space Images', 1995

Abstract

Mid-level processes in vision are understood to produce structured descriptions of images without relying on very specific semantic scene knowledge. Automatic building extraction can use geometric models to a large extent. Geometric hypotheses may be inferred from the given data in 2D or 3D and represent elementary constraints as incidence or collinearity or more specific relations as symmetries. The inferred hypothesis may lead to difficulties during spatial inference due to noise and to inconsistent and mutually dependent constraints. The paper discusses the selection of mutually not-contradicting constraints via robust estimation and the selection of a set of independent constraints as a prerequisite for an optimal estimation of the objects shape. Examples from the analysis of image and range data are given.

1 Introduction

Mid-Level Processes in Vision. Automatic image interpretation is a complex process for which no commonly accepted model exists. A three layered scheme reflecting the different levels of abstraction could follow the classical distinction of low-, mid- and high-level processes. Mid-level processes then may be characterized by two criteria: 1. The output of a mid-level process is a structural description of the image or the scene, defining the separation from image processing. 2. No or only poor semantic scene knowledge is used, in a soft way defining the separation from knowledge based vision. Typical mid-level processes are feature extraction, segmentation, grouping, or spatial reasoning.

Geometric Models for Buildings. Though many of the semantic entities of buildings, like floor, garage, passage, wall etc. are very specific for buildings, others, like roof or window are relevant also in other contexts, e. g.

when modelling cars. However, their identification heavily relies of the context, i. e. on the relations between the entities and the relations to other objects.

On the other hand, geometric entities, like boxes, prisms lying vertically or horizontally, may be identified in single images, from multiple images or from range data derived by stereo or laser range finders without very specific object modelling, but being the basis for a more semantic analysis.

Extracting man-made structures from images therefore makes heavy use of mid level processes as they show various common regularities, especially geometric ones, which are common to many types of objects. The regularity of their form puts a stress on geometric processing. The geometric richness of buildings refers to very simple constraints like lines to be vertical of horizontal, to be collinear or parallel or to more specific constraints like repetitions or symmetries.

When analysing perspective images these 3D–constraints may be transferred to the image domain. In case the exterior orientation of the camera is known and/or aerial images are taken this transfer is easy as the nadir point is known and/or the weak perspective can be well approximated by a linear projection therefore allowing a simple grouping of parallel lines and detection of affine symmetries.

Problems during Constraint Processing. There is a dichotomy in representing the geometry of an image or a scene 1. using coordinates as attributes, which eases spatial access, or 2. using datum parameters together with a form description, either directly with observable parmeters or indirectly with constraints or invariants. Whereas the representation with coordinates is unique it does not show the inherent structure of the object, e. g. two planes of a polyhedron being parallel. On the other hand, form parameters or invariants allow a direct link between observables and object, but are not unique, as new invariants may easily be derived.

Data driven processes therefore usually rely on observables or invariants, e. g. in matching or grouping. On the other hand the non-uniqueness of local features for recognition, matching or grouping easily leads to a plethora of hypotheses. This leads to the following problems. Hypotheses depend on noisy data, therefore

- the derived constraints may be mutually inconsistent
- the derived constraints may be actually wrong. Moreover,
- the derived constraints may be mutually dependent.

Depending on the number of unknown parameters and the number of used hypotheses, and depending on whether one can rely on the correctness of the hypotheses or not, different modelling may lead to estimation procedures of quite different algorithmic complexity. Therefore we have to solve two tasks which are discussed in the following section:

1. Selection of non-contradicting and/or independent hypotheses, and
2. selection of the appropriate model for estimating the objects shape.

Examples from the analysis of images and range data show the versatility of the approach.

2 Constraints for Geometric Reconstruction

In the following we assume that the geometry of the object can be described by the parameters of a set of features in image or object space together with a set of constraints between these parameters. We may assume these parameters in the most simple case to be the coordinates of points being directly or indirectly observed by some image analysis process, e. g. by direct point detection or by line intersection. These coordinates are at the same treated as observations $\boldsymbol{y}$ with a covariance matrix $\boldsymbol{\Sigma}_{yy}$ describing their uncertainty and as unknown parameters $\boldsymbol{\beta}$, which are to be determined using the geometric contraints

$$\boldsymbol{H}^T\boldsymbol{\beta} = \boldsymbol{c} \tag{1}$$

inferred by some grouping process.

We first present the different ways of modelling including the worst case complexity of the adjoint estimation process. The selection of non-contradicting hypotheses may be performed using a robust estimation with weak constraints. The selection of an independent subset of hypothesis can make use of a greedy algorithm derived from matroid theory.

2.1 Estimation with Constraints

Let the linear or the linearized model be described by the first and second moments of the n observed quantities $\boldsymbol{y}$: namely $E(\boldsymbol{y})$ and $D(\boldsymbol{y}) = \boldsymbol{\Sigma}_{yy}$. Then we may distinguish the following three models, which differ in the handling of constraints. All proof can be taken from KOCH 1988.

Gauss-Markov-Model. The observation process is made explicit and a *minimum* number u of parameters $\boldsymbol{\beta}$ is to be estimated, implicitely taking the geometric constraints into account:

$$E(\boldsymbol{y}) = \boldsymbol{X}\boldsymbol{\beta} \tag{2}$$

The worst case complexity of the estimation process is $O(c_1u^2n + c_2u^3)$, the second term usually dominating.

This model may also be used for the case where *all* unknown parameters are treated as unknowns and the hypotheses are assumed to be weak, leading to two groups of observations $\boldsymbol{y} = (\boldsymbol{y}_1, \boldsymbol{y}_2)$ with $\boldsymbol{y}_1$ being the observed coordinates and $\boldsymbol{y}_2$ the observed contradictions $\boldsymbol{c}$. The covariance matrix

$$D(\boldsymbol{c}) = \boldsymbol{\Sigma}_{cc} \tag{3}$$

of the $\boldsymbol{y}_2$ reflects the uncertainty of the assumed constraints (cf. 2.2).

Gauss-Markov-Model with constraints. The observation process as well as the constraints are made explicit:

$$E(\boldsymbol{y}) = \boldsymbol{X}\boldsymbol{\beta}, \quad \boldsymbol{H}^T\boldsymbol{\beta} = \boldsymbol{c} \tag{4}$$

where $\boldsymbol{\beta}$ carries *all* unknown parameters. The $u \times h$ matrix $\boldsymbol{X}$ must have full rank. The complexity of the estimation is $O(c_1 u^2 n + c_2 (u+h)^3)$.

Special Gauss-Helmert-Model. By observing the Jacobian in (4) to be $\boldsymbol{X} = \boldsymbol{I}$ we also can write the special Gauss-Helmert model

$$\boldsymbol{H}^T E(\boldsymbol{y}) = \boldsymbol{c} \tag{5}$$

where only a fit of the observations to the constraints is expressed. Again $\boldsymbol{H}$ must have full rank. The algorithmic complexity of the estimation process is $O(c_1 n h^2 + c_2 h^3)$, again the last term usually dominating.

An important equivalence. The two models (2) and (5) are equivalent if

$$\boldsymbol{H}^T \boldsymbol{X} = \boldsymbol{o} \tag{6}$$

and both matrices have full rank.

2.2 Selecting Non-contradicting Hypotheses

When automatically finding geometric hypotheses we cannot guarantee them to be non-contradicting. The way of finding contradicting hypotheses depends on whether they are crisp, as in (4) and (5), which cannot occur if the Jacobians $\boldsymbol{X}$ and $\boldsymbol{H}$ have full rank, or on whether they are weak, as in (3). In the last case no crisp criterion exists for detecting contradicting constraints. But a robust estimation of the unknown parameters may be performed, where the weights $p_{c_i} = 1/\sigma^2_{c_i}$ of the (statistically independent) constraints are reduced depending on the differences $\boldsymbol{h}_i^T \boldsymbol{\beta} - c_i$. Low weights indicate globally inconsistent constraints, which may be eliminated before further processing. Of course constraints with still high weights need not be true in reality.

2.3 Finding Linearily Independent Constraints

Let us assume a grouping process has found a large number h^* of geometric hypotheses which are linear. E. g. for the 6 different x-coordinates x_1, x_2, x_3, x_4, x_7, and x_8, of the 12 points in Fig. 1 the following linear relations have been found by comparing differences and second differences of x-coordinates:

$$\boldsymbol{H}^* = \left(\frac{\boldsymbol{H}_1}{\boldsymbol{H}_2}\right) = \left(\begin{array}{cccccc} 2 & & -1 & & & -1 \\ 1 & -1 & -1 & & 1 & \\ -1 & 1 & & -1 & & 1 \\ \hline -1 & -1 & & & 1 & 1 \\ & 2 & & -1 & -1 & \\ & & -1 & -1 & 1 & 1 \\ -1 & -1 & & & 1 & 1 \\ 1 & -1 & -1 & & -1 & \\ & & -1 & -1 & 1 & 1 \\ 1 & 1 & -1 & & -1 & \\ -1 & 1 & & -1 & & 1 \\ 1 & 1 & -1 & -1 & & \end{array}\right) \tag{7}$$

Figure 1: groundplan of a building showing regularities

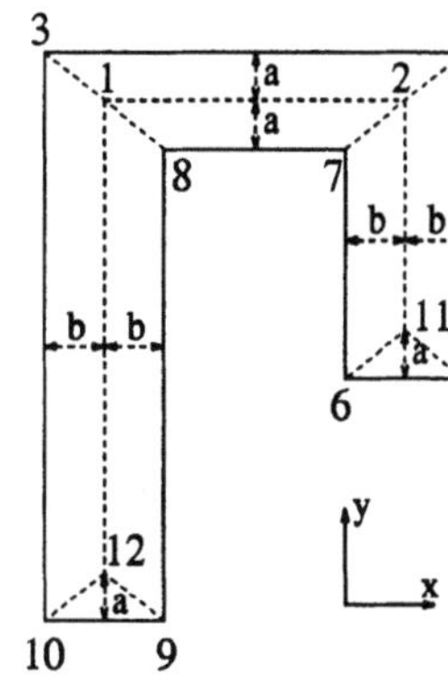

A similar matrix is found for the y-coordinates with also 3 independent constraints. But, also the not intended identity of the coordinate differences: $x_7 - x_8 = y_6 - y_{12}$ had been found.

The first row e. g. represents the symmetry of points 3 and 8 points with respect to point 1: $x_8 - x_1 = x_1 - x_3$, whereas the second row represents the two left halfs of the two building wings to be of the same width: $x_2 - x_7 = x_1 - x_3$. Obviously several constraints have been found twice. But only the first three constraints form an independent set; the others are linearily dependent on these three. The rows of such an automatically derived integer valued matrix can be interpreted as elements of a vector matroid (EDMONDS 1971), where a set of elements (vectors) is labelled independent if the matrix composed of these elements has full rank. Then a simple greedy algorithm can be defined to select an (arbitrary) set of independent vecors. This greedy algorithm sequentially selects one row vector after the other and puts it into the minimum set if it is independent on those vectors which are already in that set. Checking the rank of a matrix cannot be performed on reals, which is reasonable as the decision whether a matrix is singular is a binary decision. Therefore a Gaussian elimination with integer arithmetic may be used (cf. GLOCK 1990). The submatrix $\boldsymbol{H}_1$ therefore may be used in a Gauss-Markov model with constraints according to (4).

2.4 Minimal Parametrization

In case the reduced number h of constraints found this way still is large it may be desirable to use the normal Gauss-Markov-Model (2). Then the number of unknowns has to be reduced by h which may be performed directly which involves the inversion of $\boldsymbol{H}^T\boldsymbol{H}$. Instead, the orthogonality relation may be used for selecting a subset of the coordinates, again using a greedy algorithm, i. e. sequentially select parameters which can not be solved by the previously selected ones and using the given constraints (cf. GLOCK 1990). This way the Jacobian X can be set up fully automatically from the reduced matrix of constraints.

3 Examples

3.1 Finding Non-conflicting Constraints in Segmentation

The following example is using the Ascona Workshop test images. It shows the result of the feature extraction and hypotheses selection (cf. Fuchs and Förstner 1995) on a subpart of the image 5889. The polymorphic feature extraction (cf. Förstner 1994) extracts points, straight line segments and homogeneous regions. The analysis of the exoskeleton is used to build up a feature adjacency graph from which direct and indirect neighbours are derived. Using the DEM-analysis all blobs overlapping with building areas (after a projection into the image, cf. Fig. 2a) are used to trigger the selection of the points and line segments, shown in Fig. 2b, to be analysed for geometric relations, such as incidence, collinearity or parallelity. Only neighbouring segments are analysed and the distance between them was restricted to appr. 10 pixels, in order not to find too many wrong hypotheses. All these geometric hypotheses are checked for global consistency in a robust estimation according to the first paragraph in section 2.1. Obviously quite some links of neighbouring segments have been found and correctly closed. Also some neighbouring segments are straightened. But some short segments have been erroneously prolongated, indicating that the global check cannot guarantee the accepted hypotheses actually to be true.

3.2 Recovering the Groundplan of Buildings from Range Data

The following example shows the use of the constraint estimation onto the recovery of the groundplan of buildings from airborne laser scanner data (cf. Brunn *et al.* 1995). A simple segmentation based on the heights of the buildings with respect to their surrounding leads to a binary image. Vectorization of the boundary of three of the the connected components (cf. Fig. 4) establishing *global* geometric hypotheses and again perfoming a robust estimation leads to the optimal estimates for the ground plans of the buildings in Fig. 5, which nicely show all types of regularities included in the hypothesis generation process, namely collinearities, parallelities and also orthogonalities. The overlay with the original data reveals small differences which need to be analysed.

3.3 Using 3D-Constraints for Shape Recovery

The shape recovery also can use 3D-hypotheses. Fig. 7 shows a gray scale image of a hand drawn polyhedron and the automatically extracted features

Figure 2: a. Section of Avanches test image 5889, b. DHM-selected features and c. result of robust estimation including all non-contradicting features.

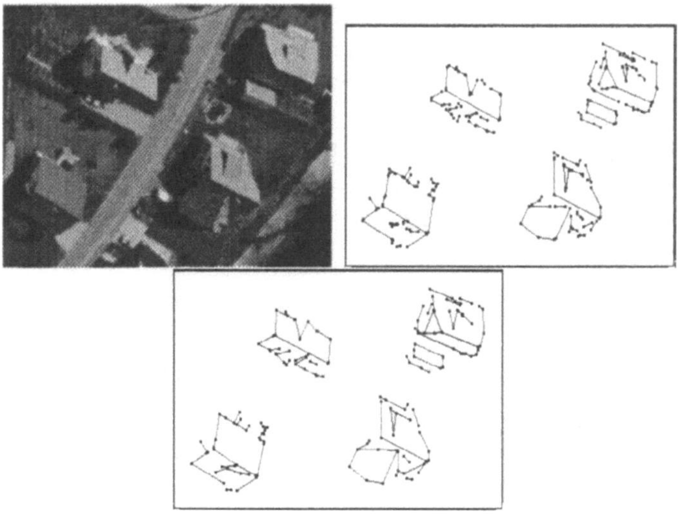

Figure 3: Result of feature extraction

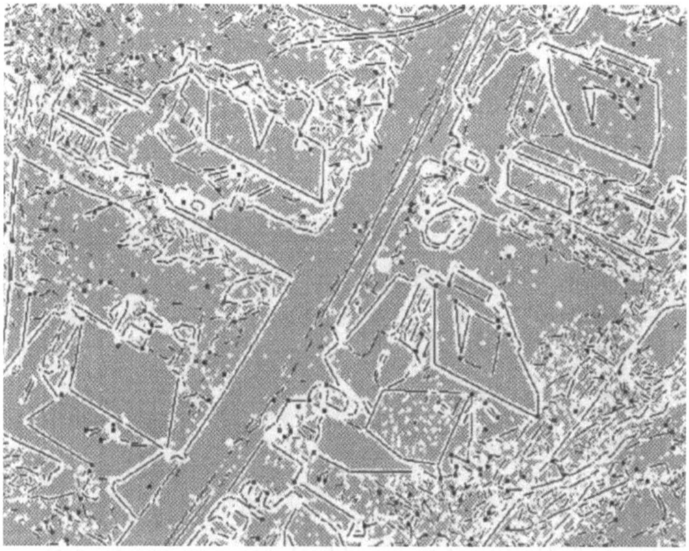

Figure 4: Range data, original boundary, DEM segments

Figure 5: Examples for range data, recovered final shape

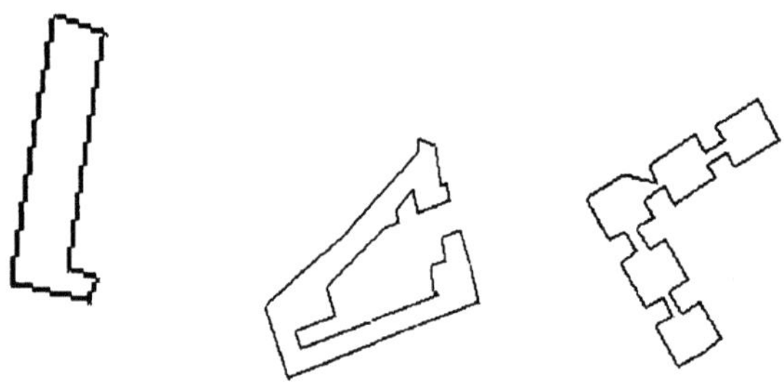

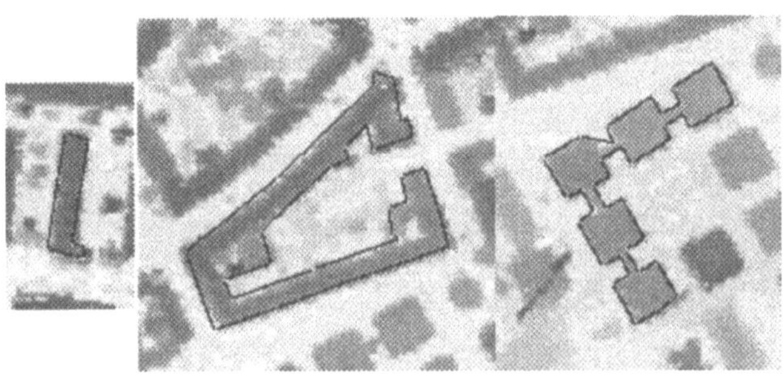

Figure 6: Overlay of recovered final shapes on original data

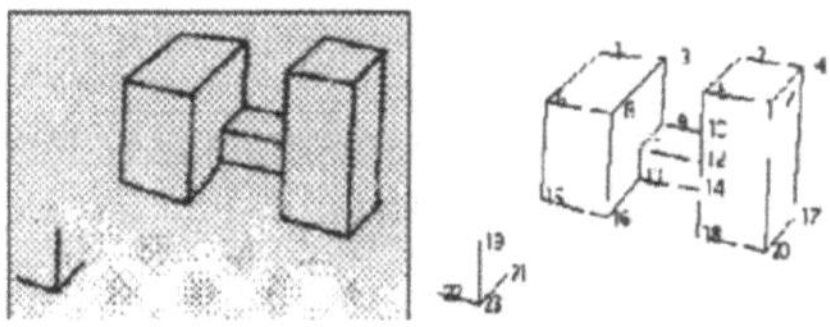

Figure 8: a. and b.: Reconstructed in two views, c. spatially corrected

using the same feature extraction module as for the aerial images. The interpretation of the sketch (cf. BRAUN 1994) using the tripod as indicator for the major axes, the planarity of the faces as only knowledge, and assuming a scale leads to a set of 3D-coordinates for the corners of the polyhedron shown in Fig. 8a and b in two different perspectives. The detection of spatial relations in 3D and the use of these constraints leads to corrected 3D-coordinates (cf. Fig. 8c).

3.4 Symmetries of Planar Shapes

The groundplan shown in Fig. 1 can be parametrized by only 6 coordinates, if reference to a rectangular coordinate system is made. This is because the only 13 different coordinates are linked by the 7 constraints mentioned above. The jacobian $\boldsymbol{X}$ automatically derived from the sketch is

$$\boldsymbol{X}^T = \left(\begin{array}{cccccccccccc} 1 & 1 & 1 & 1 & 1 & 1 & 1 & 1 & 1 & 1 & 1 & 1 \\ 1 & 1 & & 2 & 2 & & & 2 & 2 & & 1 & 1 \\ -1 & & & -1 & -1 & 1 & 1 & -2 & -2 & & & -1 \\ \hline & & & & & & & & & & & \\ & & & & & & & & & & & \end{array} \right. \tag{8}$$

$$\left. \begin{array}{cccccccccccc} & & & & -2 & -2 & & & & & -2 & \\ & & & & 3 & 3 & & & & & 3 & \\ \hline 1 & 1 & 1 & 1 & 1 & 1 & 1 & 1 & 1 & 1 & 1 & 1 \\ 1 & 1 & 2 & 2 & 1 & 1 & & & & & 2 & 1 \\ & & -1 & -1 & -1 & -1 & 1 & 1 & & & -2 & -1 \end{array} \right) \tag{9}$$

referring to the vector $\beta^T = (x_{10}, x_{11}, x_7, y_{10}, y_2, y_8)^T$ of the unknowns (rows) and to the observed x- and the y-coordinates (columns). Though the setup of the Jacobian also takes time, in an iterative e. g. robust estimation the saving at each iteration is obvious when compared to the otherwise 12×2+3=27 unknowns in an estimation following the Gauss-Markov-Model with constraints (4). Observe that this analysis can also be used for affine symmetries.

4 Conclusions

A few examples from mid-level processing for extracting polyhedral structures from intensity and range images have shown the potential of integrating hypothesis generation with robust estimation for hypothesis selection and optimal reconstruction. The reduction of linearily dependent hypotheses to an independent set can be used to advantage for a final evaluation. The procedures may be used in all contexts where parameters are sufficient to describe the essential part of the scene.

Extensions are manyfold: Nonlinear geometric constraints need to be handled in a similar way, which can be expected to be at least one order of magnitude harder. The weak information contained in the segments should be exploited. The link between symmetry detection and the use of invariants needs further analysis in order to ease the interpretation whithin single images. Finaly, all types of grouping processes are based on decisions on geometric relations, which requires an integration of hypothesis generation and selection with more model driven and knowledge based techniques.

Acknowledgments: I thank Claudia Fuchs for preparing the examples in section 3.1. The work has been partly supported by the German Science Foundation (Fo 131/4 and SFB 350) which is highly appreciated.

References

BRAUN, C. (1994): *Interpretation von Einzelbildern zur Gebäudeerfassung.* PhD thesis, Institut für Photogrammetrie, Universität Bonn, 1994.

BRUNN, A.; WEIDNER, U.; FÖRSTNER, W. (1995): Model-based 2D-Shape recovery. In: *submitted to DAGM '95*, 1995.

EDMONDS, J. (1971): Matroids and the Greedy-Algorithm. *Math. Programming*, pages 125–136, 1971.

FÖRSTNER, W. (1994): A Framework for Low Level Feature Extraction. In: EKLUNDH, J.-O. (Ed.), *Computer Vision - ECCV 94, Vol. II, Proceedings*, pages 383–394, 1994.

FUCHS, C.; FÖRSTNER, W. (1995): Polymorphic Grouping for Image Segmentation. *5th ICCV '95, Boston*, 1995.

GLOCK, C. (1990): Erkennung von Symmetrieen in Skizzen. Master's thesis, Institut für Photogrammetrie, Universität Bonn, 1990.

KOCH, K.-R. (1988): *Parameter estimation and hypothesis testing in linear models.* Springer, 1988.

Automatic Extraction of Man-Made Objects
from Aerial and Space Images
Monte Verità, © 1995 Birkhäuser Verlag Basel

Geometric versus Texture Detail in 3-D Models of Real World Buildings

Michael Gruber, Marko Pasko, Franz Leberl

Institut für Computerunterstützte Geometrie und Graphik
Technische Universität Graz
Münzgrabenstraße 11
A-8010 Graz, Austria
Ph.: ++43-316-84 17 66, Fax: ++43-316-84 63 04
e-mail: name@icg.tu-graz.ac.at

Abstract

The creation of three-dimensional CAD models of real world objects and their rendering using photorealistic texture are current topics of investigation. In the case of buildings and other objects of urban environments the need for this technology is evident from various applications. The primary source material consists of images which serve for both the reconstruction of the object's geometry and the creation of texture.

Depending on the scale of the digital source images and on the specifics of an application one may require more or less detail of the geometry. The use of photorealistic texture enhances the perceived detail even in the absence of a detailed geometric model; however one needs to overcome correspondence problems between texture and geometry, which partly may be caused by issues of illumination.

We present in this paper a progress report on one aspect of urban models, namely the reconstruction of roofs. To improve the automation of datacollection we propose the fusion of digital map data and aerial images. We discuss an affine matching procedure. In addition we illustrate the correspondence problem between geometry and photo-texture.

Our initial experiences are based on results from two test sites, one in Vienna and the other in the city of Graz.

1 Introduction

Photo-realistic rendering of CAD models taken from real-world objects is a very current topic since a high degree of naturalism of a computer model is highly desireable. Such "naturalism" is needed to create broad appeal for digital 3D graphics solutions. Powerful computing environments exist which can process large amounts of textural data. In the case of urban environments the need for so-called "photo-realistic" city models is evident from numerous applications such as urban planning, architecture, entertainment, disaster preparedness etc. A primary source for realistic rendering is of course the photograph itself. Therefore it is useful to investigate methods to create and merge phototexture and CAD models of city scenes.

In the case of building roofs the exploitation of aerial images appears applicable to both, the geometric as well as textural information. The photo-texture of facades of buildings has to be derived from terrestrial photographs. The benefit of this method is the high degree of detail which is carried in the texture data and may not be present in the geometric data set. This is obvious when considering the structure and color of roof tiles, the texture and color of windows on facades and similar examples.

Succesfull creation of vast repositories of urban 3D models will depend on one's ability to reduce the cost of data when compared with conventional stereophotogrammetric methods. This can be accomplished by means of automation. We show in this paper that the existing two-dimensional digital map, as part of the urban geographic information system, can envolve into a 3D model. This leads to a correspondence problem between map, digital image and real world object. We report on first experiences combining a 2D GIS with aerial photographs to create a detailed 3D model of buildings, as well as to render them with a high degree of realism. An item of concern is the trade-off between geometric detail and realism.

A severe correspondence problem may disturb the quality of the CAD model, when textural detail is not mapped correctly, perhaps because of missing geometric detail. In the case of roofs this may concern the chimneys and skylights; facades show baywindows, entrances and other architectural details, which are not part of the CAD model. In this case, the phototexture will be mapped incorrectly and reduce the quality of the rendering result.

Our initial experiences derive from two test-sites in Vienna and Graz. Aerial images, digital map data and partially terrestrial photographs have been available to create the photo-textured rendering of a CAD model of the sites. The test-site Vienna presents a simplified geometry model without detail and a moderate resolution of texture data; in Graz we focus on the modeling and rendering of roofs and include chimneys into the geometry of the model. We also report about a fusion experiment between digital map data and the aerial image information in an effort to automate the creation of a sufficiently detailed geometric city model.

2 Test Sites

2.1 Vienna

The test-site Vienna is situated in one of the down-town districts of Austria´s capital and contains a city block of 58 buildings (Fig.1). True color aerial photographs at a scale of 1 : 7000 were available for the roofs´ phototexture. The facades´ texture was extracted from terrestrial images at a scale of 1 : 700. The analog film was scanned at a pixel size of 20 µm with 24 bit color resolution. The digital image data were resampled to a suitable resolution for an online application. The three-dimensional CAD model of the site was compiled from the digital map and a photogrammetric acquisition of the roofs. A low level of detail was available. The contrast between a flat shaded rendering

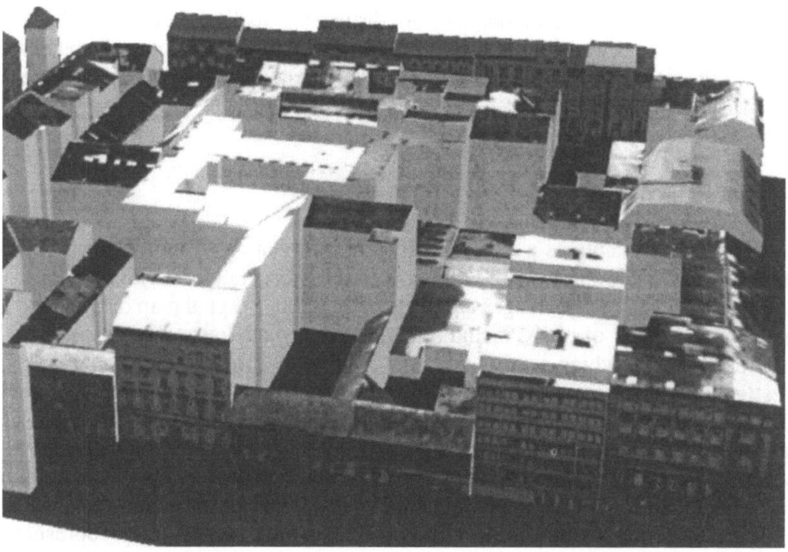

Fig. 1: The Vienna test–site, a city block of 58 buildings. Roofs and facades facing the streets are textured from aerial and terrestrial photographs

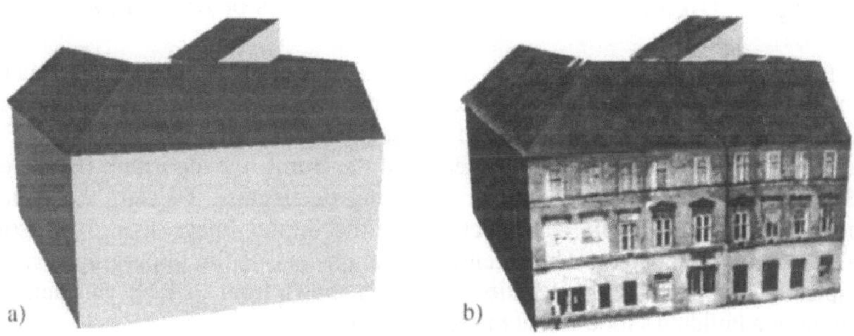

Fig. 2: A simple building from the Vienna test–site, a) flat shaded and b) rendered with phototexture

of the building shapes with a photo-textured visualization shows that photo-texture is a useful addition to the geometric data. The color and quality of the buildings, number of levels and situtation of windows and entrances is documented (cf. Fig.2). The missing detail within the geometric model does not disturb the quality of the model because of the moderate scale and resolution. In other words, the level of geometric detail and of the texture resolution supports the correspondence between CAD model and photographic reality.

2.2 Graz

The city of Graz maintains a 2.5 D geographic information system in the format of a Siemens-SICAD-GIS. The entire city is covered by recent color aerial photographs. A set of five true color aerial images at a scale of 1 : 3500, a digital map data set and a 2.5 dimensional data set of roofs, including chimneys and skylights was thus selected for the Graz test-site, which is situated in the downtown of the city (cf. Fig.3). Our investigation initially focuses on the modeling and photo-realistic rendering of the roofs, including architectural details. Such detail is dictated by the quality and resolution of the aerial photographs, which do show elements like chimneys and skylights, but also eaves and even the texture of the rooftiles (cf. Fig. 4). In addition to these object details the images contain shadows and different brightnesses as an effect of the illumination by sunlight. The two-dimensional digital map represents the planimetry of the buildings´ footprint; this does generally not correspond with the shape of the roofs.

Our work serves to understand the problem of automatically creating a geometric model of a roof, and of creating photorealistically textured representations of these roofs with the many complications arising from various aspects of the problem domain.

3 Using the Digital Map to obtain the Rooflines of Buildings

The fusion of the digital map and the digital images is based on a new affine matching algorithm [Pinz et al., submitted] which compares two token sets, the shape of one building from the map and the detected line segments in the digital images (cf. Fig.5). These line segments have been obtained from the Burns line algorithm [Burns et al., 1986], which allows a large number of tuning parameters. To start the matching procedure, the map detail is projected into the digital image using the exterior orientation of the camera. The orientation parameters of the photographs must be complemented by the terrain elevations, which is known from a DEM. This defines the position of a building's footprint at the ground level.

The transfer of these data to the level of the roofs requires an estimate of the height of the buildings to calculate the shift in the image plane. The estimate can be less accurate the more a building is situated near the center of the photograph. The direction of the shift is radial from the known nadir point of the photograph. The result of an automated matching procedure is shown in Fig.6. Each of the five edge images is matched with the digital map of one building. The correspondence between image and map varies due to several reasons. We have to accept that the digital map represents the "foot print" of

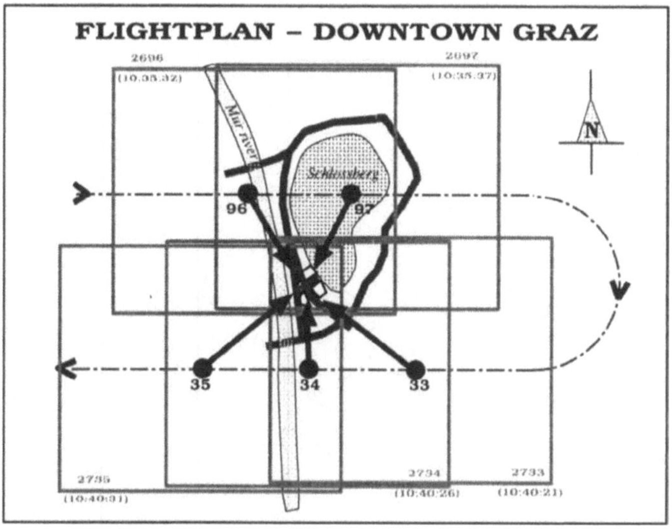

Fig. 3: The Graz test–site. A set of 5 true colour aerial images was selected to cover a quintuple overlap in the historic district. The buildings are on the foot of the Schlossberg, adjacent to the Mur–river

Fig. 4: Detail of one aerial photograph, showing the roof with chimneys, skylights and even the structure of the roof tiles

Fig. 5: Detected line segments in the digital images

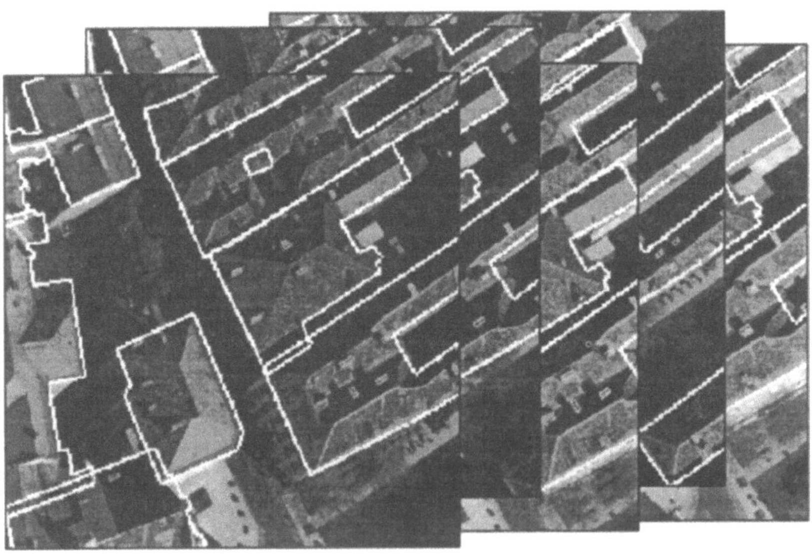

Fig. 6: Result of the affine matching procedure: the 2D digital map combined with the image data

the building and not the shape of the roof-line. This produces a shape that differs from that shown in the image. However, both shapes will find entry into a city-model.

In addition the roof exhibits details such as chimneys and skylights, which cause shadows and occlusions. This may confuse a matching algorithm. The roofline and the building's footprint describe the "building box"; however, the shape of the roof must still be computed, either by stereopsis, by shape-from-shading, by symbolic computing, using the linear features on the roof.

4 Texture versus Geometry

The work with the Vienna test-site addressed the broad problem domain of 3D city models [Gruber et al., 1995], [Leberl et al., 1994]. The next step is to focus on specific details. In this paper we deal with a high resolution representation of the medieval tile roofs of downtown Graz. The digital images describe the tile structure of the roof, the shape of small skylights and the texture of the eaves. This may be included in the textured CAD model. A much larger data set is obtained compared with a standard textured presentation (cf. Fig.7a and 8a). The part of the roof seen from the point of view chosen for Fig.8a does not contain object details which are of geometric importance. The correspondence of the CAD model and the photo-texture seems to be free of errors, at least at this level of detail and resolution. However, moving the viewing position produces the example of Fig. 8b; the geometric detail is inadequate for the quality of the model because contradictions exist between the CAD model and the photo-texture: the chimneys are not included within the geometry, but they are existing in the photorealistic texture.

This issue is resolved by a refinement of the CAD model. The chimneys are included in the scene, resulting in the presentation of Fig.9a. Geometry and phototexture correspond now to one another at the new level of detail. But occlusions and shadows in the image still disturb the visual impression. This defect has to be removed in a postprocessing step. The result is shown in Fig.9b to indicate that a solution exists that will satisfy the visual requirements. Shadows could now be computed for the chosen position of the sun.

While methods for the automated refinement of the geometric model have some tradition in photogrammetry, such tools need to be created for the texture process. A cost effective system for data acquisition must offer a high degree of machine support. It is therefore important that automation for texture processing be developed.

5 Design of a Building and Roof Modeling Module

The entire suite of modules to create 3D models of cities in the computer will have to combine an element addressing the basic shapes of buildings (= "building boxes") and roof shapes. We denote this as "building box and roof modeling" which needs to be augmented by modules for facades, vegetations, suspended structures in streets etc. to represent a complete system to create a 3D GIS. Fig.10 is the outline of a process flow which encompasses the geometric modeling procedure, the texture processing and the merging of both, CAD model and phototexture.

a) b)

Fig. 7: Rendering result of a roof with standard texture using a) simplified geometry and b) enhanced geometry including chimneys

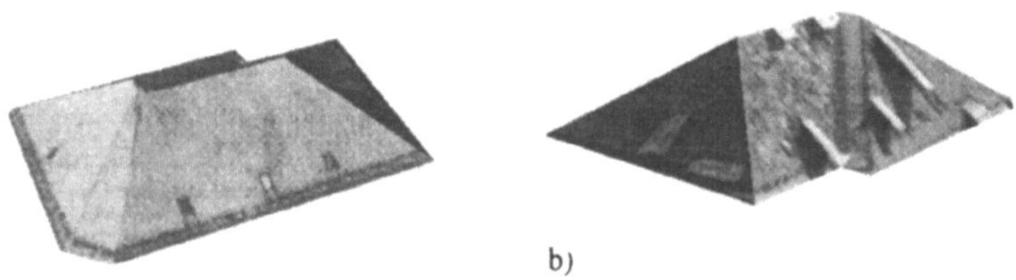

a) b)

Fig. 8: Rendering result of a roof with photorealistic texture, where a) details are visible and mapped correctly (skylights) and b) details are visible but mapped incorrectly due to the missing geometric representation

a) b)

Fig. 9: Rendering result of the same roof with photorealistic texture and enhanced geometry, where a) shadows and occlusions are visible and b) shadows and occlusions are cleaned

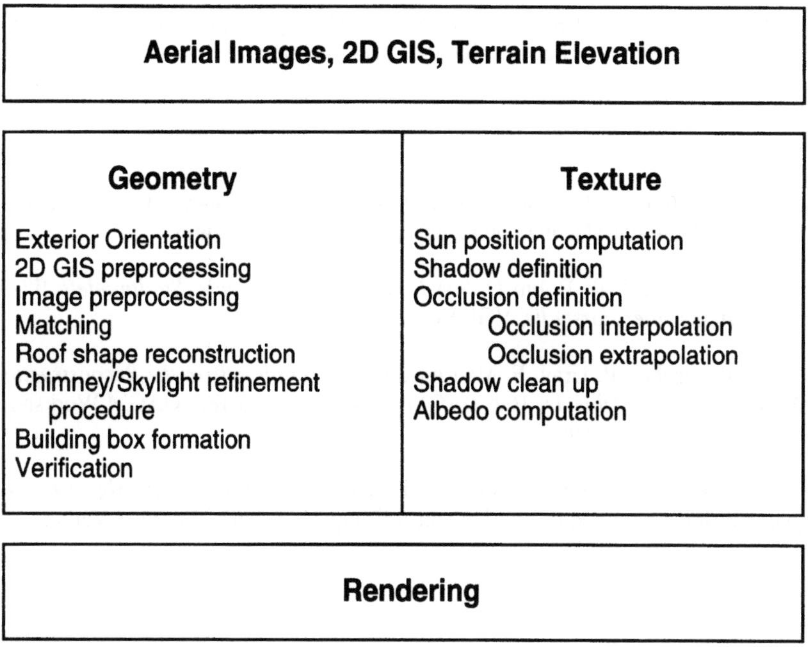

Fig. 10: Flow diagram of the "building box and roof modeling" procedure for geometric modeling and texture processing

6 Conclusions

We have presented a concept for the joint use of 2D GIS data and aerial photography to produce a detailed model of roofs. We demonstrate the importance of sufficient detail with chimneys and other structures. Our approach begins with a novel method of affine matching using so-called "token sets" based on edges obtained by the algorithm by Burns.

The use of photographic texture to render three-dimensional CAD-models of buildings is an important element in the transition from the 2D GIS to the 3D cityscape. We show the pitfalls caused by inappropriate geometric detail, shadows and occlusions.

Therefore a 3D GIS requires close correspondence between geometric and textural details. Lack of correspondence will result in a loss of quality. Currently a large amount of manual postprocessing is needed to create a texture with satisfactory results. We argue that automated texture processing methods are needed as urgently as one needs geometric reconstruction methods. Texture must closely interact with the geometric model so that satisfactory 3D geographic information systems are obtained.

References

Burns B., A. Hanson, E. Riseman (1986) *Extracting Straight Lines,* IEEE Transactions on Pattern Analysis and Machine Intelligence, Vol. 8/4 Los Alamitos

Gruber M., S. Meissl, R. Böhm (1995) *Das dreidimensionale Stadtmodell Wien, Erfahrungen aus einer Vorstudie,* VGI, Heft 1+2, Wien

Leberl F., M. Gruber, P. Uray, F. Madritsch (1994) *Trade-offs in the Reconstruction and Rendering of 3-D Objects,* 16th DAGM Symposium and 18th ÖAGM Workshop, Informatik XPRESS 5, Wien

Pinz A., M. Prantl, H. Ganster (submitted) *A new Affine Matching Algorithm for Computer Vision*, submitted to J. of Universal Computer Science, Springer Verlag

Use of DTMs/DSMs and Orthoimages to Support Building Extraction

Emmanuel Baltsavias, Scott Mason and Dirk Stallmann

Institute of Geodesy and Photogrammetry
Swiss Federal Institute of Technology (ETH)
8093 Zurich, Switzerland
{manos, mason, dirk}@p.igp.ethz.ch

1. Introduction

The acquisition of 3D models of buildings and other man-made objects is currently an issue of high importance to many users of geoinformation, including planners, geographers, architects, etc. Aerial imagery has proven to be a valuable data source for these models. The project AMOBE[1], a joint research effort between the photogrammetry and image sciences groups at ETH, aims firstly at developing practical algorithms to support the semi-automatic reconstruction of man-made objects from aerial imagery and secondly, at developing improved techniques for automatic digital terrain and surface model generation. In the latter case it is important to differentiate between terrain models (DTMs), which model the terrain, and surface models (DSMs), which model all 3D objects. In this paper, we explore the roles of both DTMs and DSMs and their derived products orthoimages in supporting the extraction of buildings from aerial imagery. In Section 3 the quality of commercially available automatic DTM/DSM generation software is investigated. In Section 4 a number of methods for automatically detecting buildings in DSMs are presented and evaluated. In Section 5, the use of DSMs in deriving coarse building models is described. Applications of orthoimages are discussed in Section 6. First, however, the fundamental assumptions and strategy employed in AMOBE is outlined. Note finally that the techniques and tests described in this paper constitute preliminary investigations. Promising directions, as noted, are the subject of ongoing research.

2. Project AMOBE

Primary attention in AMOBE is focussed on the extraction of buildings as one of the more predominant and frequently occurring 3D man-made structures. The employed imagery is assumed to be digitised photogrammetric photography of the type typically used for topographic mapping. In this nadir imagery primarily building roofs, and not walls,

1. Support for this research by ETH under project 13-1993-4 is gratefully acknowledged. The project is entitled "Automation of Digital Terrain Model Generation and Man-Made Object Extraction from Aerial Images" (AMOBE).

can be extracted. Given assumptions, e.g. from the context, or estimates of the overhang from terrestrial field measurements, the full volume structure of a building can be inferred from the reconstructed roof. The main features of the strategy we are following are illustrated in Figure 1 and include:

- Automatic generation of a DSM exploiting *all* overlapping images.
- The DSM and the extracted 2D features should be exploited to automatically derive a coarse building model that can be employed in the following steps of feature matching and model-based building reconstruction.
- As 3D object reconstruction can best proceed using 3D features, 3D information should be derived as early as possible by *multi-image matching* of 2D image features, e.g. straight lines, exploiting their attributes and graph relations obtained from low level segmentation (cf. Henricsson, 1995). The DSM is exploited along with the epipolar constraint for the determination of an approximate search space.
- The matching step produces a "cloud" of 3D linear features. Generic object models are then used to structure the 3D linear features into roof structures. The form of these models is currently being investigated. The analysis of the DSM blob for each building (see Section 5) may provide for automatic generic model selection. Information derived from 2D grouping (cf. Henricsson, 1995) will be exploited in the matching process.
- Due to the high complexity and range of building (roof) shapes, generic models will be needed to support the reconstruction process. Our concept, as shown in Figure 2, is to decompose a building's roof into 3D surface primitives, e.g. planar entities such as rectangles, trapezoids, and triangles. Models of these primitives are then used to automatically extract the roof components which, when combined, constitute the reconstructed roof. For simple buildings, parametric volume models may be applicable.
- It is presumed that operator interaction will be required to support automatic extraction procedures. The type and level of this interaction will depend on the quality of the results produced by the automated steps.

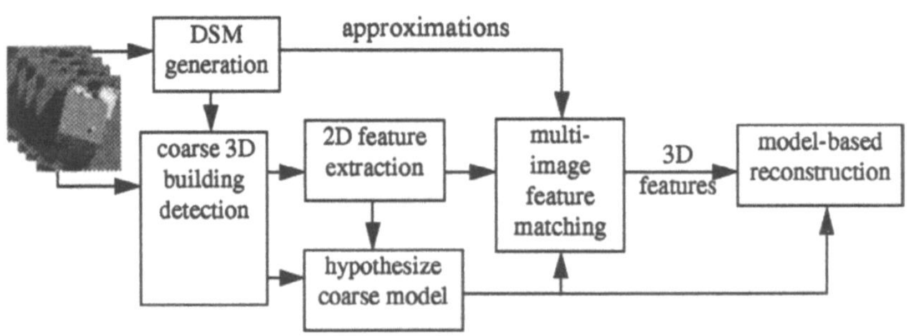

Figure 1: Strategy employed in AMOBE for building extraction from aerial imagery.

Figure 1 clearly shows that DSMs play an important role at all stages of the building extraction strategy. In the following sections, this role is described and demonstrated. The results are derived from the "Avenches" dataset[1] (Mason et al., 1994). This dataset consists of a residential and industrial scene and fulfills the aforementioned imagery assumptions: large image scale, color imagery, four-way image overlap, precise orientation data and high resolution scanning (15μm).

Figure 2: Model-based building extraction based on volumetric and surface primitives.

3. DSM Generation

The purpose in DSM generation is the derivation of a complete 3D model of the visible surface with the highest possible accuracy and taking into account terrain discontinuities. State-of-the-art automatic algorithms based on stereo imagery are not able to distinguish between the terrain surface and objects on and above this surface, e.g. man-made structures such as buildings or trees. Their output is consequently a 2.5D DSM as opposed to the generally desired DTM. For building extraction from aerial imagery, however, DSMs are well-suited offering a number of possible uses:

- Approximate detection of buildings. These results can be used to guide 2D image feature extraction, thereby speeding up processing time, support 2D feature grouping, and reduce the number of candidates in feature matching.
- Separation of building from other non-DTM objects, e.g. trees.
- Approximations for feature matching.
- Selection of a model or reduction of hypotheses in model-based building reconstruction (see Section 5), support of 3D grouping.
- Orthoimage generation (see Section 6).
- Derivation of coarse building models (e.g. building outline and maximum height) as a final product.

1. This dataset is available via ftp by contacting the authors.

3.1. DSM Generation by Stereo Correlation

In this section we report on the quality of DSM generation using the commercial state-of-the-art approach supported by the Helava DPW 770. This software employs hierarchical, stereo-based correlation. A number of strategies are offered amounting to parameter selections appropriate to different topographic classes and accuracy requirements. These selections are experience-based. We investigated different strategies as well as the sensitivity of the results with respect to the choice of the Ground Sample Distance (GSD), the sampling density of the DSM in object space. For the residential scene the strategy "steep_1" for high accuracy extraction under steep terrain conditions (here, the buildings and trees) offered the best results.

Figure 3 shows that, even though it will not guarantee close modelling of a building, the choice of a small GSD is a requirement; a coarse GSD can, in isolated cases, lead to the building being poorly modelled or not at all.

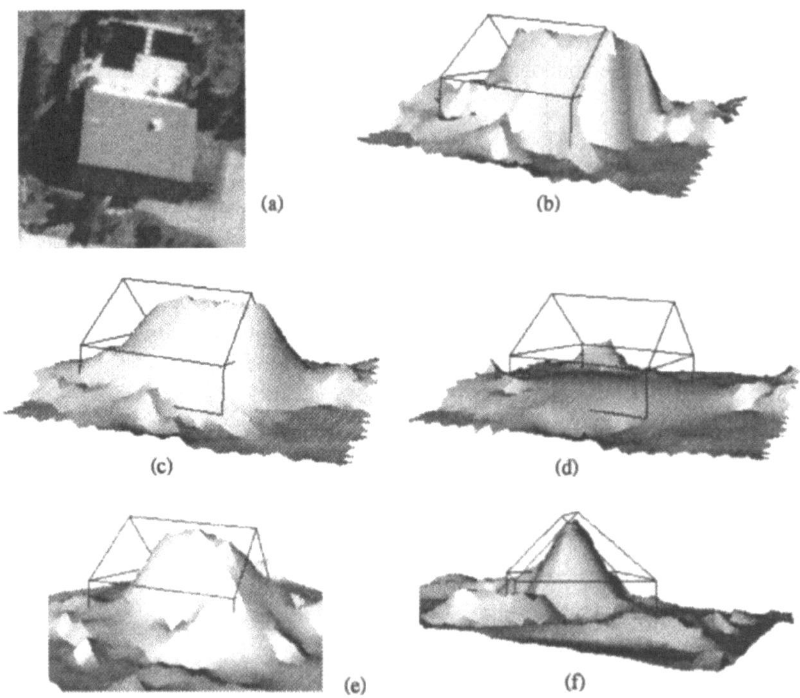

Figure 3: DSM results for varying ground sample distances: (a) image of the building; (b) 0.1m; (c) 0.25m; (d) 1.0m; (e) and (f) illustrate the result of merging multiple DSMs derived from pairwise matching of six images (GSD 0.25m).

For the dataset, a GSD of 0.25m was determined to be a good comprise between modelling accuracy and processing requirements (cf. Table 1). Figure 3 illustrates the "worst case" result achieved by the Helava DPW 770, i.e. with a GSD of 1m the 12 x 10 x 6 m house was not modelled. All other buildings in the dataset were modelled for GSD = 1m.

Table 1: DSM generation by stereo correlation

Test	GSD (m)	Number[a]	Time[b]
1	0.1	49216	23.2
2	0.25	7925	3.3
3	1.0	505	1.0

a. Number of acceptably matched points
b. Processing time relative to Test 3

3.2. Alternatives

The results above demonstrate some of the limitations of image matching procedures for DSM generation including:

- The measurement density may not be sufficient to model discontinuities such as buildings, buildings close to each other etc. Buildings may even totally disappear.
- Discontinuities are smoothed.
- Many matching errors occur around buildings due to occlusions, trees and shadows.

We are currently developing a multi-image approach based on the Multiphoto Geometrically Constrained matching (MPGC) (Baltsavias, 1991). In this approach, the points to be matched are selected in one image by an interest operator, basically a direction sensitive edge extractor, using a variable-size window. We believe that the inclusion of points on the discontinuities in the matching will lead to better modelling. Results will be reported in due course. Sensors such airborne laser range scanners and, to a lesser extent, interferometric SAR are also expected to provide viable alternatives to and/or support for DSM generation by image-based techniques.

3.3. Merging DSMs from Multiple Stereopairs

Inherent occlusions associated with 3D objects in stereo matching may, to some degree, be reduced by a multi-image approach. Moreover, it may be anticipated that the accuracy of the resultant DSM would be improved by this approach. An alternative to simultaneous multi-image approaches, such as in MPGC, is to merge the results of multiple correlations[1]. For example, if four-way stereo overlap is available, a total of six separate DSMs can be produced. We conducted an experiment in which six such DSMs were automatically generated using the Helava DPW 770 for the residential scene in the "Avenches dataset". The DSMs were then merged on the basis of a "figure of merit" (FOM) priority. The FOM is an estimate of the quality of each correlated point and therefore of the resultant height. Figure 3e and Figure 3f illustrates the results of merging the

1. This approach is clearly suboptimal in terms of efficiency and the statistical testing afforded by the simultaneous approach.

six DSMs. Clearly the modelling of the house remains unsatisfactory, because the FOM is not a reliable measure of the matching accuracy.

4. Use of DSMs in Building Detection

Because a DSM ideally models the man-made objects as well as the terrain, it may be applied to automatically *detect*, for example, buildings in the scene and thus, by projecting the positions of the detected buildings into the images, provide initial windows for building extraction schemes. A number of difficulties immediately arise: (1) buildings will be at best modelled as smooth "lumps" by matching approaches and thus will appear similar to other 3D objects such as trees, small mounds, etc.; (2) closely situated 3D objects will often be "melted" together; (3) in residential areas, trees are often found close to buildings; (4) in some cases, the matching process will fail to model a building (see Figure 3). In particular due to (4), the results of detection should not be relied upon but used as a starting point.

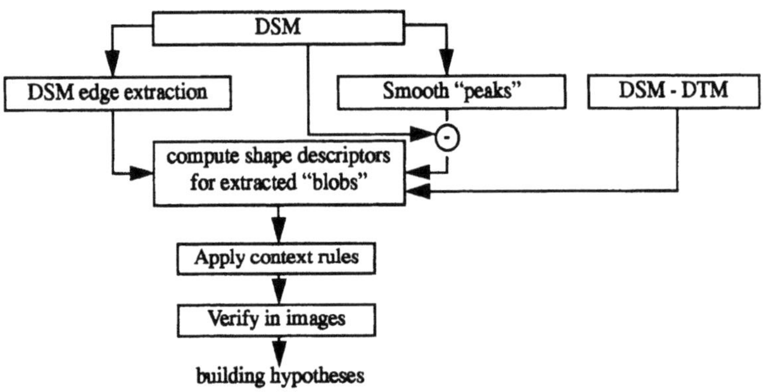

Figure 4: DSM-based building detection schemes.

Three schemes for building detection from DSMs are presented in Figure 4. The middle scheme is based on applying the morphological operation "opening" to smooth the DSM. Subtracting the processed DSM from the original DSM results in the extraction of the smoothed peaks. By thresholding this new "image" according to the expected building height range for the scene, the individual peaks can be simply segmented, e.g., using a component labelling algorithm. Shape descriptors, e.g. the compactness, area, etc., can be applied to judge whether or not an individual component is hypothesised as a building. Limits for these descriptors can be obtained from the context, e.g. residential buildings have an area greater than 50 m^2. There are clearly a number of weaknesses in this approach. First, it is sensitive to the choice of structuring element size, particularly in dense urban areas. Second, the opening operation shows no discretion when a building is closely situated to other DSM "lumps", e.g. steep terrain changes such as retaining walls might be extracted with the building leading to gross errors in the shape descriptors. Figure 5 illustrates some of these problems.

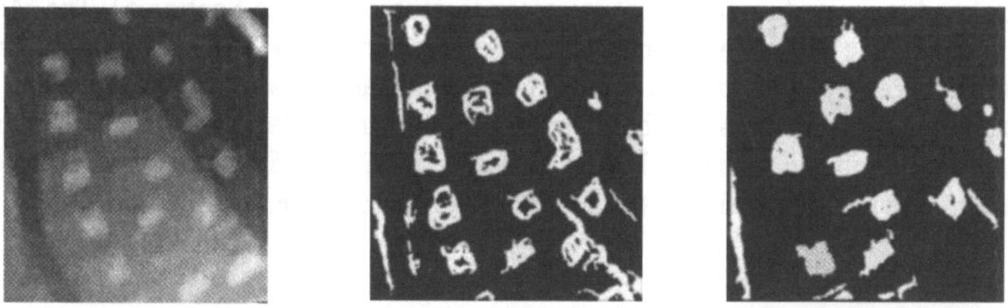

(a) (b) (c)

Figure 5: Results of morphological building detection: (a) DSM generated by DPW 770 for the residential dataset; (b) and (c) illustrate the effect of varying the structuring element size from 3 m to 5 m on the hypothesised building components.

A more promising approach to building detection is to segment the DSM with a 3D edge detector (left-hand scheme in Figure 4). Clearly, shape descriptors will still need to be used to form hypotheses but the problematic choice of the structuring element is avoided. Neighbouring "lumps" in the DSM will still be discernible, as illustrated in Figure 6a. This approach is being investigated further.

If a DTM is available, and for many developed countries this is the case, a subtraction of the DTM from the DSM (right-hand scheme in Figure 4) can provide a useful clue in building detection. The quality of the detection results using this method (see Figure 6d) may be a little optimistic, as in this case the DTM was derived to a higher accuracy and at a higher density than that of the Swiss[1] national 25m DTM (DHM25).

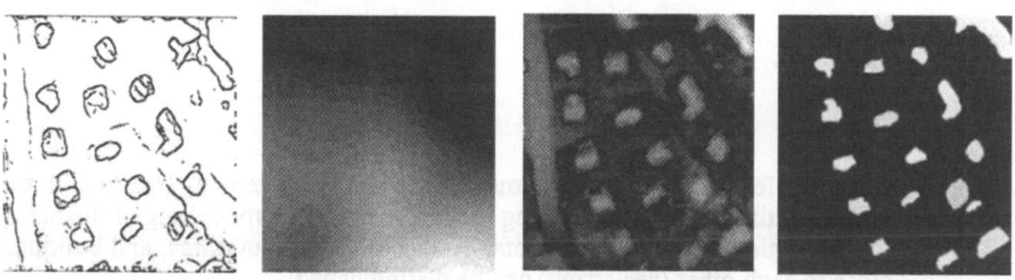

(a) (b) (c) (d)

Figure 6: (a) segmentation of DSM with a range image edge operator (Boulanger, 1994); (b) DTM; (c) DTM subtracted from the DSM; (d) building detection based on DSM - DTM.

An alternative to the above building blob detection methods is the use of multiple height bins (MHB). In the MHB method the DSM heights are grouped into consecutive bins (height ranges) of a certain size. This results in segmentation of the DSM in relatively few regions that are always closed and are easy to extract (see Figure 7). The method is

1. The DTM for the dataset was measured as a 10m grid to an accuracy of about 20 cm. The Swiss DHM25 has a 25m grid spacing and an accuracy of 2-4 meters for this region.

applied hierarchically whereby the large bins are used to detect possible buildings, and the small ones to verify and refine the coarse detection, find the approximate building model and separate buildings that are close to each other (see Section 5). The MHB method is simple and fast and can be applied globally or locally. The maximum and minimum bin sizes are determined from the known height accuracy of the DSM (e.g. 0.5 - 1 m) and the estimated minimum building height in the image (e.g. 3 - 4 m). The method can be also combined with the 3D edge extractor (see Figure 6a) to help bridge contour gaps and provide information in the interior of the contours, i.e. building roofs.

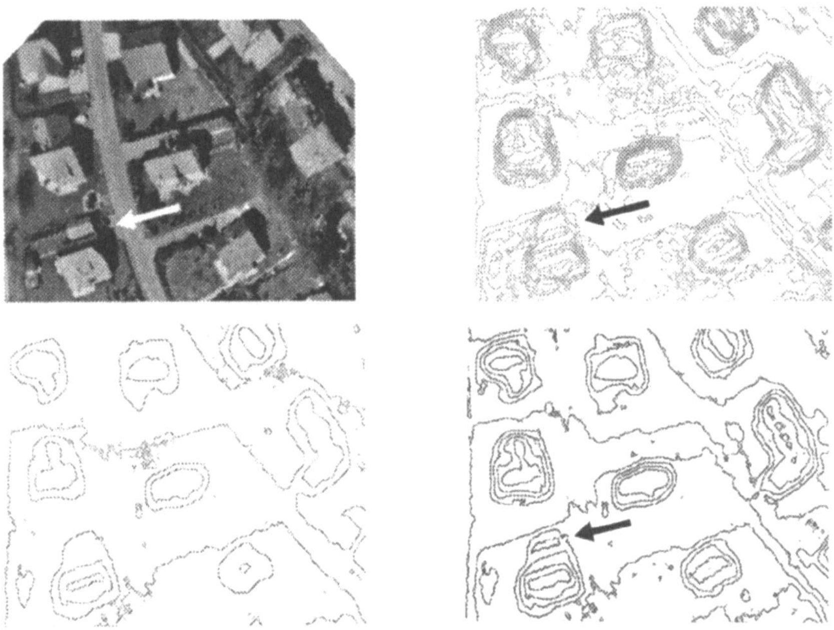

Figure 7: From top left clockwise: Orthoimage, DSM height bins with 1m, 2m, 3m bin size (quantization). By decreasing the bin size a better modelling of the buildings is achieved, and gabled roofs, T- and L-shaped buildings, and buildings close to each other (see arrow) can be distinguished.

4.1. Blob classification: seeing the buildings from the trees (and other objects)

The blobs detected in a DSM can be attributed to many different 3D objects. To reduce the false alarm rate in building detection we need to be able to separate those belonging to buildings and those belonging to other objects, e.g. trees. Here we assume that context knowledge, such as shape bounds (height, size, elongation, etc.) on buildings in a given area, has already been applied to preclassify the blobs and rely now on information in the images. A simple procedure for distinguishing buildings from trees is illustrated in Figure 8. A histogram of the gradient orientations in the range [0°, 180°] at all pixels belonging to strong gradients in an image region associated with a DSM blob is computed.

Histograms belonging to buildings will contain significant peaks 90° apart (for regularly shaped buildings), with additional peaks (usually one or two) being detected for more angular buildings. On the contrary, histograms belonging to trees are flat. Additional classification cues that can be used to distinguish buildings from trees and other 3D objects include the number and length of long straight lines, texture, colour including infrared for vegetation segmentation, motion and context.

An important product of classifying the DSM blobs is the opportunity to generate a DTM. Those blobs which belong to non-terrain features can be subtracted from the DSM, and in the blob regions heights can be interpolated from the surrounding terrain or in case of forests the DSM heights can be reduced by an estimation of the tree heights, to derive a DTM.

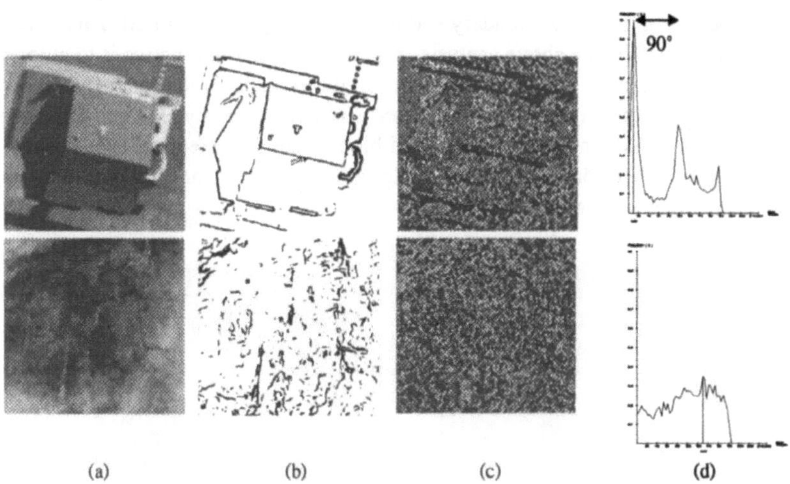

Figure 8: DSM blob classification: distinguishing buildings from trees: (a) original images; (b) strong gradients obtained from thresholding; (c) gradient orientation images; (d) histogram of orientations of strong gradients.

5. Estimation of Approximate Building Model

An analysis of the blobs detected in a DSM can, when the DSM is dense and relatively accurate, provide in combination with the 2D extracted features an approximation of the building model. As an example, small elongated regions in the middle of a DSM blob (see top right in Figure 7) that include long straight lines in the image form a very strong hypothesis that there is a gable on the roof. The approximate model can then be used to (i) support grouping of 2D features, and (ii) automatically select the appropriate model or restrict the possible hypotheses in the 3D reconstruction stage. The aims of the analysis for approximate building detection may be to:

- Determine approximate building size (height, length, width).

- Distinguish between flat and peaked roofs. In the latter case, distinguish whether the roof plane(s) intersect at a highest point or at a gable.
- Distinguish between rectangular, T, L, U and + shaped buildings.
- Determine the number and form of the major roof planes.
- Detect buildings that are close together (see top right in Figure 7).
- Exclude small disturbances on roofs (skylights, chimneys).
- Detect smaller buildings on building roofs and group them.

The above analysis makes use of certain properties (e.g. symmetry) and rules (e.g. a gable is the intersection of either two roof planes or one roof plane and a wall).

In certain applications, e.g. telecommunications, flight simulators etc., only a coarse building model, e. g. building boundary and maximum height, is required. This information can be provided by the above analysis. Another attractive alternative is to automatically derive the outline of the buildings from digitised maps (cf. Carosio and Nebiker, 1995) and the height from the DSM. The latter is accomplished by estimating the maximum building height using the nearest DSM blob to the projected building outline, as shown in Figure 9. This procedure can lead to rapid establishment of 3D building databases, a very important practical application.

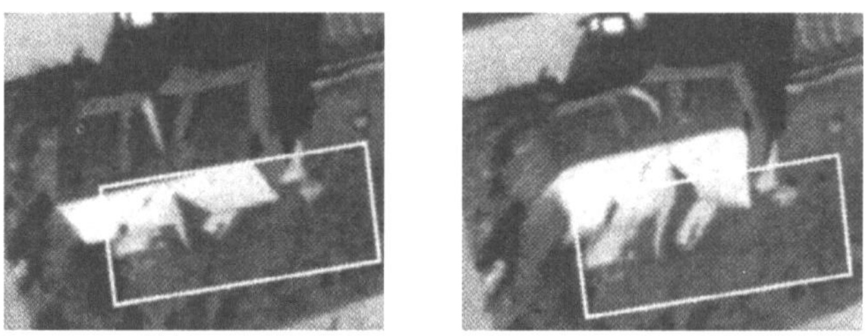

Figure 9: Coarse building modelling. Building outline digitised from a 1: 25000 map overlaid on an orthoimage from a DTM (left) and a DSM (right).

6. Orthoimages and Orthorectified Stereo Images

Orthoimages are potentially useful in building extraction:

- Geocoded information such as DTMs, DSMs, roads or utility lines from, e.g. a GIS or a map, can be directly combined with the orthoimage.
- 3D measurements can be made by detecting corresponding points in orthorectified stereo images. These 3D measurements are correct even if the DTM or DSM used for the orthoimage generation is totally wrong (Baltsavias, 1993). Matching by use of multiple orthoimages and geometric constraints is possible. Matching in orthoimages is easier due to good approximations and small geometric distortions.

There are, however, a number of problems with their use. First, buildings in an orthoimage produced using a DSM will be deformed due to DSM errors. These deformations cause problems in the extraction of straight linear features and matching between orthorectified stereo images (see Figure 10). In orthoimages produced from a DTM the buildings lie on the ground and are partly out of the detected DSM blobs. However, straight lines remain straight and can be extracted and matched. Orthorectified stereo images generated from a DTM can also be used for building reconstruction by detecting corresponding characteristic roof points in a semi-automatic approach.

(a) (b) (c)

Figure 10: Orthophotos generated using (a) DTM, (b) DTM and 3D building model, and (c) DSM.

Ideally, 3D non-terrain objects could also be detected by differencing orthorectified stereo images produced by a DTM. However, there will be no differences in the overlapping areas of the non-DTM objects, and there will be differences due to radiometric differences of the images and geometric misregistration (see Figure 11). The latter problem can be avoided by subtracting images in a higher pyramid level. More useful is the subtraction of orthorectified stereo images produced by a DSM (see Figure 12b) for quality control of the DSM and manual or automated corrections. In regions of large differences either DSM errors or radiometric differences will exist. This is a very simple and practical way of controlling the DSM quality in the absence of ground truth.

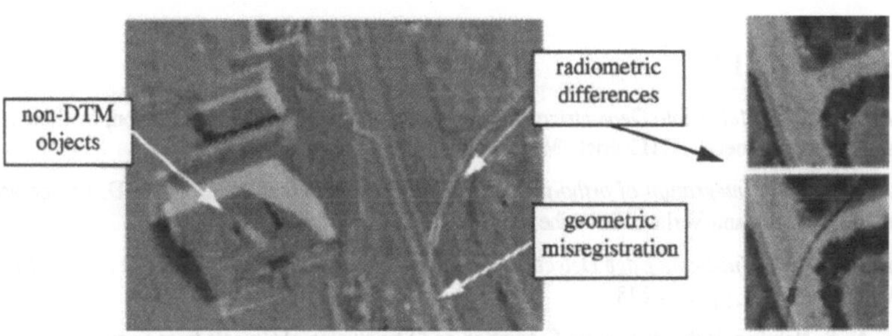

Figure 11: Left: differencing orthorectified stereo images. Not all differences can be attributed to non-DTM objects. Right: differences due to a hair (noise) on one of the scanned films.

Figure 12: Differences of orthorectified stereo images: (a) using DTM, (b) using DSM

7. CONCLUSIONS

DSMs provide a very useful source of information in extracting buildings from aerial imagery. They can be employed to automatically detect buildings as blobs and to provide approximations in matching of features in overlapping images. The detected buildings blobs and the extracted 2D features, in turn, can be analysed to hypothesize a coarse model of the building. This model can be used in 2D and 3D feature grouping and 3D roof reconstruction. Alternatively, for certain applications the coarse building may be the final product. Clearly these methods require that the DSM models the buildings to a sufficient extent. To this end it has been shown that in automatically generating DSMs state-of-the-art matching procedures require a dense sampling rate. It has also been shown that the quality of a DSM can be assessed by subtracting orthorectified stereo images produced with it. Finally, by classifying and removing non-terrain objects from a DSM, a DTM may be generated.

8. REFERENCES

Baltsavias, E., 1991. *Multiphoto Geometrically Constrained Matching*, Ph. D. dissertation, Institute of Geodesy and Photogrammetry, ETH Zurich, Mitteilungen No. 49, 221 p.

Baltsavias, E., 1993. *Integration of ortho-images in GIS*, Photogrammetric Week '93, D. Fritsch and D. Hobbie (Eds.), Wichmann Verlag, Karlsruhe, pp. 261 - 272.

Boulanger, P., 1994. *Multiscale Edge Detection Based on a New Geometrically Intrinsic Filter*, SPIE Vol. 2350, Videometrics III, pp. 264-278.

Carosio, A., S. Nebiker, 1995. *Automatic Extraction and Structuring of Objects from Scanned Topographic Maps - An Alternative to the Extraction from Aerial and Space Images?*, Workshop on Automatic Extraction of Man-Made Objects from Aerial and Space Images, Ascona.

Henricsson, O., 1995. *Inferring Homogeneous Regions from Rich Image Attributes*, Workshop on Automatic Extraction of Man-Made Objects from Aerial and Space Images, Ascona.

Mason, S., E. Baltsavias, D. Stallmann, 1994. *High Precision Photogrammetric Data Set for Building Reconstruction and Terrain Modelling*, Institute of Geodesy and Photogrammetry, ETH Zurich, Internal Report.

Automatic Extraction of Man-Made Objects
from Aerial and Space Images
Monte Verità, © 1995 Birkhäuser Verlag Basel

Data fusion for the detection and reconstruction of buildings

Norbert Haala and **Michael Hahn**
Institute of Photogrammetry, Stuttgart University
Keplerstraße 11, 70174 Stuttgart
Phone: 0711-121-3384
e-mail: norbert.haala@ifp.uni-stuttgart.de

Abstract

Research in detection and reconstruction of man-made objects from aerial images has made significant progress in the past two or three years. Two important reasons for that are: (1) data fusion of different sources provides more information for the ill-posed image analysis processes and (2) more sophisticated algorithms are developed which apply grouping and reasoning processes using a model of the object class of interest.

This paper presents an algorithm for automatic detection and reconstruction of buildings using height and image data. A given Digital Height Model (DHM), in the experiments computed by automatic stereo matching, is used to focus attention on regions where buildings are expected. Detection relies on the heuristic that buildings are represented in a DHM by regions with local height maxima. Object contours of buildings can be modelled by straight lines. Therefore, three-dimensional line segments are extracted from the image pair by stereo matching of grey-value edges. Again, the DHM is used to provide approximate parallaxes for the line segments. A building can be reconstructed by matching these observed three-dimensional lines to the lines of a model of the building. Position and shape of a building is estimated by minimizing the distances between the observed lines and the corresponding lines of a parameterized building model. The resulting error of least squares estimation provides a measure on how good the observed lines fit to the model. Thus it can be used to evaluate the result of the reconstruction. To have a second quality check the extracted roof regions of a building in a stereo image pair are matched by area based matching.

1 Introduction

The growing number of applications using 3-D urban models promises to become a considerable market. Examples like 3-D simulations for planning telecommunication networks demand for urban models which are complete and up-to-date. Most significant objects in urban model data bases are buildings and roads. This paper describes a procedure which aims on automatic reconstruction of buildings using height and aerial image data.

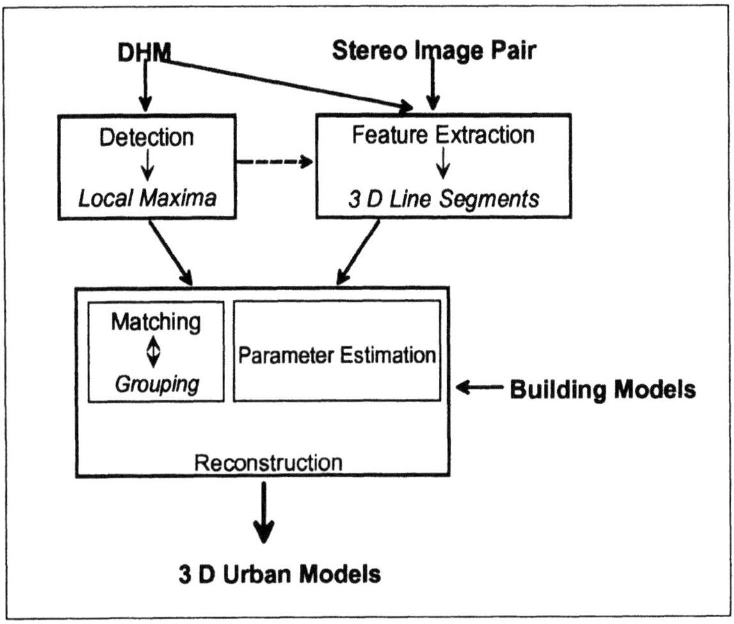

Figure 1: Work-flow diagram

The work-flow of the proposed algorithm is shown in figure 1. A stereo image pair with known exterior orientation and a Digital Height Model (DHM), computed automatically by stereo image matching, are used as input data. In this context the term DHM is used for a surface description in which objects like trees or buildings are represented together with the terrain.

The reconstruction step is preceded by two other steps, the detection of 3-D object regions and feature extraction, in particular linear segments. The basic assumption for the first step is that buildings rising from the terrain surface appear as lump in the DHM. The detection of local height maxima in the DHM is used to define the regions of interest for the following steps. All further processes can be restricted to these 3-D object regions. In the second step line segments are extracted from grey value edges in both images. Using the parallaxes given by the DHM corresponding lines are determined and 3-D line segments are adjusted. In summary, the result of the first two steps are linear segments of 3-D lines which are likely to be roof break-lines, other object borders, but also shadow or texture lines.

Further analysis of these line segments is done within a reconstruction module. Reconstruction includes a matching of observed data with object models. This implies that suitable building models have to be provided. We use primitive instancing to represent rectangular buildings with gabled roofs. This basic type of buildings is taken as an example, because it occurs quite frequently in suburban areas. A modification of the building model does not change the overall process. If a library of different object models is taken into account this implies that model matching has to select the best fitting building type. The reconstruction consists of a two-step process. First corresponding pairs of extracted linear segments and lines of the building model have to be established. Without any prior knowledge on shape and position of a building each observed line could be a potential matching partner of a model line. The search space is cut down by combining the linear segments to rectangular structures applying relations like proximity, parallelity and rectangularity. The grouping leads to more meaningful structures which are combinations of linear segments and provides information on shape and position of a building. This information is utilized to estimate parameters of the building model. The estimation process is a least squares adjustment minimizing the distance between the model lines and the observed three-dimensional linear segments. The least squares adjustment provides a measure on how good the lines fit to the selected model. This is used for evaluation of the obtained result. For a further evaluation the roof regions of both images are matched using area based matching. If area matching is successful the mapping between corresponding regions indicates consistency with an inclined plane in space.

In the following some more details on the approach are outlined. Results of the processing chain applied to a height and image data of a scene are visualized to illustrate the algorithm.

2 Extraction of 3-D line segments

The starting point for the analysis are height data of the imaged scene and a stereo image pair. The exterior orientation of the images is given, thus height and image data are explicitly related to each other. This is a prerequisite for a reasonable fusion of information extracted from both data sets. Figure 2 shows a DHM of a built-up area, computed by a commercial program for surface reconstruction applying automatic stereo matching (Ackermann & Krzystek 1991).

The first step in image analysis is to focus attention on regions where 3-D objects are present. Extraction of those regions is done by detection of local maxima in the height model and thresholding. If the heights within local maxima (3-D object regions) are eliminated, a surface model can be computed representing more or less

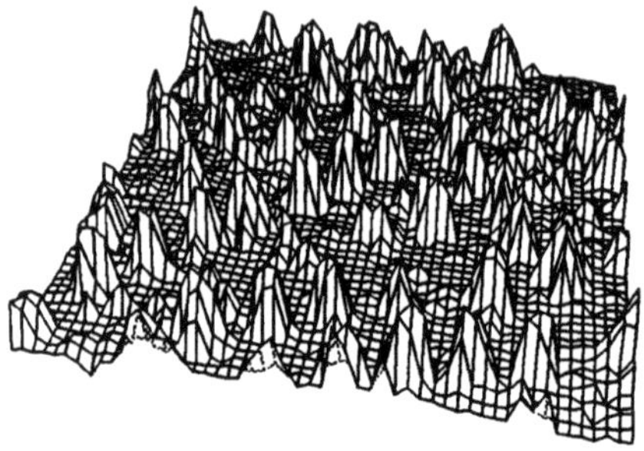

Figure 2: DHM, computed by stereo image matching

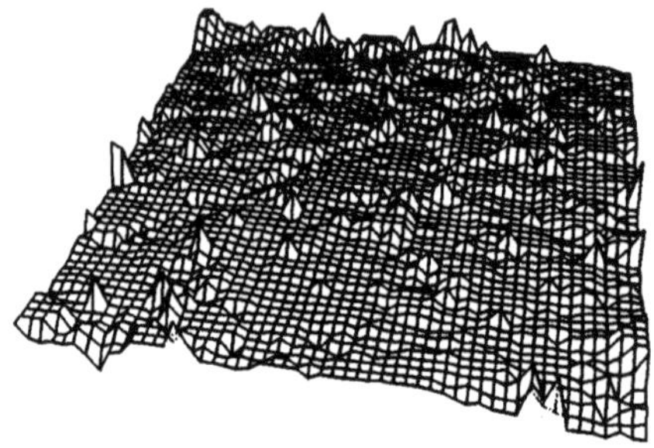

Figure 3: Approximate DTM, computed by elimination of local maxima

the terrain surface. The resulting Digital Terrain Model (DTM) without the blobs caused by buildings is shown in figure 3.

A more complex task than 3-D region extraction is the extraction of 3-D line segments within these regions. The reason for relying on straight lines is that contours of buildings and other man-made objects give a very significant characteristic of the object. Corresponding lines can be reasonably extracted from the grey level images. Edge detection, edge linking and straight line approximation are standard tools of digital image processing. The lines are determined in three-dimensional object space by stereo matching. For this purpose the DHM is used to provide approximate values for the parallaxes of corresponding line pairs. The height of a line above the terrain is used to classify lines which are likely to represent roof break-lines and lines which are more likely to belong to linear segments on the ground level.

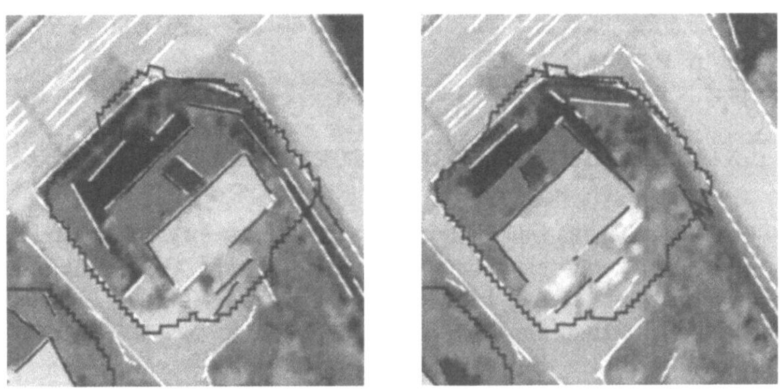

Figure 4: Stereo image pair with extracted grey-value edges (white), detected 3-D region (black border line), selected three-dimensional line segments (black)

Figure 4 presents results of detected 3-D object regions and feature extraction, illustrated by a small part of the stereo image pair. Grey-value edges approximated to straight line segments are extracted by an algorithm proposed by Burns, Hansen & Riseman (1986). These lines are superimposed as white lines to the grey level images. The border line of the 3-D region extracted from the DHM is represented by a black polygon. The black straight line segments are a subset of the white lines. These lines could be determined three-dimensional by stereo line matching. The matching restricts to lines which are located in 3-D object regions, where buildings are supposed. For reconstruction only the black 3-D line segments are used. A more detailed description on the processing steps which result in three-dimensional line segments is presented in Haala (1994).

3 Reconstruction

If certain objects have to be geometrically reconstructed from image and height data, they first have to be recognized in the data. Recognition, in general, presumes knowledge about the percepted objects dealing with any kind of object model. For the geometrical reconstruction it is desirable to have an appropriate geometric object model.

The *boundary representation* is probably the most widespread 3-D model and already many algorithms are available for computing physical properties from that representation. In *constructive solid geometry* (CSG) simple primitives are combined by means of regularized Boolean set operators. We use *primitive instancing* which is very suitable for representing several objects of the same type. A group of similar objects is described by a parametric model with a fixed set of parameters. The individual objects of this group differ only in the values of the parameter.

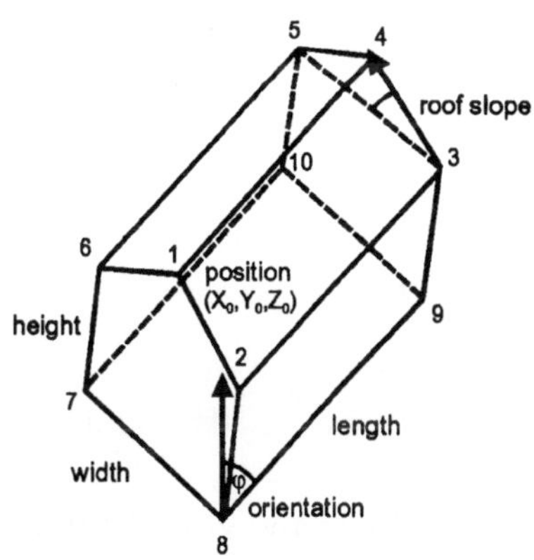

Figure 5: Parametric model of a rectangular building with gabled roof

The shape of a basic type of a building is plotted in figure 5. Shape parameters of this object model are roof slope, width and length of the building. The position of the building is defined by the coordinates X_0, Y_0, Z_0 of a selected point, the orientation of the building corresponds to the angle φ of the ridge. A parametrization of other building types, e.g. L-shaped buildings or buildings with hipped roof can be set up in a similar way.

Due to perspective projection and occlusion, vertical walls of buildings in general are not visible in an aerial image. Despite from occlusion by other objects like trees or larger buildings this is not a problem for the roof of a building. Therefore, the linear 3-D segments extracted in the preceding steps are matched against the roof lines of the building model. If the roof of the building is reconstructed its height can be determined in a second step by computing the differences between the eaves and terrain heights shown in figure 3.

3.1 Grouping

In order to match the observed data to the building model it is necessary to find a mapping between the set of extracted linear segments and the set of model lines. Primitives are matched if they have similar attributes, e.g. length, position or orientation. Without any a priori knowledge on the position, orientation and shape of a building, all possible mappings between observed primitives and model primitives have to be considered. In that case a matching procedure would fail in practice due to the exponential complexity of the matching problem. Aggregation to more abstract primitives is a standard way to reduce the correspondence problem to model primitives. Meaningful combinations of linear segments can be found by grouping.

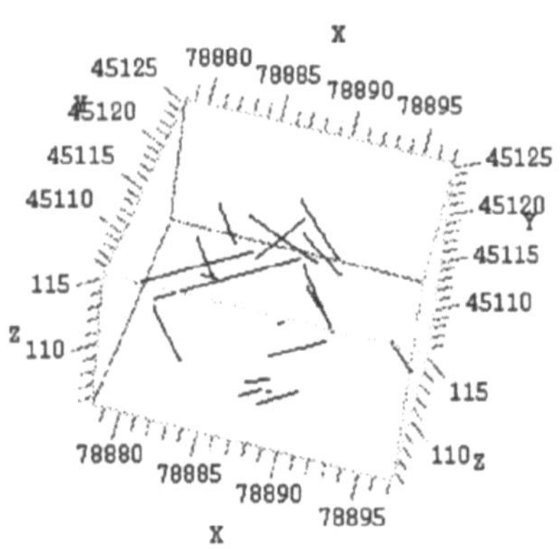

Figure 6: Perspective view of 3-D line segments.

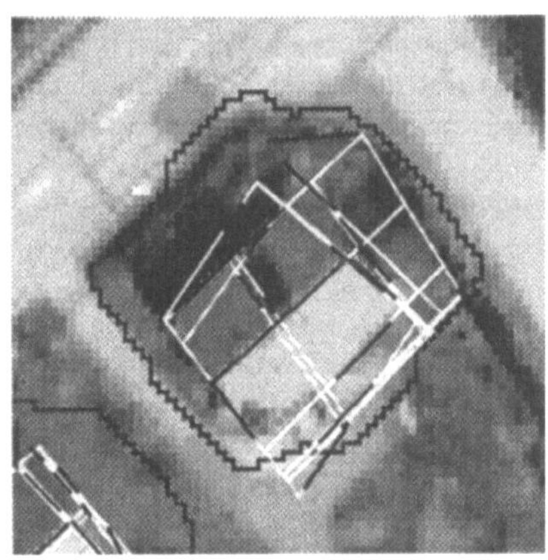

Figure 7: Part of the orthoimage, extracted line segments (black) and detected groupings (white).

Rectangular structures are detected by combining four linear segments, respectively. For that purpose relations like proximity, parallelity and rectangularity are applied. Invariance to viewpoint position is obtained by performing the grouping in object space; the topological relations of the linear segments are computed using ground coordinates. Figure 7 shows the detected rectangles and the used linear segments projected into the orthoimage. The definition up to what extent lines are parallel, rectangular or proximate is based on thresholds. In order to avoid that a correct solution might be eliminated by too strict thresholds they are set very generously. Constraints based on the height or slope of the grouped lines are not used at all. The example (figure 7) shows several grouping results in which texture lines and shadows are involved as well as the lines which represent break-lines of the roof. That the matching process creates a certain number of mismatched lines can be observed from a perspective view of all linear segments in figure 6. Mismatched lines are a reason for the multiple overlapping rectangles constructed by the grouping.

Each side of a grouped rectangle is defined by one extracted line segment. If the observed lines are border lines of the roof its vertical projection to the ground indicates the ground space of the building. Position, orientation, length and width of the building can be computed approximately and used in the next step which aims at estimating the parameters of the building model. The slope and relative height of the four grouped linear segments are essential features for establishing correspondences between model lines and observed lines. One pair of opposite lines of the four grouped lines is parallel to the ground plane. If one of these two lines is higher than the other then this line defines the ridge of the roof. The second pair fits to the front sides of the building and defines the gable of the roof. The slope of these lines indicates the inclination of the roof.

3.2 Estimation of building parameters

The parameters of the roof model (roof slope, width, length, position and orientation of the building) can be determined by least squares adjustment. As objective function minimization of the distances between the observed linear segments and the corresponding model lines is chosen. Each grouped rectangle is taken as a candidate of the estimation process. Initial parameters of the roof are estimated with four lines combined by the grouping. Sequentially all detected groupings are processed. Within this estimaion step a detected grouping may be rejected if the line combinations do not fit the roof model. Otherwise the roof parameters are used to include further line segments into the estimation process. The error of the residual sum of squares of the least squares adjustment measures totally the discrepancy between observed lines and fitted roof model. This error expresses also the quality of topological conditions between the lines initiated by grouping. The corresponding topology is implicitly covered by the building model. Among the solutions determined with all detected groupings the reconstruction with the smallest error is chosen.

Figure 8 shows a wire frame model of the reconstructed building with all lines used for reconstruction. As mentioned before the height of the building is estimated by the distance between the eaves of the roof and the terrain surface. The reconstruction process runs over all 3-D object regions sequentially.

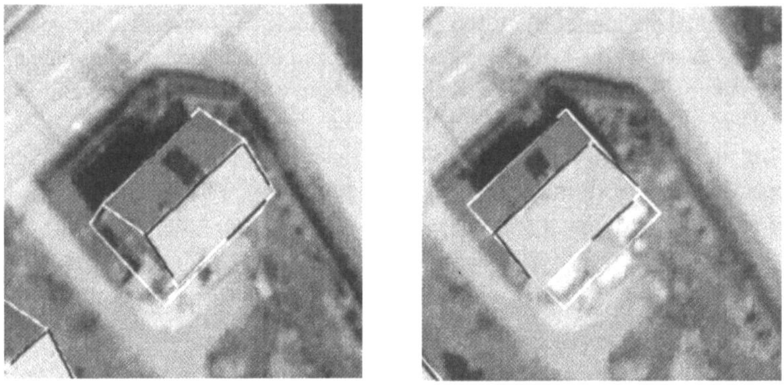

Figure 8: Wire frame model of a building with lines used for reconstruction

Figure 9: Left image with reconstructed buildings and used linear segments

3.3 Further evaluation

The reconstruction process described so far selects the best grouping among several candidates and estimates the parameters of a given building model. As a further

evaluation procedure we use classical area based matching. The aim is to confirm or reject the reconstruction by matching the roof regions localized in both images in the processes before. To restrict area based matching to those pixels which are elements of roof regions the matching algorithm is modified in such a way that it can handle arbitrary shaped image regions. Because the affine distortion between corresponding regions of a roof is often considerable (for an example look at figure 8), starting values for affine mapping are provided by the reconstruction step. The area based matching algorithm relies on similarity of image structure of the roof regions and on consistency with the mapping function. The planarity of the roof regions fits quite well to affine mapping in image space. Typical for area based matching is that if good starting values are given, the algorithm determines iteratively an optimal solution. If matching is successful, the reconstructed roof is confirmed; if not this indicates that something might be wrong or very inaccurate.
Potentially the matching result could be used to improve the reconstruction. But often image texture is poor at roof regions, thus a real improvement can not be expected in general. Experimentally in 1/4 of the reconstructed buildings (figure 9) an improvement is observed.

4 Concluding remarks

The algorithm was applied on height and image data of the *Test on Image Understanding* provided by the working group III/3 *Semantic Modelling and Image Understanding* of ISPRS (Fritsch, Sester & Schenk 1994). From this data set 80% of the visible houses have been successfully detected and reconstructed.

Figure 10: Perspective view of reconstructed 3-D urban model

Figure 9 shows the left stereo image of the test area including all the reconstructed buildings together with the lines used for reconstruction. By evaluation with area based matching 24 of the 29 reconstructed buildings are confirmed. For 7 of them the reconstructed shape of the roof could be improved significantly. In this cases often edge extraction failed to extract ridge of the roof. Two of the reconstructed models were rejected correctly by matching. For the last 3 houses matching indicated an error on one side of the roof even though the reconstruction was correct. Figure 10 shows a perspective view of the test area. The DTM and the image data were used to generate this view. In addition, the reconstructed buildings were visualized by shading.

References

Ackermann, F. & Krzystek, P. (1991), 'MATCH-T: Automatic Mensuration of Digital Elevation Models', Presented paper to the 3rd Technical Seminar of the Sociedad Espanola de Cartografia Fotogrametria y Teledeteccion.

Burns, J., Hansen, A. & Riseman, E. (1986), 'Extracting straight lines', *IEEE Transactions on Pattern Analysis and Machine Intelligence.*

Förstner, W. & Pallaske, R. (1993), 'Mustererkennung und 3D-Geoinformationssysteme', *Zeitschrift für Photogrametrie und Fernerkundung* pp. 167–177.

Fritsch, D., Sester, M. & Schenk, T. (1994), Test on image understanding, *in* 'Proc. ISPRS Congress Comm. III', München, pp. 243–248.

Haala, N. (1994), Detection of buildings by fusion of range and image data, *in* 'Proc. ISPRS Congress Comm. III', München, pp. 341–346.

McKeown, D. & McGlone, J. (1993), Integration of Photogrammetric Cues into Cartographic Feature Extraction, *in* 'SPIE Conference on Integrating Photogrammetric Techniques with Scene Analysis and Machine Vision', Orlando, Florida, pp. 2–15.

Mohan, R. & Nevatia, R. (1992), 'Perceptual Organization for Scene Segmentation and Description', *IEEE Transactions on Pattern Analysis and Machine Intelligence.*

Building Extraction and Verification from Spaceborne and Aerial Imagery using Image Understanding Fusion Techniques

Taejung Kim, Jan-Peter Muller

Department of Photogrammetry and Surveying
University College London
Gower St., London WC1E 6BT, UK
Telephone No: 44-171-380 7227, Fax No: 44-171-380 0453
INTERNET address: kim@ps.ucl.ac.uk, jpmuller@ps.ucl.ac.uk

Abstract

For automatic building extraction, the need of the fusion of several image understanding cues has been emphasised. In this paper, the fusion of a stereoscopic (pyramidal matching) and a monoscopic technique (building detection) was investigated. In the fusion technique proposed, building detection output is used as boundaries of height interpolation and pyramidal matching output as the source of height information. After height interpolation, buildings are reconstructed in 3D space. This fusion technique was tested with several examples and proved to be very powerful. The results of this paper support the role of fusion in more complete image understanding.

1 Introduction

One of the fundamental tasks in image understanding is the automated extraction of man-made objects from aerial or spaceborne imagery. In particular, buildings are an important feature of urban areas and techniques to extract them from an urban scene have numerous applications in urban mapping, urban planning, and Geo-Information Engineering. However, building extraction remains difficult as it requires not only good low-level vision techniques such as edge or line extraction but also good middle-level or high-level vision techniques such as cognition and interpretation.

There have been several techniques suggested for building extraction. Pyramidal stereo matching was suggested to extract building height from an urban scene where many height discontinuities are likely to be present (Kim and Muller, 1994a, 1995). Other approaches have proposed building detection by extracting lines from edge or corner analysis and grouping them based on perceptual grouping (Mohan and Nevatia, 1989, Dang et al., 1989, Sakar and Boyer, 1993) or using a line relation graph (Kim and Muller, 1994b). The use of shadow analysis (Huertas and Nevatia, 1988, Irvin and McKeown, 1989) and perspective geometry (Herman and Kanade, 1986) for building detection have also been reported. However, it was shown that a single technique is not enough to solve all the problems of building extraction. The fusion of image understanding cues has been suggested by many authors (Kim and Muller, 1995, Shufelt and McKeown, 1993) to overcome this single technique restriction.

The aim of this paper is to apply the fusion of image understanding cues for building extraction and to highlight the importance of fusion. In particular, the fusion is carried out combining stereoscopic and monoscopic techniques developed at University College London. The next section briefly describes these stereoscopic and monoscopic techniques. Section 3 describes fusion techniques with the two Ascona95 several examples. Section 4 draws some conclusions.

2 Stereoscopic and monoscopic techniques

2.1 Stereoscopic technique

Urban area imagery have many distinctive characteristics such as height discontinuities, occlusions, and shadows. Their effects on the region growing variant of Gruen's adaptive least squares correlation (ALSC) stereo matching algorithm ("Gotcha") (Gruen, 1985, Otto and Chau, 1988) have been investigated. It has been reported that buildings create isolated regions for region growing matching (Kim and Muller, 1995).

Arising from these studies, a fully automated pyramidal matching technique was developed to solve these problems. It solves the problem of isolated regions by utilising results of lower resolution matching as initial seed points for higher resolution matching. In this way, it facilitates well-distributed initial seed points to cover as many isolated regions as possible. It also reduces the magnitude of height discontinuities and the number of isolated regions by averaging down the image resolution. At the lowest (coarsest) level, matching can be performed successfully and at the higher level, matching can be enhanced by utilising the results of lower level matching. Initial seed points at the lowest level are generated automatically by assuming a constant disparity, applying pre-matching, and selecting those points with high accuracy. This method, which enables the process fully automated, was shown to be very effective (Kim and Muller, 1995).

In order to reduce the problem of blunder propagation, a tile-based selection strategy was adopted as a control strategy between levels. This can achieve well-distributed seed points by defining tiles spread over the image space and selecting one match point with the best accuracy within a tile. The maximum eigenvalue of the variance-covariance matrix of estimation is used as a measure of accuracy, which should be zero for an ideal match.

Output points from pyramidal matching are converted into the ground coordinates by applying camera model. Height information of the object space is then stored in a Digital Elevation Model (DEM).

2.2 A monoscopic technique

A monoscopic building detection technique based on a line relation graph has been developed (Kim and Muller, 1994b). It constructs a line relation graph with lines (as nodes) and relations between lines (as arcs) to generate building hypotheses (BHs) by finding closed loops in the graph. It has the following procedure.

- **Line detection**
- **Line relation search**
- **Building Hypotheses generation**
- **Building Hypotheses verification**

Lines are detected in an image using several steps: edge elements are detected by a Canny-Petrou-Kittler (CPK) filter; a connected edge labelling algorithm is applied; two end points of a linear elements are searched with end point templates. A line is defined with two end points. Small lines broken from a long line and parallel lines located closely are merged.

After detecting lines, the relations between lines are searched. The line relations are classified as *positive*, *negative*, *neutral* , or *parallel connection*. Lines and line relations are stored in a graph. The nodes of the graph consist of lines. Arcs represent the relations between lines.

The system finds building hypotheses (BHs) by searching for closed loops in the line relation graph. To traverse a graph, a depth-first search algorithm is used. In order to make sensible building hypotheses, the search is limited by some connection rules (Kim and Muller, 1994b) and the number of nodes it visits.

BHs then undergo a verification process. Firstly, similar BHs are merged. Similarity is defined when BHs all have identical components except one in their search chain. Secondly, false BHs, which are made from ground-level lines, are removed. In our current implementation, this is done by using shadow analysis and perspective geometry (Herman and Kanade, 1986). However these are optional processes as they require knowledge of the direction of illumination and the position of the principal point. After verification, each BH indicates a building or a part of a building.

3 Fusion

One of the weakest application areas of monoscopic approach is the extraction of height or depth information. Our building detection system suffers the same drawback. Although some indication of building height is possible by shadow analysis and perspective geometry, the system cannot provide accurate information on building height. If oblique perspective geometry were available, this situation could be improved.

Stereo matching results usually give accurate height information. However height is assigned only on grid-points and it is difficult to obtain the height of objects such as building roofs without further processing. Also there is a problem of interpolation of height for urban areas. In such areas, the surface does not vary in a geo-statistical manner (Davis, 1986). Large areas of the surface are very flat or linearly varying and there are many cliffs due to building boundaries. Hence "kriging", one of the common techniques for height interpolation, cannot be applied. Height interpolation requires external guidance.

The fusion of stereoscopic and monoscopic cues can potentially solve these problems. Rich height information from the stereoscopic process can be used to assign height onto buildings. BHs detected by the monoscopic process can provide the external guidance of height interpolation so that interpolation can take place within the boundaries of a BH. Figure 1 shows a schematic diagram of a procedure developed to test this idea.

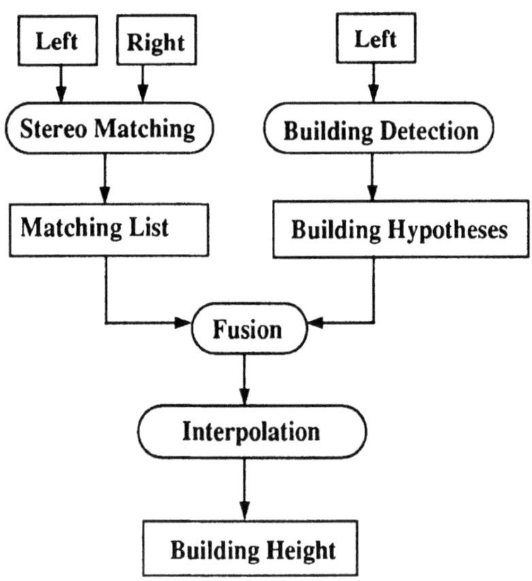

Figure 1: A schematic diagram for extracting building height

A matching list and BHs are produced by the stereoscopic and monoscopic ap-

Figure 2: A test stereo pair

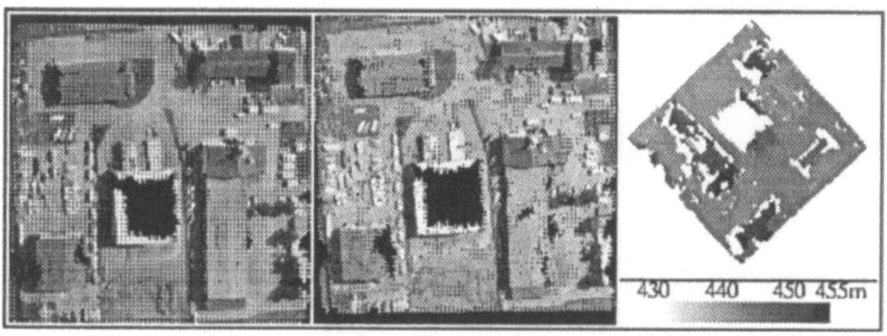

Figure 3: Left and Middle: matched areas in the stereo pair. Right: a DEM derived from matching output at 1 m grid spacing.

proaches respectively. Each BH provides interpolation boundaries and an interpolation is performed by finding coefficients of a surface model of a BH using the matching points in the matching list which fall within the BH. For a surface model, a planar or apex model is used in our current implementation. The coefficients of each model are calculated using the matching points through least squares estimation. The estimation errors of both models are compared and the model with less error is chosen as a proper surface model. Interpolation is done simply by calculating the corresponding height using the surface model equation.

This algorithm was tested using the images shown in figure 2 provided by ETH Zurich. Pyramidal matching was applied with 121 automatically generated seed points and a 5-level image pyramid. 16123 matched points were obtained and these were converted into ground coordinates to produce a DEM. Matched areas in the left and right images and the DEM are shown in figure 3. The large building in the middle of the image space was not matched very well as the left and right images had very different (noise-like) texture on that building roof.

The right image of figure 2 was used for building detection. After applying a CPK filter with a size of 3×3 pixels followed by a connected edge labelling and

Figure 4: Left: 1120 lines. Middle: 257 lines after merging. Right: 15 BHs overlaid with the original image.

Figure 5: A perspective view of the original right image (No height exaggeration).

finding end points, 1120 lines were detected (see the left image in Figure 4). After merging them and removing small lines, 257 lines remained finally (see the middle image in Figure 4). After BH generation and verification process, 15 BHs were produced. They are shown in the right of figure 4 overlaid with the original image. The buildings appear to have been detected very successfully.

The fusion of stereoscopic cues and monoscopic cues was then carried out. The building detection output in figure 4 were used as boundaries of height interpolation. The height information from pyramidal matching was used to model the surface. Among 15 BHs, 9 were modelled as planar and 3 apex. 3 BHs were not modelled as there were not enough matched points for them. As a result, height is assigned to buildings. Figure 5 shows a perspective view of buildings in the image space.

Another stereo pair supplied by ETH Zurich was tested. Figure 6 shows the images. Pyramidal matching with automatically generated seed points was applied to the images. 16538 points were matched and a DEM was created from the matching points. The left image in figure 7 shows the matched areas in the left image.

Figure 6: Another stereo image pair

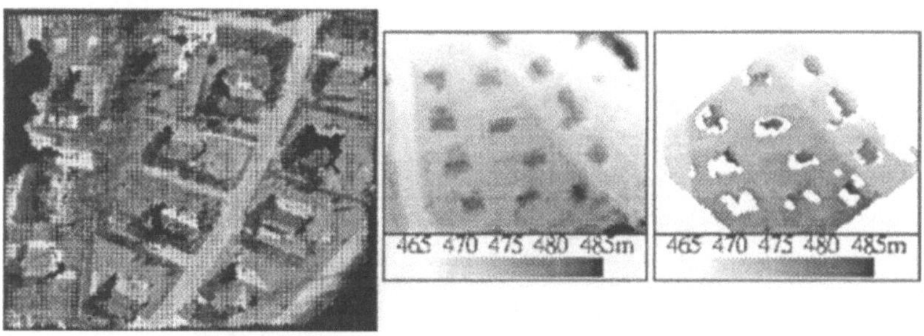

Figure 7: Left: matched areas. Middle: the reference DEM at 1 m grid spacing. Right: a DEM derived from matching output at 1 m grid spacing.

The middle image shows the reference DEM (from ETH Zurich) and the right the resultant DEM. This resultant DEM was compared with the reference DEM and the RMS height error was 0.986 m. Table 1 summarises the statistics of elevation errors.

Table 1: Statistics of the elevation error.

min	max	mean	RMS	St. Dev (σ)	blunder ($> 3\sigma$)
-9,7m	14.6 m	0.126m	0.986 m	0.977m	2.3 %

Building detection was carried out with the right image in figure 6. 2355 lines were detected initially and 338 lines were left after merging (see the right and middle images in figure 8). With 338 lines, BHs were generated and 37 BHs were finally verified. The 37 BHs were shown in the right of figure 8 overlaid with the original right image. In this case, 8 buildings out of 12 were successfully detected and 3 partly detected. There are some false BHs.

Fusion was performed as before. 37 BHs in figure 8 and the height information in figure 7 were combined for height interpolation. 28 BHs were modelled as planar

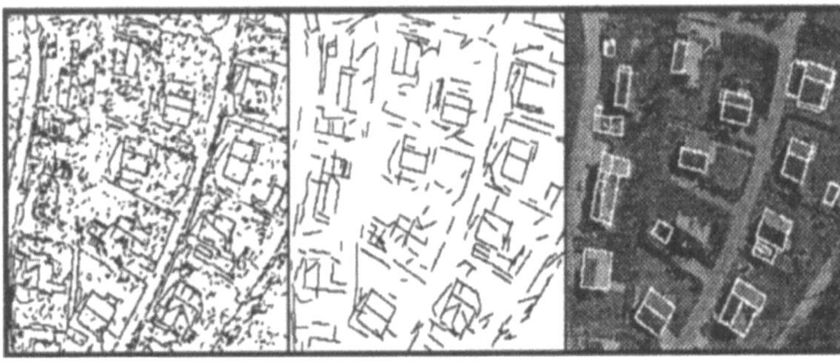

Figure 8: Left: 2355 lines. Middle: 338 lines after merging. Right: 37 BHs overlaid with the original image.

Figure 9: A perspective view of buildings (Height exaggeration factor 1.5).

and 9 BHs were not modelled [1]. A perspective view of buildings in the image space is shown in figure 9.

The left image in figure 10 shows a spaceborne DD5 ($2m$ resolution) image over Berlin airport. 437 lines from the scene and building detection output are shown in figure 10. Due to the limited availability of spaceborne high resolution stereo imagery, the fusion technique could not be tested. However, the results in figure 10 indicate that our monoscopic system, together with the fusion technique once stereoscopic data is available, should be capable of extracting large buildings from 2 m resolution spaceborne imagery.

[1] Apex model was not used in this case as matching points were not distributed evenly within a BH and hence apex model created high errors.

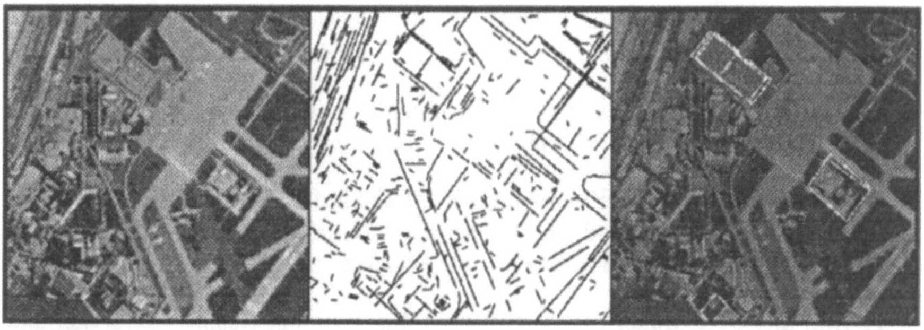

Figure 10: Left: A DD5 image over Berlin airport. Middle: 437 lines. Right: Building detection output.

4 Conclusions

In this paper, the fusion of monoscopic and stereoscopic process was demonstrated in order to extract 3D buildings in urban area imagery. The proposed fusion technique combines a building detection output and a pyramidal matching output for the height interpolation of the areas on building roofs. The results shown prove that this fusion technique is a very powerful way of reconstructing 3D building structures. They also support the role of fusion in more complete and better image understanding.

Acknowledgement

We would like to thank SaTReC (Satellite Technology Research Centre) for partly supporting the work described here and WIB for the use of the DD5 image. We thank ETH Zurich for the supply of the test data-sets shown here.

References

1 T. Dang, O. Jamet, and H. Maitre (1994) "Applying Perceptual Grouping and Surface Models to the Detection and Stereo Reconstruction of Building in Aerial Imagery", *Proc. ISPRS Commission 3 Symposium on Spatial Information from Digital Photogrammetry and Computer Vision*, **30(3/1):165-172**.

2 J.C. Davis (1986) "Statistical Data Analysis in Geology", Second Edition, Pub. Wiley.

3 A.W. Gruen (1985) "Adaptive Least Squares Correlation: A Powerful Image Matching Technique", *S. Afr. Journal of Photogrammetry, Remote Sensing and Cartography*, **14(3):175-187**.

4 M. Herman and T. Kanade (1986) "The 3D Mosaic Scene Understanding System", *From Pixels to Predicates edited by A.P. Pentland*, **pp.322-358**.

5 A. Huertas and R. Nevatia (1988) "Detecting Buildings in Aerial Images", *Computer Vision, Graphics, and Image Processing*, **41:131-152**.

6 R.B. Irvin and D.M. McKeown (1989) "Methods for Exploiting the Relationship Between Buildings and Their Shadows in Aerial Imagery", *IEEE Trans. on System, Man, and Cybernetics*, **19(6):1564-1575**, Nov./Dem.

7 T. Kim and J-P. Muller (1994a) "Automated Building Height Extraction and Building Detection from High Resolution Aerial and Space Imagery", *The 4th IAPR International Workshop on Machine Vision Applications*, pp. 364-367, Kawasaki, Japan, 13-15 December.

8 T. Kim and J-P. Muller (1994b) "A New Algorithm for Building Detection: A Graph-based Approach", *IEEE Trans. on Pattern Analysis and Machine Intelligence*, submitted.

9 T. Kim and J-P. Muller (1995) "Automated Urban Area Building Extraction from High Resolution Stereo Imagery", *Image and Vision Computing*, (in press).

10 R. Mohan and R. Nevatia (1989) "Using Perceptual Organization to Extract 3D structures", *IEEE Trans. on Pattern Analysis and Machine Intelligence*, **11(11):1121-1139**, Nov.

11 G.P. Otto and T.K. Chau (1988) "A Region-Growing Algorithm for Matching of Terrain Images", *Image and Vision Computing*, **7(2):83-94**.

12 S. Sarkar and K.L. Boyer (1993) "Integration, Inference, and Management of Spatial Information Using Bayesian Networks: Perceptual Organization", *IEEE Trans. on Pattern Analysis and Machine Intelligence*, **15(3):256-274**.

13 J. Shufelt and D.M. McKeown (1993) "Fusion of Monocular Cues to detect Man-made Structure in Aerial Imagery", *Computer Vision, Graphics, and Image Processing: Image Understanding*, **57(3):307-330**, May.

Automatic Extraction of Man-Made Objects
from Aerial and Space Images
Monte Verità, © 1995 Birkhäuser Verlag Basel

Building Extraction from Stereo Pairs of Aerial Images: Accuracy and Productivity Constraint of a Topographic Production Line

Olivier JAMET, Olivier DISSARD, Sylvain AIRAULT

Institut Géographique National
Direction Technique, Laboratoire MATIS
2, avenue Pasteur ,94 160 Saint-Mandé, France
Ph.: +33-1-43988069, Fax: +33-1-43988581
e-mail: jamet, dissard, airault @ matis.ign.fr

Abstract

This paper presents the research that has been lead for a few years at IGN on automatic building extraction from aerial photographs. A description of the operational context and of the topographic application that is aimed at introduces the main constraints to be overcome. We give then a quick sketch of the building extraction process that has been worked out up to now, and develop its quality assessment on a test site. The analysis of the results underlines the part played by the planar approximation of disparities resulting of an area-based correlation process in the fulfilment of the accuracy requirements. Further research will focus on reliability improvement and on more complex shape handling.

1. Operational context

1.1. The Topographic Database

As a national mapping agency, the French National Geographic Institute (IGN) is mainly dedicated to the acquisition and updating of the national reference geographic data, including the geodetic network, a periodic medium scale photographic coverage of the territory, and the topographic mapping.

Till 1984, the reference topographic data has been the set of the 2000 1:25000 maps covering France, stored as paper map sheets as well as original films. In 1984, facing the increasing needs for geographic digital data, IGN decided to provide France with geographic databases at various scales. The smallest scale ones were to be produced from the digitisation of existing maps, while the topographic database (BDTopo®), that aimed at replacing the topographic reference, was to come from a new photogrammetric plotting of the territory.

This complete refection of our topographic data will provide its users with higher accuracy data than the former maps (1.5 m rms in XY on the average), each point being stored in 3D with a Z accuracy of 1m rms. The data model includes all information already present on the maps (contours lines, communication and hydrographic networks, buildings, vegetation...) in vector form, within a topological structure, each object supporting attributes and relations. The BDTopo® will thus provide a complete topographic data set, oriented, from its content, toward GIS applications, and dedicated to the scales 1:25000 up to 1:5000, according to its accuracy.

1.2. Evolution of the production capacity

The data input actually started in 1989, and very soon, the first analytical "integrated production lines" (LPI) were settled up. This LPI concept, gathering within a same team the photogrammetric data input, the complementary ground survey, and the GIS management, was designed in order to optimise productivity and to allow a better quality control along the process. Each LPI is composed of four analytical plotters, one data server, a 2D workstation for GIS correction and ground survey input, and a plotter for control drawings. It represents a staff of 15 people (two people per stereoplotter, topographers, controllers...), and can produce five to six 550 km^2 work units a year (a work unit corresponds to the area covered by two 1:25000 map sheets).

In march 1995, 6 LPI are currently working, and 6.5% of the territory (for 43% of the population) are available. This production capacity should extend up to 12 LPI with an average rhythm of 2 LPI a year, and will then correspond to an updating capacity of whole France every 8 years (on the average, the most evolutive areas being updated every 3 or 4 years). Within this production frame, the complete coverage of France will require 25 years, and represent 2 millions hours of stereoplotting. To meet faster the user needs, IGN aims at reducing this delay down to 15 years. If the main way to reach this ambition keeps promoting the use of the data, so as to fund part of the plotting task within private companies on commercial income, IGN also decided to invest in digital techniques. In 1991, a research program was started up in image analysis to provide semi-automatic helps to the plotting, while more recently (in 1994), an evaluation of the available softcopy stereoplotters was undertaken, and lead to the choice of digital plotters for the equipment of the forthcoming LPI.

1.3. Data input

The data acquisition is performed on 24 cm panchromatic aerial stereo pairs at the scale 1:30 000 in rural areas, and 1: 20 000 in urban areas. The focal lengths were chosen to fulfil the Z accuracy requirements (152 mm in rural areas and 210 mm in urban ones). The photographs are taken by strips with a 60% overlap (and 10% overlap between the strips). On each working unit, the LPI are supplied with the photographs and with the external orientations computed by a triangulation process. For the softcopy plotters, the photographs will be digitised at a step of 15 to 20 microns, and a quantification in 256 grey levels.

1.4. Research aims

The image processing research program that was settled up in 1991 aims at providing softcopy stereoplotters with semi-automatic tools in order to speed up the data input. The part to be played by IGN in this field corresponds to leading our own academic research as well as ensuring an interface between external research teams and our needs. It stays however within the research field (extended to validation studies): operational developments will be left to contracted industrial partners.

In first analysis, focus was given to three major topics: contour line production from automatic correlation techniques, road extraction, and building extraction. Starting from a small team in October 1991, the activity was to increase up to 15 people in 1995 within the laboratory "Méthodes d'Analyse et de Traitement d'Images pour la Stereorestitution" (MATIS, created in 1993), with a goal of 18 people in 1997. Keeping the same main topics of interest, the work is currently shared between long term research (5 PhD studies) and short term activity (evaluation of current algorithms in accordance with the production staff).

2. Constraints analysis

2.1. Data cost

The first constraint of this applicative context concerns of course the data cost. Within the frame of the BDTopo® initial acquisition, the *a priori* available information is restricted to the currently used aerial photographs. Any additional information input has to be confronted to a comparison between its cost and the productivity it brings. On the average, taking only into account the photographic and the digitisation processes, a single image represents from 2.5 to 5% of the cost of the photogrammetric plotting of one stereo pair. This means that doubling the image scale (using 4 times more images), for instance, would only justify under a productivity increase of 10 to 20%.

In the long term, expecting the use of digital cameras, this over-cost will be much lower - so we can't utterly ignore this kind of possibility. But in the short term, that is within the frame of initial acquisition, the likeliness for such improvements of the productivity to rely on changes in the used imagery is weak: the main limits of the road extraction or contour line extraction algorithms from stereo pairs at the scale 1:30000 rely more on the lack of high level understanding of the scene (interpretation of complex landscapes) than on the lack of the data available to simple algorithms; the only topic which would maybe request larger scale imagery would be the building extraction.

2.2. Work flow management

We can't expect to modify the general work flow within a LPI. Splitting the operator work into several phases is likely to lower the general efficiency in a greater proportion than all the improvement that is to be expected from automatism. This implies:

- either to offer preprocessing fully automatic tools, for a further interactive exploitation of the results;
- or to propose real-time semi-automatic extraction processes (under the hypothesis of the current CPU power of commercial workstations).

In both cases, the main issue remains the auto-diagnostic ability of the automatic techniques. In the case of preprocessing, a complete scanning and correction of the data is generally less efficient, and yields more errors, than manual plotting. The operator intervention should be guided by the internal confidence in the results, i.e. by a confidence level given by the automatic process itself. In the second alternative, the help provided by the machine should preferably be able to require on line operator intervention in case of possible error, rather than impose an *a posteriori* correction of the result.

2.3. Reliability vs exhaustivity

The need for auto-diagnostic is emphasised by the specificity of the cartographic application, that doesn't aim at detecting some particular feature, but at giving an exhaustive model of the scene. While we aren't able to handle all the cartographic objects, an operator will have to perform a complete scanning of his working area. This means that the efficiency of a semi-automatic technique will much more depend on its reliability (predictable behaviour, low sensitivity to small changes in the parameters, preference to no result than to a wrong one...) than on its exhaustivity (an exhaustive visual control being performed).

3. Building extraction

The previous remarks lead us to start the studies on building extraction on the aerial stereo pairs at the scale 1:20 000 that are actually used for manual plotting. The chosen approach aims at extracting the building roofs and relies theoretically on a rather generic building model (polygonal footprint with orthogonal angles, plane facets), with a (huge) restriction of the goal to isolated buildings with planar roofs.

The developed method inherits of previous works by H.Maître and W.Luo (1992), who proposed the use of surface models to improve disparity maps on building areas. The basic idea relies on a fusion of the monoscopic information given by a segmentation process and of the stereoscopic information yielded by an automatic correlation process. Assuming that the main weakness of correlation techniques is a misbehaviour on places where disparities are discontinuous (i.e. where occlusions occur), and that homogeneous areas, from a radiometric point of view, should correspond to continuous disparity areas, theses authors proposed the use of a least square polynomial approximation of the disparities inside each region of a segmentation.

This filtering of the disparity noise showed efficient even on complex buildings. The extension of this technique we worked on consists in processing a 2D shape recognition

on the hypothesised buildings, based on the grouping techniques of Mohan and Nevatia (1988):
- first, in order to filter noise on the buildings boundaries; on the first hand, geometrical shapes have to be identified to describe topographic objects; on the other hand, the boundaries of the roofs cannot always result from the segmentation process (in presence of low contrast with the context);
- secondly, to give more clues for the acceptance or rejection of the building hypothesis.

Applied first on rectangular buildings with horizontal roofs, the method was complemented with a mixed surface approximation technique, and an edge selection before the grouping (Dang 1994). Further improvements were added by Olivier Dissard (1995), particularly through a study of the behaviour of the correlation process and of the grouping, according to the content of the scene (orientation of the shadows, relative position of the occlusions...).

A scheme illustrating the current process is presented below (Fig. 1).

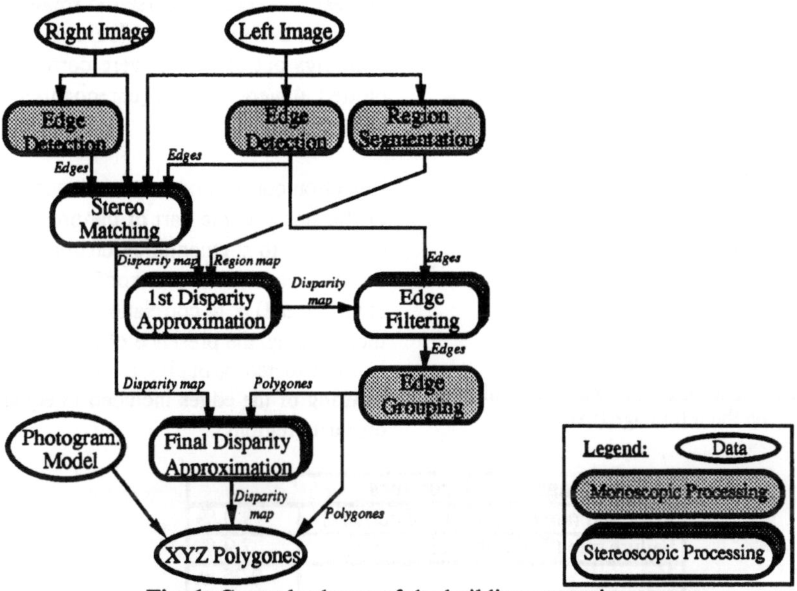

Fig. 1: General scheme of the building extraction process

4. Assessment of the process

4.1. Quality assessment

The method was evaluated on a test site located in the suburb of Paris. The photographs used for the BDTopo® acquisition were digitised at a step of 20 microns (40 cm ground resolution). The outer orientation was provided by the production department, and the vector data issued of the manual plotting was used as reference data. Forty one buildings of large size and rectangular shape were selected as test set. The results of the automatic extraction are illustrated by the examples provided on figure 2.

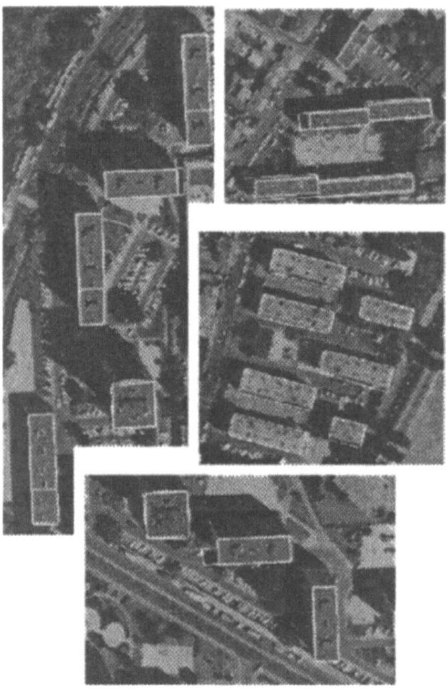

Fig. 2: Projection of the extracted buildings on the left image (extracts)

Accuracy measurements were performed on the corners of the detected buildings (according to the BDTopo® quality requirements), in XY on the monoscopic extraction in the image geometry, as well as in XYZ on the final results in geographic coordinates. This evaluation is summed up in the table 1.

As far as the exhaustivity is concerned, the test set keeps of course too small to provide definitive conclusions. 30 buildings out of the 41 were correctly plotted, 4 were rejected (unrecognised) and 7 failed.

Rejection correspond to misbehaviours of the stereoscopic part of the process, either due to segmentation errors (low contrast between the roof and its environment) or to correlation errors. In theses cases, the presence of a building is not detected (grouping rejection after filtering of the edges included in equal disparity regions).

41 Buildings / 178 corners	
Rejection rate (% number of corners)	9.0 %
Failure rate (% number of corners)	7.9 %
X rms	0.75 m
Y rms	0.71 m
XY rms	1.02 m
Z rms	0.60 m

Table 1: Quality measurements

We call failures extracted buildings presenting shape errors. Theses errors are mainly due to the fusion of a visible vertical face with the roof during the grouping process (enlargement of the roof: see the top right extract on fig.2), and affect generally only one side (even sometimes only one corner) of the building. The same effect can be observed

in presence of vegetation in the near neighbourhood of the building (see the bottom extract on fig.2). More seldom, some building can be shortened, equally during the grouping phase (closure of a rectangle with pieces of edges within the roof).
The accuracy results provided on table 1 were computed after exclusion of the failures (in fact, we only excluded the misinterpreted corners of the failed extractions). The 3D accuracy shows compatible with BDTopo® requirements:
- the XY accuracy keep low (2.5 pixels rms), but satisfies the 1.2 m that was aimed at;
- the Z accuracy is pretty satisfying (less than 1 disparity pixel rms).

4.2. Comments on the exhaustivity of the process

This study for an automatic process for building extraction was first of all motivated by the need to evaluate the performances of the algorithms before any analysis of what could be the real operational use of them (that is the user interface).

We can however consider this first method as a possible preprocessing doing part of the plotting, the results of which would be presented to the operator during the interactive plotting phase. If the accuracy results quite fulfil the BDTopo® requirements, exhaustivity keeps rather low. The rejection rate isn't by itself the main weakness of the method, since exhaustive extraction isn't presupposed. Failures may cause much more supplementary work: the interactive correction of objects as simple as rectangular buildings will probably cost more time than their entire manual plotting, and their simple deletion can even be heavy if they are connected to neighbouring objects.

A reduction of the failure rate (even to the "benefit" of a higher rejection rate) which could be reached by working out a confidence measure from complementary image processing (for instance by using the building shadows as a way to verify the consistency of the extraction, as proposed by Irvin and McKeown in 1989), would, in our application context, bring a serious improvement to the method.

4.3. General conclusions on building extraction techniques

The proposed methods will not, of course, be implemented "as is" on our softcopy plotters. First of all, the range of handled building shapes is still narrow, and the consequent low productivity improvement doesn't make software developments worthwhile. Secondly, as discussed in the previous section, progress is still to be done as far as the automatic control of the results is concerned. This work brings however some interesting pieces of information for the orientation of our further works.

Area-based vs feature-based matching. The first conclusions concern the use of area based matching techniques on this kind of imagery. The scale constraint, which bounds the resolution to values upper than 30 cm (40 cm on our test site), associated with a rather low B/H ratio (0.46 in the case of our test site), make the aimed Z accuracy a real challenge. Many authors have proposed building reconstruction techniques based on feature matching techniques. On our test site, supposing the feature location errors

independent on both images, a one metre Z accuracy would suppose a feature position accuracy better than 0.7 pixel. Knowing that the features of interest are *in fine* the corners of the buildings, such a quality seems difficult to reach (even visually, the correct position of a building corner can appear ambiguous). This very Z accuracy was reached by an estimate of the elevation based on area-based correlation and surface models.
As a confirmation of this analysis, we performed two complementary tests. First, we computed the 3D building rectangles issued of the two monocular groupings on the left and the right image. Despite the imperfection of the process (the double monoscopic grouping is of course less exhaustive than a 3D grouping procedure), the accuracy of the obtained building is representative of the accuracy we can expect of a feature matching process (table 2): the difficulty for locating correctly the edges of the building comes mainly from the need for grouping neighbouring parallel edges to cope with unavoidable ambiguities in presence of multiple detection (as in the case of buildings bordered by a thin white wall on fig.2), or of perturbation in the surrounding (especially vegetation).

Estimation technique	Z rms
Matched rectangles (17 buildings)	2.98 m
Surface model on manual plotting	0.54 m
Surface model on automatic plotting	0.60 m

Table 2: Z accuracy according to the estimation technique

Secondly, we computed the elevation resulting from the surface model approximation within the manually plotted rectangles, in order to have an estimation of the sensitivity of the approximation to an incorrect delineation of the buildings. The accuracy of this new Z estimate is slightly better (table 2), but the low difference with the accuracy obtained with the automatic rectangle extraction shows the robustness of the surface model technique.

Error propagation handling in the correlation phase. The correlation algorithm makes use of the same classical geometric constraints as many others: matches are performed along the epipolar lines (which are known with a subpixelar accuracy under the production conditions), with an order constraint, and a figural continuity constraint based on edges (Hoff 1989). The behaviour of this kind of algorithm on the building roofs highly depends on the direction of the computation (right to left or reverse along the epipolar line), due to the propagation effect induced by the order constraint (the match for a point is to be looked for after the match of its predecessor).

This is illustrated by the figure 3: one direction shows more favourable to high elevation surfaces (Fig.3-d), while the other one to the ground and the low elevation surfaces (Fig. 3-c). This behaviour is rather deterministic. The correlation matches quite well the shadows (and thus the ground) when they are first examined, and jumps more easily on the roofs otherwise.

Up to now, we cope with this problem by merging both results (Fig. 3-e). From an operational point of view, this default keeps bearable on large building, if the results we obtained on our test site are confirmed on larger areas. Small buildings seem however out of reach of this correlation process: even after the combination of the two results, the noise keeps too high to handle small surfaces. This limitation of the process can raise up the interest we may have in feature matching techniques.

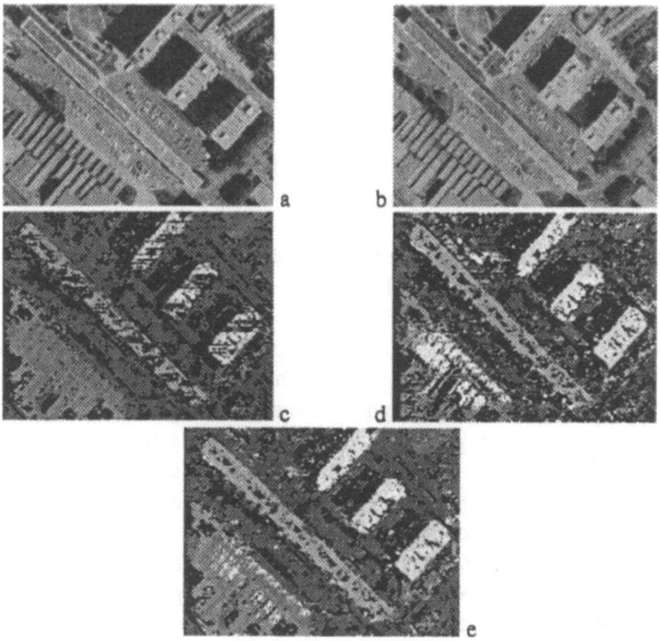

Fig. 3: Behaviour of the correlation process according to the propagation direction
a,b: aerial stereo pair; c,d: correlation results; e: merging of c and d

Non linearity in the disparity estimation. The median filter proposed by T.Dang (1994) in the disparity approximation phase proved to be efficient. In the current implementation of the method, surface model approximation is first tried by least square adjustment, in order to allow non horizontal plane handling; median approximation is tested afterwards, in case of failure. On the test data, median filtering was finally used most of the time, due to its lesser sensitivity to non gaussian noise. During the first disparity approximation, segmentation errors on the delimitation of the regions induce non gaussian noise in the disparity map within a region. Besides, even after the edge grouping, non gaussian noise can still occur within the roofs (due to the presence of staircase accesses to the roof, of chimneys...).

This latter effect is inherent to any object recognition problem. The object model is necessarily a simplification of the real world, suppressing all the details that will be too small, at the imagery resolution, to be correctly extracted, but that can affect any basic detector (such as an automatic correlation process) at the pixel level. A research effort keeps needed for the generalisation of non linear decision making procedures to more complex buildings shapes.

5. Conclusions

The topographic application we are aiming at places the real bottlenecks of automatic building extraction techniques in their reliability and geometrical accuracy. In the context of exhaustive stereoplotting of the working area, the need for correcting too often unreliable automatic extraction results is likely to make the process more tedious than a fully manual one. As for geometric accuracy, while the productivity increase keeps low, we are bound to using medium scale photographs, bringing a need for a subpixellar disparity estimation.

The research we have lead for a few years on building extraction showed that the accuracy requirements could be fulfilled by automatic extraction on flat roof buildings, under the use of a median Z estimation from a disparity map. Further research should answer the following issues. First, within an application to large isolated buildings, the results have to be generalised to more complex shapes, while a better reliability should be achieved. Theses improvements suppose a better handling of the polygon recognition process (we are currently integrating additional constraints, such as elevation and gradient direction), additional automatic evaluations of the results (such as a shadow consistency check), and an adaptation of the non linear surface model computation to oblique roofs. Secondly, the limits of the proposed techniques have to be more precisely defined, as far as the size and the shape of the buildings are concerned. The operational use of building extraction techniques supposes to handle a larger proportion of objects. A better compromise between the used techniques (correlation and feature matching) and the imagery scale will probably have to be defined, on the base of a refined estimation of the possible productivity that automatism can bring.

References

Dang T., Maître H., Dissard O., Jamet O. (1994) *Using stereo matching and perceptual grouping to detect buildings in aerial images*, SPIE, Vol. 2315, pp. 64-77

Dissard O., Jamet O. (1995) *3D-reconstruction of buildings from stereo images using both monocular analysis and stereo-matching: an assessment within the context of cartographic production*, SPIE, Vol. 2486, to be published

Hoff W., Ahuja N. (1989), Surface from stereo: integrating feature matching, disparity estimation and contour detection, IEEE-PAMI, Vol.11, No.2, pp. 121-136

Irvin B.R., McKeown D. (1989) Methods for exploiting the relationship between buildings and their shadows in aerial imagery, IEEE Trans. on Syst. Man and Cybernet., Vol. 19, No. 6, pp. 1564-1586

Maître H., Luo W. (1992) *Using models to improve stereo reconstruction*, IEEE-PAMI, Vol. 14, No. 2, pp. 269-277

Mohan R., Nevatia R. (1988) Using perceptual grouping to extract 3D structures, IEEE-PAMI, Vol. 11, No. 11, pp. 1121-1138

Road Extraction

Automatic Extraction of Man-Made Objects
from Aerial and Space Images
Monte Verità, © 1995 Birkhäuser Verlag Basel

Tracking Roads in Satellite Images by Playing Twenty Questions *

Bruno Jedynak
INRIA - Rocquencourt
Domaine de Voluceau, B.P. 105
78153 Le Chesnay Cedex, France
Email: Bruno.Jedynak@inria.fr,
and
Jean-Philippe Rozé
INRIA - Rocquencourt
Domaine de Voluceau, B.P. 105
78153 Le Chesnay Cedex, France
Email: Jean-Philippe.Roze@inria.fr

April 15, 1995

Abstract

We present new experiments on tracking roads from SPOT satellite images. The principle of the algorithm is as follows: we choose "tests" (matched filters for short road segments) one at a time in order to remove as much uncertainty as possible about the road position given the results of the previous tests. The tests are chosen based on a statistical model for the joint distribution of tests and road positions. On-line, we alternate between data collection and optimization: at each iteration new image data is examined and a minimization problem is solved, resulting in a new image location to inspect, and so forth. We report experiments using panchromatic SPOT satellite imagery with a ground resolution of ten meters: given a starting point and starting direction, we are able to track in real time mountain highways in southern France over large distances without manual intervention.

*This work was partially supported by the CNES under Marché 833/CNES/93/0972/00.

1 Introduction

We present an algorithm for tracking major roads from pan-chromatic SPOT satellite imagery with a ground resolution of ten meters. Mathematical foundations, computer vision related methods, and detailed implementation already appear in Geman et al. (1995) and Jedynak (1995). Here, we present new experiments, and we show that with proper training, our algorithm is able to track not only regular highways but also mountain highways. The setting of the problem together with previous work on road extraction is presented in section §2. Detailed geometric and statistical models appear in section §3. The principle of our algorithm is presented in section §4 and section §5 is devoted to experiments.

2 Previous Work

The satellite SPOT provides images of the entire globe with high resolution ranging from 10 to 20 meters. A single image covers an area of about 4,000 km^2. Examples of sub-images are presented in Figures 3 and 5. For cartography, urban planning or resource management, the row data must be converted into meaningfull (semantic) information. This task is usually long and tedious. Our work is aimed at providing assistance in the automatic extraction of major roads. Whereas these networks can usually be rather easily identified by people at a large scale (say on the order of several kilometers), the problem is far more difficult at small scale. (see the discussion in Serendero (1989) and in Boggess (1993) for Landsat TM images.) In other words, the problem of road extraction is more *global* than it might at first appear.

Most of the algorithm previously developed contain two major components, one dealing with low-level processing and one with higher level algorithms. The low-level algorithms are devoted to the identification of local, road-like structures, often based on local filtering; see Duda et al. (1973), Neviata (1982), Wang (1988), Estival et al. (1986) and others. We also employ local operators, see §3; however, in contrast to most of the work cited above, we have chosen *invariant* operators, based only on *relative* brightness, returning the same result under any linear transformation of the grey levels. The second component consists in *global* methods. The aim is to organize local detections using geometric information (eg. curvature) about the shape of actual roads. The formalism usually leads to an optimization problem solved by combinatorial optimization as in Fischler et al. (1981), Geman et al. (1991) and Merlet al. (1993), variational methods, or others. In Serendero (1989), the focus is on *tracking*. The algorithm is iterative. At each step, a certain number of paths emanating from the seed are inspected, some of which are kept and extended, and other discarded. These choices are based on a cost function motivated by heuristics about road properties, such as brightness, width, curvature, and the presence of edges. The algorithm we propose next is also a tracking algorithm. In Geman et al. (1995) we compare it to a simpler version of the one in Serendero (1989).

3 Modeling the Road Problem

A major road in a SPOT image is modeled as a discretization of a smooth, planar curve whose curvature is bounded by a known value. We use a piecewise linear approximation; each knot corresponds to a pixel and the segments (which we call "arcs") are digitized lines of approximatively the same length. Such an object is completely determined by an ordered list of regularly spaced pixels (the knots) and an algorithm which constructs a digitized line between any two pixels (we use Bresenham algorithm). The curvature constraint is expressed by setting the length of the arcs and limiting the angle between successive arcs. Here we stick on allowing only three angles labeled $\{0, 1, -1\}$, corresponding to "no turn" between successive arcs and the next two smallest ones at the level of discretization of the image. We will play with the common length of the arcs to model different types of roads.

We suppose from now on that the first arc is given. This is the seed from which we are going to track the road. Therefore each road candidate is identified with a sequence of digits in $\{0, 1, -1\}$. The road in the image is described as a realization of a stochastic process $X = X_1, X_2, \ldots$ where each $X_m, m \geq 1$, is a random variable in $\{0, 1, -1\}$ representing the angles between successive arcs. The simplest model would be to consider the X_m independent and equally uniformly distributed. However, this is not realistic for major highways since curvature tends to be maintained: there are long straight segments and roads tend to continue turning at the same rate. So, we model X by a low-order Markov chain. The transition matrix is estimated using a sample of roads extracted manually.

Let $\mathcal{A}$ denote the set of all arcs contained in some road candidate; (If all roads contain exactly M arcs then $\sharp\mathcal{A} = \frac{3}{2}(M-1)$.) There is a test associated with each arc $a \in \mathcal{A}$. Performing one of these test consists in evaluating a filter response. The filtering algorithm is designed to identify short, linear segments that are likely to lie on major roads. The basic assumption that is utilized is that two pixels in close proximity and both on the road should have a smaller intensity difference than that of two pixels, one of which lies on the road and the other off the road. Since the major roads appear in SPOT images as structures of width between two and three pixels, the filter is based on a pair of adjacent, digitized lines (see Figure 1). The overall response of the filter is the aggregate along the strip of the individual, binary responses of elementary probes, each based on a pattern of six pixels $(t_1, ..., t_6)$ arranged as shown in Figure 1. Let $I(t)$ denote the image intensity at pixel t. Each such pattern is assigned a value 0 or 1; the value is 1 if $|I(t_1) - I(t_2)| < \min\{|I(t_3) - I(t_1)|, |I(t_5) - I(t_1)|, |I(t_4) - I(t_2)|, |I(t_6) - I(t_2)|\}$ and 0 otherwise. The test result is the average value along the strip, which is renormalized to obtain a value in $\{0, ..., J\}$. ($J = 10$ for example)

Let $\{Y_a, a \in \mathcal{A}\}$ denote this family of tests. We assume these are *conditionally independent* given X. This is not true for the functionals we use, but an approximation that leads to very good experimental results. Another assumption is that the marginal (conditional) distributions depend only on whether or not $a \in x$, meaning that arc a is contained in road x. Let us denote these distributions by p_1 and p_0

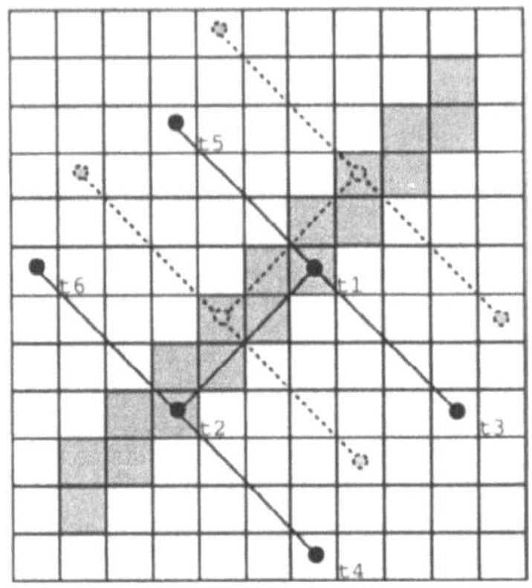

Figure 1: Local Filtering.

respectively. Thus,

$$P(Y_a = y_a, a \in \mathcal{A}|X = x) = \prod_{a \in x} p_1(y_a) \prod_{a \notin x} p_0(y_a).$$

The distributions p_0 and p_1 quantify the performances of the filtering algorithm. They are estimated from sample data. See Figure 2.

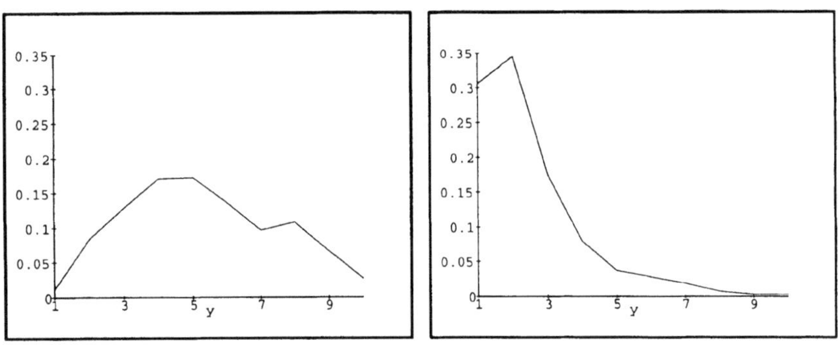

Figure 2: Estimated Distribution of the Filter. **Left:** On Roads; **Right:** In Background

4 Algorithm

The tracking problem being modeled as described, in principle we can imagine doing all the tests and computing the maximum a posteriori (MAP) estimator

$$\hat{X}^{MAP} = \arg\max_{x \in \mathcal{X}} P(X = x|Y_1, ..., Y_N)$$

$$= \arg\max_{x \in \mathcal{X}} \; P(X = x)P(Y_1, ..., Y_N | X = x).$$

where $x = x_1, \ldots, x_M$, and $N = 3(M-1)/2$, for some large value of M.

We prove in Jedynak (1995) that this estimator would be consistent asymptotically for $M \to \infty$. However, it is not feasible computationally. In effect, we wish to track roads for many kilometers. So, we need to choose M in the order of 100, but then there are 3^{100} road candidates and order of 3^{100} tests. Indeed, we wish to get roughly the same performance as the MAP estimator by considering only a very small fraction of all the tests, such as order M to 100 M tests rather than 3^M. The main idea is as follow. We are going to proceed sequentially and adaptively. At each step, we are going to choose the arc that most reduce the current *expected uncertainty* about the road position. (Uncertainty is measured by Shannon entropy.) Then we perform the corresponding test and move to the next step. The distribution on the road position X is going to evolve as tests are performed. After a fixed number of iteration, we stop and estimate the road by the *mode* of this distribution. The detailed mathematical formulation and implementation is fully presented in Geman et al. (1995) and Jedynak (1995).

5 Experiments

All the result below are obtained by choosing for the process X the second order Markov chain defined as follow:
$P(X_1 = 0) = 0.688, P(X_1 = 1) = P(X_1 = -1) = 0.156.$
$P(X_2 = x_2/X_1 = x_1)$ is given by:

$x_1 \backslash x_2$	0	1	-1
0	0.730	0.135	0.135
1	0.598	0.382	0.020
-1	0.598	0.020	0.382

$P(X_3 = x_3/X_1 = x_1, X_2 = x_2)$ is given:

$x_1 x_2 \backslash x_3$	0	1	-1
00	0.736	0.132	0.132
01	0.703	0.274	0.023
0-1	0.703	0.023	0.274
10	0.712	0.217	0.071
11	0.431	0.550	0.019
1-1	0.650	0.000	0.350
-10	0.712	0.071	0.217
-11	0.650	0.350	0.000
-1-1	0.431	0.019	0.550

This distribution was estimated using a sample of roads extracted manually. The distributions p_1 presented in Figure 2 was estimated using the same data. p_0 was

Figure 3: **Left**: Sophia, France. **Right**: Output of the tracking algorithm for the upper half of the image

Figure 4: Output of the tracking algorithm for the lower half of the Sophia image

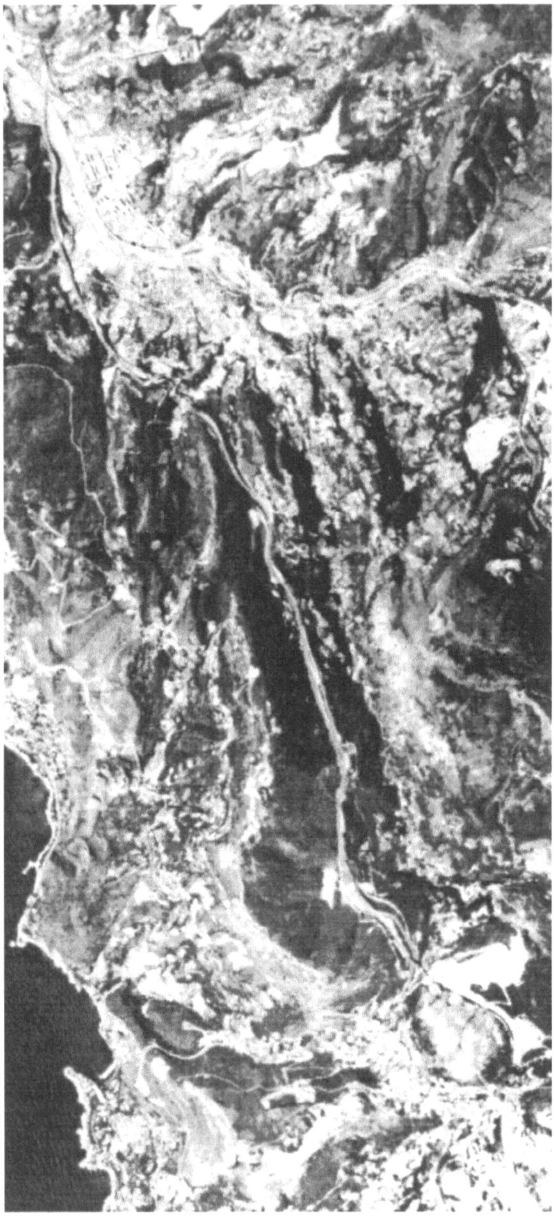

Figure 5: Nice, France.

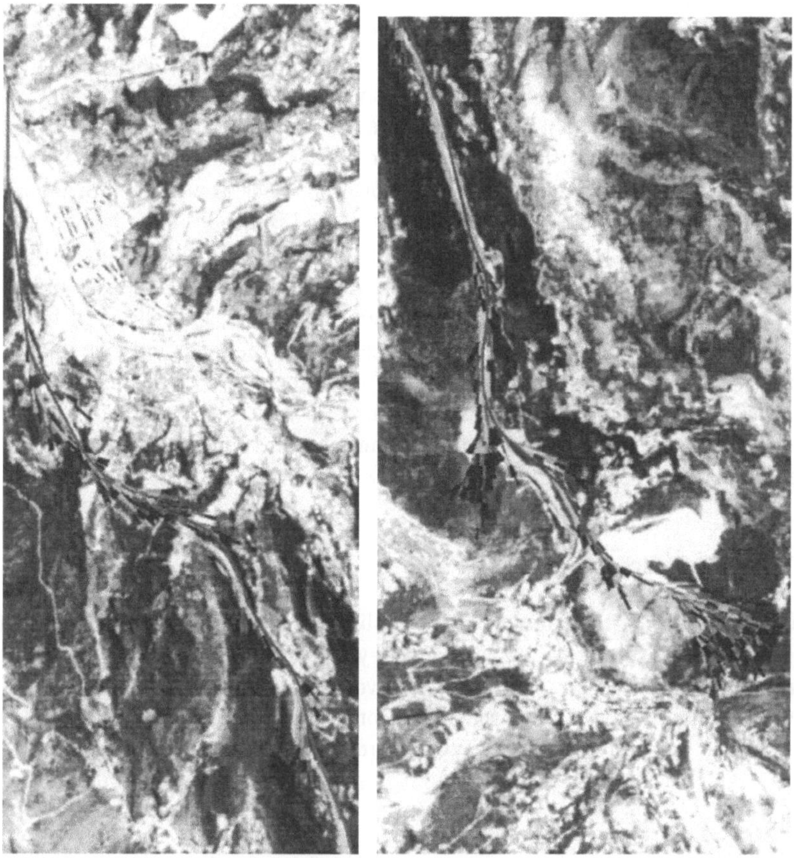

Figure 6: Output of the tracking algorithm **Left**: upper half of the Nice image. **Right**: lower half of the Nice image.

obtained by selecting arcs off major roads but containing eventually small roads. These two distributions are the same for all the experiments presented here and also for all the ones in Geman et al. (1995). The common length of each arc is set to 8. This is the required value for mountain highways in France. In Geman et al. (1995), we present experiment for regular highways where the length 12 is sufficient.

Figure 3 is a 1000x1000 image of the area of Sophia, in South France. The arcs selected by the algorithm for the upper side of the Sophia image are shown on the right hand side of the same Figure. The color, from black to white is proportional to the likelihood ratio p_1/p_0. The estimate of the road is also visible in grey. Tracking for the lower part of the image is presented in Figure 4. The all tracking of the road took about 20 seconds on a Spark 10 computer. Figure 5 is a 1000x1000 image of the area of Nice, in South France. The main road is here specially hard to follow. Tracking results appear in Figure 6. Notice that in the left hand side image, the tracker is stock due to a tunnel. Tracking took about the same time as the Nice image. Notice that the estimate of the road is close to the true road, less than 5 pixels but not right on it. This estimate can be used to initialize an algorithm for fine localization.

6 Conclusion

We contribute a partial solution to the problem of extracting road networks, from high resolution (10 meters) satellite imagery. Given a starting point and starting direction, we are able to track highways over large distances, without manual intervention. Morever, we prove that the algorithm can be set up to track mountain highways as well as regular highway with proper training.

References

[1] Boggess, J. E., "Identification of roads in satellite imagery using artificial neural networks: a contextual approach," Technical Report, Mississippi St. Univ., August, 1993.

[2] Estival, I. and Le Men, H., "Detection of linear networks on satellite images", Proc. Conf. on Pattern Recognition, 856-858, October, 1986.

[3] Duda, R.O. and Hart, P.E., *Pattern Classification and Scene Analysis*, John Wiley, New York, 1973.

[4] Fischler, M., Tenenbaum, J. and Wolf, H., "Detection of roads and linear structures in low-resolution aerial imagery using a multisource knowledge integration technique," Computer Graphics and Image Processing, 15, 201-223, 1981.

[5] Geman, D. and Jedynak, B., "Detection of roads in SPOT satellite images," Proc. IGRASS 91, Helsinski, 1991.

[6] Geman, D. and Jedynak, B., "Shape recognition and twenty questions," Technical Report No. 2155, INRIA-Rocquencourt, November, 1993.

[7] Geman, D. and Jedynak, B., "An Active Testing Model for Tracking Roads in Satellite Images," to appear in IEEE Trans. on PAMI, 1995.

[8] Jedynak B., "Modèles stochastiques et méthodes déterministes pour extraire les routes des images de la Terre vues du ciel" Thèse de Doctorat, Modèlisation Stochastique, Université Paris Sud, 1995.

[9] Merlet, N. and Zerubia, J., "A curvature-dependent energy function for detecting lines in satellite images," 8th SCIA Conference, May, 1993.

[10] Neviata, R., "Locating structures in aerial images," IEEE Trans. PAMI, 476-484, September, 1982.

[11] Serendero, M. A., "Extraction d'informations symboliques en imagerie SPOT : réseaux de communication et aglomations," Thèse de Doctorat, Université de Nice, 1989.

[12] Wang, F. and Newkirk, R. , "A knowledge-based system for highway network extraction," IEEE Trans. Geoscience and Remote Sensing, 26, September, 1988.

New Geometric Stochastic Technology for Finding and Recognizing Roads and Their Features in Aerial Images.

Meir Barzohar

Computer Vision Group
RAFAEL Dep. 39
Haifa P.O. Box 31021, ISRAEL
e-mail: mb@lems.brown.edu

David. B. Cooper

Laboratory for Engineering
Brown University
Providence, RI 02912
e-mail: cooper@lems.brown.edu

Abstract

This paper presents an *automated* approach to finding main roads in aerial images. The approach is to build geometric-probabilistic models for road image generation. We use Gibbs Distributions. Then, given an image, roads are found by map (maximum aposteriori probability) estimation. The map estimation is handled by partitioning an image into windows, realizing the estimation in each window through the use of dynamic programming, and then, starting with the windows containing high confidence estimates, using dynamic programming again to obtain *optimal* global estimates of the roads present. The approach is model-based from the outset and is completely different than those appearing in the published literature. It produces two boundaries for each road, or four boundary when a mid road barrier is present.

1 Introduction

In this paper we introduce an approach to building models for and estimating main roads in aerial images, which is an extended version of our 1993 CVPR paper [M.Barzohar and D.B.Cooper, 1993a],[M.Barzohar and D.B.Cooper, 1993b]. The goal is a completely automated system. The idea is that this approach could then be extended to finding other types of objects in areal images of the ground.
In recent years a number of papers have appeared in the published literature dealing with *semi-automatic* extraction of roads from aerial photos. In general a human operator gives the road starting points and the road directions at the starting points. This is a *huge* help to the road finding algorithm. This interaction has been necessary because road images can be very complicated. Examination of our images in this paper, Figuers 6, 7, 8, and 9, reveals that image intensity across a road can vary

noticeably. There may be a barrier running along the center of the road. Road-width can vary appreciably, especially when a barrier is present. Image intensity edges along a road boundary may disappear, especially when there is a building entrance with a very small parking area alongside the road. The image intensity structure at road intersections can be very complicated. There may be cars or markings on roads, or shadows cast by buildings or trees, etc., etc.. Three major types of road finders were used: edge linkers, correlation trackers, region based followers.

Edge linkers, based on an edge operator output for magnitude and angle for each point in the image and then linking the edges according to some criteria, were used first in [Nevatia and Babu, 1980] and later in [Z.Aviad and P.D.Carnine, 1988] and in [McKeown and L.Denlinger, 1988].

Correlation trackers based on the assumption that there exists a pattern or texture on the road surface was used first by [Quam, 1978], and later by [McKeown and L.Denlinger, 1988] in combination with edge linkers.

In [Cooper, Oct 1979], a Bayesian approach to low-level boundary estimation and object recognition was introduced. The problem considered in this paper is the automatic extraction of main roads when road curvature, width, image intensity and edge strength can vary considerably and when a barrier along the road center may or may not be present. The approach is general, and we feel that it can be extended to handle the full range of road image variability. The approach is to build geometric-stochastic models for representing road images, and then use maximum aposteriori probability estimation for estimating the road boundaries (and other important features) in an image. The modeling approach forces the designer to model all significant phenomena, and the model is generative so that its representation power can be assessed. The map estimation provides for the most accurate road finding. Global map estimation is realised in a computationally reasonable way by using dynamic programming to search small windows to obtain initial road candidates, and then dynamic programming again with small windows in order to obtain global estimates.

2 Road Generation

2.1 Road Geometry And Internal Grey Level Model

We build a geometric-stochastic road model based on the following assumptions:
1) Road width variance is small and road width change is likely to be slow.
2)Road direction changes are likely to be slow.
3)Road grey level is likely to vary only slowly.
4) Grey level variation between road and background is likely to be large.
5) Roads are unlikely to be short.

A stochastic process model is built exhibiting the preceding behavior. Specifically, autoregressive processes are designed to model road center line, road width, grey level within the road, edge strength at the road boundary,and regions outside the

roads and adjacent to the boundaries. We refer to these regions as background. Note that the road geometry processes are hidden, i.e, they are not observed directly in the data. For this purpose, consideration is restricted to an $N \times N$ pixel window. The stochastic processess are function of a discrete parameter i which can be thought of as time or distance, in pixels, along a horrizontal or a vertical axis. As an example, consider Fig 1. The i axis here is horrizontal. The road center line at i is x_i which takes values in the vertical direction. This variable is not quantized, it takes arbitrary real values. The $\{x_i\}$ process is given by (1), where ε_{x_i} is a zero mean, white, Gaussian driving noise. This process is designed to generate a straight line if the driving noise is zero.

$$x_i = 2x_{i-1} - x_{i-2} + \varepsilon_{x_i} \tag{1}$$

$$d_i = \frac{1}{2}d_{i-1} + \frac{1}{2}d_{i-2} + \varepsilon_{d_i} \tag{2}$$

Equation (2) describes road width, where d_i is the perpendicular road width through the unquantized center point $[x_i$,i]. The perpendicular direction is measured as perpendicular to the line segment from the point $[x_{i-2}, i-2]$ to the point $[x_{i-1}, i-1]$. The stochastic processes ε_{x_i} and ε_{d_i} are independent Gaussian white noise sequences with zero means and variances σ_{x_i} and σ_{d_i}, respectively. The road obtained for the unforced solution ($\varepsilon x_i = 0, \varepsilon d_i = 0$) will be two parallel line as seen in Fig 1. Road boundary locations are uniquely detemined by the x_i and d_i. The road boundary locations on the grid locations are determined as in Figure 1; where $\tilde{x}_{i,1}$ denotes the upper unquantized boundary and $\tilde{x}_{i,2}$ denotes the lower unquantized boundary.

We use a second order Markov Process to model the mean intensity of the image

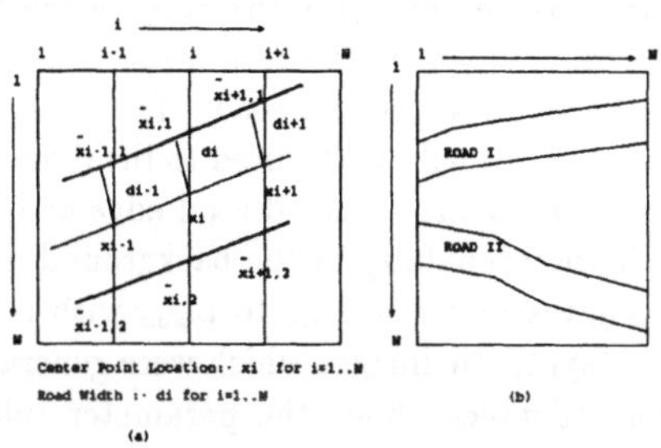

Figure 1: (a)A road in a window for unforced solution. (b) Two Roads in a Window

data in sequential vertical strips of the road to be consistent with feature 3 of our road model.

$$u_i = \frac{1}{2}u_{i-1} + \frac{1}{2}u_{i-2} + \varepsilon_{u_i} \tag{3}$$

The variable u_i is the mean intensity in column i of the window, and ε_{u_i} is a Gaussian white noise sequences with zero mean and variance σ_{u_i}, and is independent of the processes ε_{x_i} and ε_{d_i}. The intensity varies across the vertical direction of the

road too. We model this by adding an additive white noise.Therefore, the observed picture function at the (j,i)th pixel (jth row, ith column) in the road is y_{ji}, given in (4), where n_{ji} is Gaussian white noise with zero mean and variance σ.

$$y_{j,i} = u_i + n_{j,i}, \quad i = 1..N, \quad j = \bar{x}_{i1}..\bar{x}_{i2} \tag{4}$$

Assume for now that the we deal with a step edge (we also deal with the blur edge but this is more involved). The image intensities $y_{x_{i2}+1,i}$ and $y_{x_{i1}-1,i}$ immediately outside the lower and upper road boundaries, respectivly, are modeled by :

$$y_{x_{i2}+1,i} = u_i + r_2 * \theta_{l_i}, \quad i = 1....N \tag{5}$$

$$y_{x_{i1}-1,i} = u_i + r_1 * \theta_{u_i}, \quad i = 1....N \tag{6}$$

where r_2 and r_1 are random variables, taking values ± 1 with equal probabilities, θ_{l_i} and θ_{u_i} are a white Gaussian iid sequence having some nonzero mean. The purpose of r_2 and r_1 are to model whether each of the background grey level is lighter or darker than that in the road.

Assume that the image intensities in the background regions above and below the road boundaries are modeled by different Markov processes. These are causal Gaussian AR (autoregressive) models. The lower background AR model is:

$$y_{j,i} = u_{j,i} + \sum_{k=0}^{1}\sum_{l=0}^{1} \beta_{kl} \times (y_{j-k,i-l} - \mu_{j-k,i-l}) + \varepsilon_{y_{j,i}} \tag{7}$$

where β_{kl} are the model parameters, u_{ij} is the mean value function, and $\varepsilon_{y_{j,i}}$ is a Gaussian white noise driving sequence with zero mean and variance $\sigma_{y_{i,j}}$. A similar model generates the background above the road. The reason for using causal AR models is that they are computatioally well suited to the dynamic programming road estimation algorithm used. With the specified road edge and background models, it is now possible to specify the probabilty of the background data. We are expecting to detect roads having lengths between L_{min} to L_{max} with uniform distribution. In fig 2 are displayed various synthetic images which were generated by the road model described in this section, (detailes about the parameter values used are given in [M.Barzohar and D.B.Cooper, 1993b]).

3 Road Finding As Map Estimation Problem

Our general framework for viewing road finding is to estimate the geometry of the road by formulating the problem as map estimation. We look for that road for which $P(hypothesized\ road\ model|image\ data)$ is a maximum . Since this posterior likelihood of a hypothesized road model given the image data can be written as

$$\frac{P(hypothesized\ road\ model,\ image\ data)}{P(image\ data)} \tag{8}$$

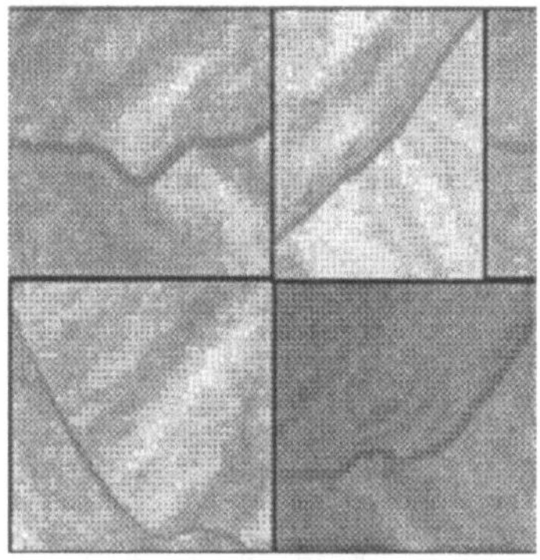

Figure 2: Various synthetic images generated by our road model.

and the denominator is not a function of the hypothesized road model , the road model estimate can be found as that which maximizes the numerator, i.e, the joint likelihood of a hypothesized road model and the image data. The map estimation is handled by partitioning the image into windows, realizing the estimation parameters in each window through the use of dynamic programing (more detailes are given in [M.Barzohar and D.B.Cooper, 1993b]). Our approach also includes of decision rules for starting points of a road in a window, and stopping points of a road in a window.

3.1 Probability Of Road Geometry And Image Data:Recursive Computation

By road image data we will now mean all the image data inside and in the first pixel layer immediately outside a hypothesized road boundary. The road geometry at i is denoted X_i, $X_i = (x_i, d_i)^t$. Denote by $\underline{X}$ the vector having all the x_i, d_i in a road as its components and by $\underline{Y}$ the vector having as its components the image intensities $y_{j,k}$ at all the pixels (j,k) in the road region. Then the system seeks the road geometry maximizing (9)

$$\max_{\underline{X}}(P(\underline{Y}|\underline{X}) \times P(\underline{X}) \tag{9}$$

Note, the geometry $\underline{X}$ is the hidden structure in this model. The pixels in the road region are determined by the geometry vector $\underline{X}$. Hence $\underline{X}$ specifies the data to be used, so that the probability of $\underline{Y}$ is conditional on $\underline{X}$. Let $\tilde{L}_i(X_i,\ X_{i-1})$ denote the joint probability of road geometry and road image data from column 1 through road geometry X_{i-1} and X_i in columns i-1 and i *for the best road.* Let $D_i(X_i, X_{i-1}, X_{i-2})$ denote the summation of a transition probability with an observation probability, (detials are given in [M.Barzohar and D.B.Cooper, 1993b]). Then for any acceptable geometry X_{i-2}, X_{i-1}, X_i, we have

$$\tilde{L}_i(X_i, X_{i-1}) = \max_{X_{i-2}}\{\tilde{L}_{i-1}(X_{i-1}, X_{i-2}) + D_i(X_i, X_{i-1}, X_{i-2})\} \tag{10}$$

Hence if we know $\tilde{L}_{i-1}(X_{i-1}, X_{i-2})$ for all geometry states X_{i-1}, X_{i-2} at stage i-1, we can compute $\tilde{L}_i(X_i, X_{i-1})$ from (10), (more detials are given in [M.Barzohar and D.B.Cooper, 1993b]).

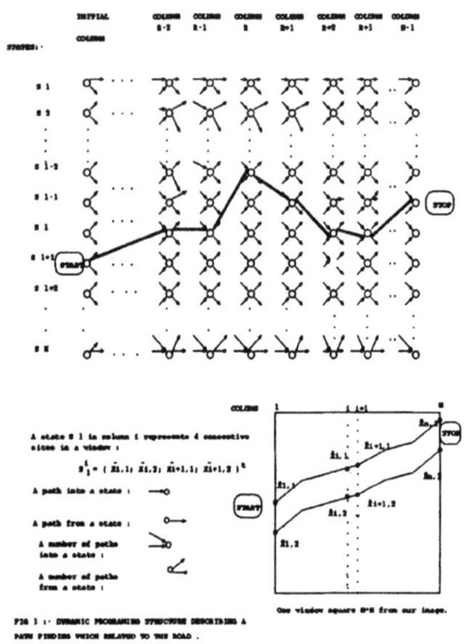

Figure 3: Dynamic Programming Structure describing a path finding which related to the road.

3.2 The Complete Low Level Processing

In the low level processing we are searching for seeds of roads. The image is divided into square windows, N x N pixels in each window, and the system searches for road candidates that fit the road model with high probability by using the dynamic programming structure. We run the road search four times, with a separate search starting at each of the four sides of the window, to pass over all the possibilities of road geometries. In Fig 3a we represent those examples. There may be more than one candidate road within a window. This is handled by searching a window for the best road, and then repeating a window search for the next best road. We do not want the second-best road to be a variation of the best road. This is handled by not permitting boundary points for the best road also to be boundary points for roads in subsequent searches. In this way, the system will find a pair of roads such as in figure 1b. The procedure is repeated until all road candidates are found. Fig 3b illustrate another example, road junction where a main road forks into two roads. In a first search starting on, the right side of the window road1 between points H, K, E, F will be found, then road2 between points L, M, F, G will be found, and road3 between points A, B, C, D will be found, by a search starting at the left side of the window. The splitting area represented by points C, D, E, F will not be found at this level . The hidden structure for every road candidate is found and observed to the high- level processing.

4 Combining Road Candidates

A high level processing stage is now used to extend each road candidate produced by the low level window search in order to obtain global road estimates. This is done by using shifting windows, as illustrated in Fig 4, where an extension window is abutted to the side of the present window from which the road is exiting. The best extension of the road candidate through the extension window is estimated by using the dynamic programming algorithm. This process is repeated until the stopping criterion stops the road search or until the estimated road hits the image border. In the process of extending a road through an extension window, the dynamic programming algorithm begins by using the last estimated state of the road and the associated road image data. Upon termination of the road search, if the length of the estimated road from the initial road candidate is greater than a threshold, the estimated road is accepted. If the length is less than the threshold, the estimated road is rejected, unless there is other supporting evidence. Supporting evidence we have used in the experiments is that if three long roads enter an intersection and the short road is adjacent to the intersection and appears to be an extension of one of them, i.e., has roughly the direction and width of one of them, then accept the short road.

5 Experimentel Road Results

This section describes the results of road finding, which were carried out on four different real images called, Rad1 (Figure 6a) and Rad2 (Figure 7a) that were obtain from the Radius Program, Rad3 (Figure 8a) that was obtain from the old city Jerusalem picture, and Rad4 (Figure 9a) that was obtain from the roads in terrain picture. The goal is to find the roads in the real images using our approach. The image field is partitioned into an array of square windows (32x32). The road finder runs simultaneously within the windows to find initial road seeds in the images (section 3). It then combines all the local results to obtain the final main roads in the images by using the high-level approach. The recognized road boundary points for Rad3 and Rad4, are indicated by black dots (Figures 8b and 9b), (detailes are given in [M.Barzohar and D.B.Cooper, 1993b]). The recognized roads for Rad1 are indicated in (Figure 6b), (detailes are given in [M.Barzohar and D.B.Cooper, 1993b]). The recognized barriers and roads for Rad2 are indicated in (Figure 7b). The system starts first with recognizing the road barriers, using knowledge that the barrier intensity is lighter than the road surrounding it. The initial road barrier seeds in the image were obtain by low-level processing within the windows, and the high-level following stage combines the local results to obtain the final barrier (details are given in [M.Barzohar and D.B.Cooper, 1993b]). Each side of the barrier is bordered by a road; therefore, by knowing the boundaries of the barriers, the system also knows the corresponding boundary along one side of each main road. The other boundary along the second side of each road is estimated by using the high-level approach for road finding.

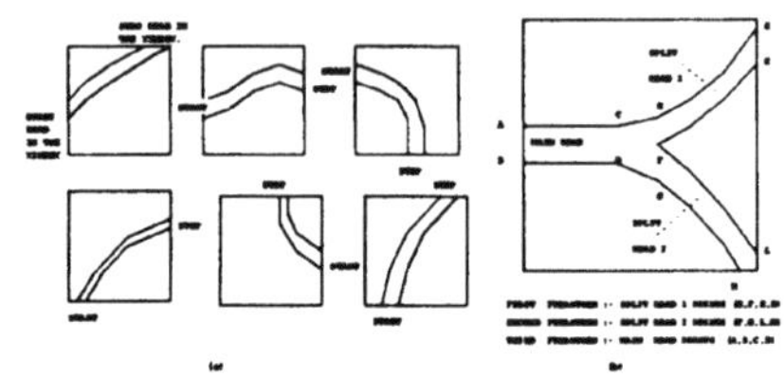

Figure 4: (a) Disparate roads with different starting and terminating side locations in a window.. (b) A Main Road split into two roads.

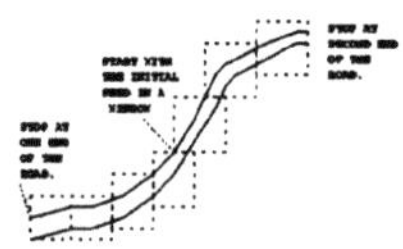

Figure 5: Following road candidate through its neighborhood shifted windows.

References

[Cooper, Oct 1979] D.B. Cooper. Maximum likelihood estimation of Markov process blob boundaries in noisy images. *IEEE PAMI*, Oct 1979.

[M.Barzohar and D.B.Cooper, 1993a] M.Barzohar and D.B.Cooper. Automatic Finding Of Main Roads In Aerial Images By Using Geometric-Stochastic Models and Estimation. *IEEE CVPR*, July 1993.

[M.Barzohar and D.B.Cooper, 1993b] M.Barzohar and D.B.Cooper. Completely Automatic Reliable Finding Of Main Roads In Aerial Images By Using Bayesian Methods. *Brown-LEMS-118*, March 1993.

[McKeown and L.Denlinger, 1988] D. M. McKeown and L.Denlinger. Cooperative methods for road tracking in aerial imagery. *CVPR*, 1988.

[Nevatia and Babu, 1980] R. Nevatia and K. R. Babu. Linear feature extraction and descripition. *IEEE CGIP*, pages 257–269, 1980.

[Quam, 1978] L. Quam. Road tracking and anomaly detection in aerial imagery. *Proceedings: Darpa Image Understanding Workshop*, pages 51–55, 1978.

[Z.Aviad and P.D.Carnine, 1988] Z.Aviad and Jr P.D.Carnine. Road finding for road network extraction. *CMU-CS-88-157*, 1988.

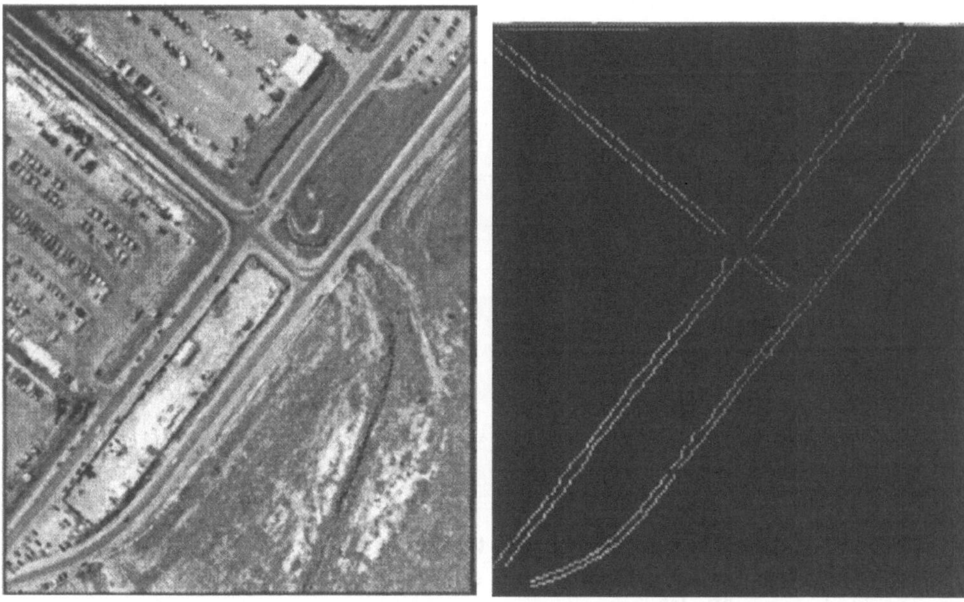

Figure 6: (a) Real Image number 1 called "rad1" taken from the Radius program. (b)The recognized and estimated roads of our road finder, in image "Rad1"

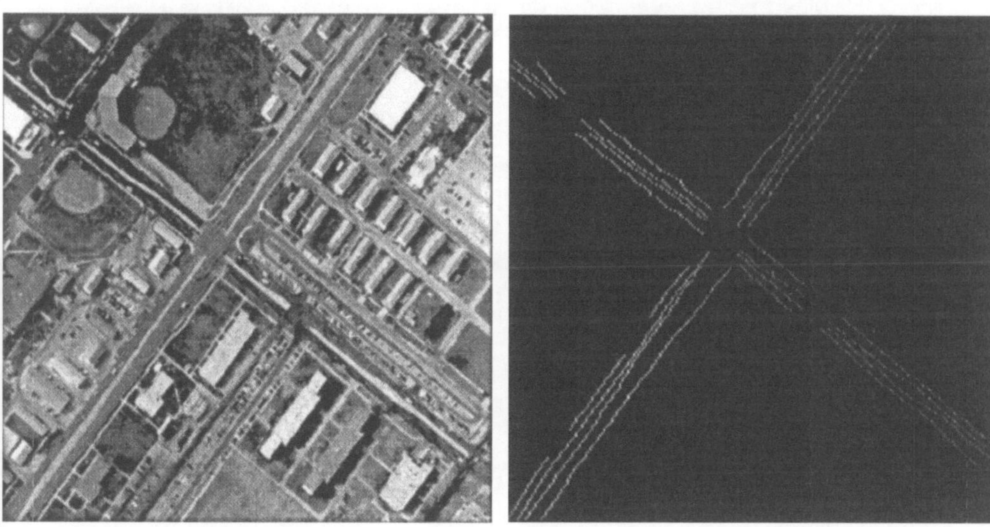

Figure 7: (a) Real Image number 2 called "Rad2" taken from the Radius program. (b) The recognized and estimated roads and barriers of our road finder, in image "Rad2"

Figure 8: (a) Real Image called "Rad3" taken from the old city Jerusalem picture. (b) The recognized and estimated roads of our road finder overlaid on the original Image, in image "Rad3".

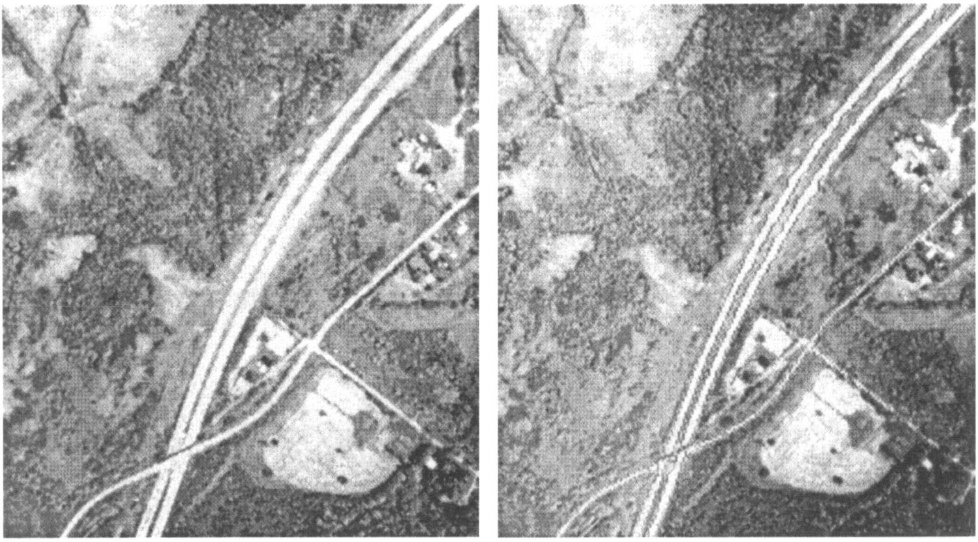

Figure 9: (a) Real Image called "Rad4" taken from the roads in terrain image. (b) The recognized and estimated roads of our road finder overlaid on the original Image, in image "Rad4".

Automatic Extraction of Man-Made Objects
from Aerial and Space Images
Monte Verità, © 1995 Birkhäuser Verlag Basel

Road tracing by profile matching and Kalman filtering

George Vosselman[1] and Jurrien de Knecht[2]

[1]Faculty of Geodetic Engineering, Thijsseweg 11, 2629 JA Delft
E-mail: vosselm@dutgs1.tudelft.nl
[2]Faculty of Applied Physics, Lorentzweg 1, 2628 CJ Delft
E-mail: jurrien@ph.tn.tudelft.nl
Delft University of Technology
The Netherlands

Abstract

Road tracing is a promising technique to increase the efficiency of road mapping. In this paper a new road tracing algorithm is presented. Road positions are computed by matching the average grey value profile of a reference road segment with profiles taken from the image. The road parameters are estimated by the recursive Kalman filter. By utilizing the prediction step of the Kalman filter the road tracer is able to continue following the road despite temporary failures of the profile matching that are due to road crossings, exits and cars.

1 Introduction

Of all parts in the process of making a map from aerial photographs, the actual mapping phase is one of the most time consuming and expensive parts. Research, like presented at this workshop, is therefore increasingly focusing on the development of efficient methods to extract topographical features like houses and roads from digital aerial photographs. As fully automatic methods for mapping are still far out of reach, semi-automatic methods for feature extraction that interact with a human operator are considered to be a good compromise between the mensuration speed of a computer algorithm and the interpretation skills of the operator.

This paper deals with road tracing, a semi-automatic approach to measuring roads in digital aerial images in which a human operator initializes the mapping process by indicating a small segment of the road to be mapped. After initialization the continuation of the road is mapped by the road tracing algorithm (Bajcsy and Tavakoli (1976), Groch (1982), McKeown and Denlinger (1988), Quam and Strat (1991)).

Another class of road mapping techniques also has to be mentioned. Whereas road tracers try to extend the initial segment, techniques from this other class use a sparse set of points distributed along the complete road to be mapped and try to interpolate between the given points with dynamic programming, relaxation or snakes (Lemmens et al. (1988), Gunst (1991), Gruen and Li (1994)). We will now return to road tracing.

Most road tracing algorithms consist of two steps: detection of road positions and concatenation of those positions. The detection step is supported by a prediction based on the previously detected road segment. Detection is performed by either an edge or line detection operator (Bajcsy and Tavakoli (1976), McKeown and Denlinger (1988)) or an analysis of the road profile (cross section) (Groch (1982), McKeown and Denlinger (1988), Quam and Strat (1991)). In most cases a straightforward connection of the detected road pixels is used to describe the road.

In this paper we present a new road tracing algorithm based on least squares profile matching and the Kalman filter. In section 2 we will outline the motivation for this using least squares matching and the Kalman filter. Section 3 describes the profile matching while section 4 contains a brief description of the Kalman filter and the model used to parameterize the shape of the road. In section 5 results of the road tracer are presented.

2 Concepts

2.1 Road characteristics

Road characteristics can be classified in five groups: geometrical, radiometrical, topological, functional and contextual characteristics (also see Garnesson et al. (1990)). Examples of these characteristics are:

- Roads are elongated (geometry)
- Roads have a maximum curvature (geometry)
- The road surface usually is homogeneous (radiometry)
- The road surface often has a good contrast with the adjacent areas (radiometry)
- Roads do not stop without a reason (topology)
- Roads intersect and build a network (topology)
- Roads connect cities (functional)
- Higher roads (fly-overs) may cast a shadow (contextual)
- Trees may occlude the road surface, but, on the other hand, an array of trees may also indicate a road (contextual)

The characteristics describe many features that are used by a human operator to recognize and map roads. Especially the functional and contextual characteristics require quite some intelligence in order to exploit them in the image interpretation process. In the developed semi-automatic road tracer we will not attempt to include these high level vision components, but instead primarily focus on radiometric and geometric characteristics. Complex situations (occlusion by trees, interpretation of shadows) will be left to the operator.

It will be shown that by examining radiometric and geometric properties the road tracer is capable of detecting indications about the topology of the road network. This capability potentially can be used to extend the road tracer to a road network tracer.

2.2 Short outline of the road tracer

The operator initializes the road tracer by measuring two points that indicate a short road segment. Between the two points grey value cross sections are computed at intervals of one pixel. The model road profile is taken as the average of these cross sections. This model profile is used as a template in the profile matching.

Based on the indicated road segment an initial estimate is made of the parameters that describe the road's position and shape. This estimate is used to predict the position of the first road profile adjacent to the indicated segment. The profile at the predicted position is matched with the model profile. The result of this match is a shift between the two profiles. This shift is used by the Kalman filter to update the parameters that describe the road's position and shape. In the following iterations, the position of the next profile is predicted, the profile at this position is matched with the model profile, and the road parameters are updated. The road tracer continues until some break-off criterion (discussed later) is fulfilled.

2.3 Why least squares profile matching?

The grey value profile is a very characteristic property of a road. It often shows a good contrast between the road surface and its vicinity. Since adjacent profiles usually have a very similar shape, profile matching algorithms should be able to detect changes in the road direction.

Least squares matching is preferred over maximizing cross correlation because least squares matching can estimate the precision of the profile shift. This precision can be used to evaluate the success of the matching algorithm and moreover is required as input for the Kalman filter.

Another advantage of least squares matching is the possibility to model the geometric (and radiometric) transformation between the two profiles. Not only the road position, but also the road width can be estimated. Thus, when the road width is changing, least squares matching can obtain good results, while cross correlation will fail. Besides, increasing road widths in a sequence of profiles can be regarded as an indication for a Y-junction.

2.4 Why Kalman filtering?

The Kalman filter is a recursive procedure to estimate the parameters of a dynamic system and has found many applications in navigation (Kalman (1960), Gelb (1974)). In the case of road tracing the parameters to be estimated are the parameters that describe the position and shape of the road. These parameters are called the state.

Strictly speaking, we do not have a dynamic system since the state is not time-dependent. However, we will treat the distance along the road as if it were the time variable. In this way, we can use the recursive estimation procedure.

The Kalman filter consists of two steps: the time update and the measurement update. In the time update an estimate of the state at time **t+dt** is made using all observations (i.e. profile matches) that have been made up to time **t**. Thus the time update predicts the state at the future epoch **t+dt**. In the measurement update the results of the profile match at time **t+dt** are combined with the prediction from the time update to obtain an optimal estimate for the state at time **t+dt**.

For several reasons the use of the Kalman filter in a road tracing procedure seems very attractive:

- The Kalman filter has a solid background in statistics and estimation theory. At each epoch it can produce an optimal estimate of the current road parameters. By using all available observations in combination with a parameterized road model, the filter result will be much better than a sequence of adjacent road pixels that are determined independently.
- The background in statistics enables the Kalman filter to detect and remove outliers in the observed profile shifts. Thus, if the profile match outputs a very unlikely shift, this observation is classified as erroneous and will not corrupt the state estimation.
- When following a road the road tracer may encounter situations like intersections and fly-overs. In those situations the template profile will not match the profile extracted from the image. Hence, there will be no observation to update the road parameters in the measurement update. However, by using one or more iterations of the time update the future states can be estimated by prediction only. After crossing the intersection the profile matching will succeed again and the time and measurement updates will again be applied alternately.
- Due to the extrapolation and uncertainty in the time update (so-called system noise) the variances of the estimated state parameters will increase very rapidly if no observations are available. This allows us to formulate a break-off criterion for the road tracer: once the variance of the estimated parameters exceeds some preset threshold it is assumed that (for some reason) the road tracer lost the road.

2.5 Optimal estimate of road position

The described road tracer is optimal in a sense that it produces the least squares estimate of the road parameters at time **t** based on all observations at and prior to time **t**. It does, however, not make use of observations after time **t**. Hence, the result of the road tracer can be improved by inserting all observations along a road segment into a single adjustment problem. The result will be a road description that is smoother than the one obtained by the described road tracer. This final adjustment will not be discussed here and has not been used for the results in this paper.

3 Profile matching

The profile matching compares the model profile with the road profile at the position predicted by the time update. The differences between the two profiles are modeled by two geometric (shift and width) and two radiometric (brightness and contrast) parameters. These parameters are estimated by minimizing the square sum of the grey value differences between the profiles (Ackermann (1983)).

After determining the optimal transformation between the profiles the matching results are evaluated by three checks:

- The cross correlation coefficient between the grey values of the two profiles after the transformation is required to be higher than 0.8.
- The estimated values of the geometric and radiometric parameters should be reasonable. E.g., if the estimated contrast parameter has a high value, say 10, the match can not be accepted. A contrast value of 10 would mean that the grey value contrast in the model profile is 10 times the contrast in the profile at the predicted position. A high contrast value therefore indicates that the latter profile hardly contains any signal and most likely does not correspond to a part of the road.
- A match is only accepted if the estimated standard deviation of the estimated shift parameter is below 1 pixel.

If for one of the above reasons the result of the matching is not accepted, the Kalman filter will not perform a measurement update but instead continue with another time update. Several consecutive rejections of the profile matching can be used as an indication for a road junction or the end of the road.

The standard deviation of the estimated profile shift is obtained by propagating the a posteriori standard deviation of the differences between the grey value profiles and based on the assumption that the grey value differences have a Gaussian distribution. While testing the road tracer, it appeared that the estimated standard deviation usually is very optimistic. For many profiles with disturbances like cars or shadows the assumption of Gaussian distributed residuals is not met. Instead, the systematic differences often cause a significant bias in the estimated profile shift. Therefore, the estimated standard deviation could not be used as input for the Kalman filter. In the Kalman filter a constant standard deviation of the profile shift of 0.3 pixel was used.

Profiles along a road often change (both gradually and abruptly). Quam and Strat (1991) therefore suggest to update the model profile by calculating a weighted average of the initial model profile and the profiles that were successfully matched. In our experiments this approach led to shifts of the road position within the model profile. Whereas the model profile was initialized with the road centre in the centre of the profile, the updating could cause the road centre to shift by several pixels. These shifts were caused by biases in the shift estimation of the profile matching due to unequal contrast at both roadsides. The results shown in section 5 were therefore produced with a constant model profile derived from the segment indicated by the operator.

4 Kalman filtering

For the parameterization of the road a constant road curvature is assumed. In each position along the road the row and column coordinate of the road position (r_t, c_t), the road direction (ϕ_t) and the change in road direction $(\dot{\phi}_t)$ are estimated. These four parameters constitute the state vector (x_t) of the Kalman filter.

In the time update the state vector at time **t+dt** is predicted by the following equation using the state vector at time **t**. Both these state vectors at time **t+dt** and time **t** are based on all observations that have been made up to time **t**. This is expressed by their indices **t+dt|t** and **t|t**.

$$\hat{x}_{t+dt|t} = \begin{pmatrix} \hat{r}_{t+dt|t} \\ \hat{c}_{t+dt|t} \\ \hat{\phi}_{t+dt|t} \\ \hat{\dot{\phi}}_{t+dt|t} \end{pmatrix} = \begin{pmatrix} \hat{r}_{t|t} + dt \cos\left(\hat{\phi}_{t|t} + \frac{\hat{\dot{\phi}}_{t|t}}{2} dt\right) \\ \hat{c}_{t|t} + dt \sin\left(\hat{\phi}_{t|t} + \frac{\hat{\dot{\phi}}_{t|t}}{2} dt\right) \\ \hat{\phi}_{t|t} + \hat{\dot{\phi}}_{t|t}\, dt \\ \hat{\dot{\phi}}_{t|t} \end{pmatrix}$$

The covariance matrix of the predicted state vector is

$$P_{t+dt|t} = \Phi_{t+dt|t} P_{t|t} \Phi^T_{t+dt|t} + Q_{t+dt}$$

where $\Phi_{t+dt|t}$ contains the coefficients of the linearized time update equation. The assumption of a constant road curvature is not quite correct. The deviation of the true road shape from the shape modelled in the time update equation is known as the system noise. The resulting uncertainty in the prediction is accounted for by the covariance matrix Q_{t+dt}.

After the time update, a profile is extracted from the image at the predicted position at time **t+dt** and perpendicular to the road's direction at that time. The match of this road profile with the model profile gives an estimated shift s and its variance σ_s^2. This shift is not directly used as an observation in the Kalman filter, but converted to a two-dimensional observation vector y_{t+dt} with covariance matrix R_{t+dt}:

$$y_{t+dt} = \begin{pmatrix} r_{t+dt} \\ c_{t+dt} \end{pmatrix} = \begin{pmatrix} r_{t+dt|t} - s\sin(\phi_{t+dt|t}) \\ c_{t+dt|t} + s\cos(\phi_{t+dt|t}) \end{pmatrix}$$

$$R_{t+dt} = \sigma_s^2 \begin{pmatrix} \sin^2(\phi_{t+dt|t}) & \sin(\phi_{t+dt|t})\cos(\phi_{t+dt|t}) \\ \sin(\phi_{t+dt|t})\cos(\phi_{t+dt|t}) & \cos^2(\phi_{t+dt|t}) \end{pmatrix}$$

Note that R_{t+dt} is singular. This implies that the Kalman filter can only adjust the observed road position perpendicular to the road direction.

With the above observation the measurement model becomes

$$E\{y_{t+dt}\} = A\,x_{t+dt} = \begin{pmatrix} 1 & 0 & 0 & 0 \\ 0 & 1 & 0 & 0 \end{pmatrix} x_{t+dt}$$

This measurement model is used in the measurement update to estimate the state vector at time **t+dt** based on the prediction and the observations at time **t+dt.**

$$\hat{x}_{t+dt|t+dt} = \hat{x}_{t+dt|t} + K_{t+dt}\,(y_{t+dt} - A\,\hat{x}_{t+dt|t})$$

with

$$K_{t+dt} = P_{t+dt|t}\,A^T\,(R_{t+dt} + A\,P_{t+dt|t}\,A^T)^{-1}$$

As can be seen from these formula the difference between the observation y_{t+dt} and the "predicted observation" $A\,\hat{x}_{t+dt|t}$ plays an important role in the Kalman filter. Since the covariance matrix of this residual can be computed, the observation can be submitted to a statistical test to decide whether it can be used to update the state vector.

5 Road tracing results

Figure 1 shows the results of the road tracer on the blue channel of a colour image of a complex road junction. The initial road segments measured by the operator are shown in black. The traced roads are shown in white (good profile matching) and black (rejected profile matching: the road position is based on the time update only).

Image size:
551 x 527 pixels
Pixel size:
1.6 m
Average road width:
10 pixels
Model profile width:
16 pixels
Min. curvature radius:
160 m
Step size **dt**:
1 pixel

Fig. 1: Road tracing results for a complex road junction.

Since the road model assumes a constant curvature, the variance of the system noise is derived from the possible curvature change. The major roads in the scene show a road curvature with a minimum radius of 160 m or 100 pixels. In the worst case (a straight road connected to an arc with a radius 100 pixels) the road tracer could encounter a change in road direction ($\dot{\phi}$) of 1/100 rad. A common approximation of the standard deviation of the system noise is 1/4 of the maximum error. Thus, the standard deviation of the system noise in the derivative of the road direction was set to 1/400 rad. The standard deviations of the system noise of the other state variables were obtained by propagating the noise in the direction change. In this way, knowledge about the road design was used to estimate the covariance matrix of the system noise.

The iterations of the road tracer were stopped after the profile matching failed for 10 consecutive profiles and the standard deviation of the predicted road position became larger than 1 pixel. With this break-off criterion all roads could be traced with exception of the two major straight roads. For the latter two roads many more predictions were necessary to cross the fly-overs. For these cases it is useful to have an interaction between the operator and the road tracer. If the Kalman filter is predicting in the right direction, the operator can overrule the break-off criterion and let the road tracer continue despite the high variances of the estimated road positions.

Figure 2 depicts the variance of the road positions of a traced road segment. The two peaks correspond to the intersecting fly-overs. At these positions the profile matching failed and the estimation of the road parameters was done by prediction only. The steep increase of the variances in these areas is caused by the extrapolation and the additive system noise.

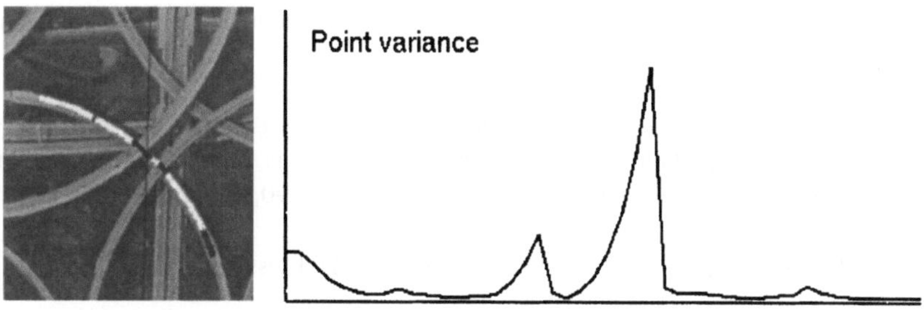

Fig. 2: Estimated road position variances along a road segment.

Further results showed that the tuning of the ratio between the variance of the profile matching and the variance of the system noise is rather sensitive in case of small distances between the profiles (step size of 1 pixel). As stated before, the estimated variance obtained from the least squares matching is very optimistic and could not be used in the Kalman filtering. The low variances make the Kalman filter react vehemently to every little disturbance and, as a consequence, obstacles like fly-overs can not be crossed if they are preceded by a small mismatch of the road profiles. On the other hand, if the variances of the profile shifts are set to a high value, the coefficients of the gain matrix K_{t+dt} may become very small. Hence, the estimation will be primarily based on the prediction and a change in road curvature may cause the road tracer to get off the road.

6 Conclusions

In this paper we presented a new road tracing algorithm which combines least squares matching of grey value profiles with the Kalman filter. Since both techniques have a solid background in statistics, decisions about the usefulness of the matching results could be based on statistical tests.

First results showed that the road tracer is able to correctly predict the road path at positions where the profile matching fails. Thus, the algorithm can cope with intersections, fly-overs and cars.

The rejected profile matches appeared to be a good indicator for intersecting roads. Together with the estimation of the road width, they can potentially be used to extend the road tracing algorithm to an algorithm for mapping road networks.

Literature

Ackermann, F. (1983): *High Precision Digital Image Correlation.* Proceedings 39th Photogrammetric Week, Stuttgart, pp. 231-243.

Bajcsy, R. and M. Tavakoli (1976): *Computer Recognition of Road from Satellite Pictures.* IEEE Transactions on Systems, Man, and Cybernetics, vol. 6, no. 9, pp. 623-637.

Garnesson, Ph., G. Giraudon, and Montesinos (1990): *An Image Analysis System, Application for Aerial Imagery Interpretation.* Proceedings 10th IAPR International Conference on Pattern Recognition, Atlantic City, June 1990, vol.1, pp. 210-212.

Gelb, A (Ed.) (1974): *Applied Optimal Estimation.* MIT Press, Cambridge, MA.

Gunst, M. de (1991): *Automatic Extraction of Roads from SPOT Images.* Technical Report, Delft University of Technology, 91 p.

Groch, W.D. (1982): *Extraction of Line Shaped Objects from Aerial Images using a Special Operator to Analyze the Profile of Functions.* Computer Graphics and Image Processing, vol. 18, pp. 347-358.

Grün, A. and H. Li (1994): *Semi-automatic Road Extraction by Dynamic Programming.* International Archives of Photogrammetry and Remote Sensing, vol. 30, part 3/1, pp. 324-332.

Kalman, R.E. (1960): *A New Approach to Linear Filtering and Prediction Problems.* ASME Journal of Basic Engineering, vol. 82D, pp. 35-45.

Knecht, J.P. de (1994): *Tracing of Roads in Aerial Photographs by Profile Matching and the Kalman-filter* (in Dutch). MSc-Thesis, Faculty of Geodetic Engineering, Delft University of Technology, 90 p.

Lemmens, M.J.P.M., E.W. Bruel and F. Fennis (1988): *Linear Feature Extraction and Road Recognition from Large Scale Digital Aerial Images.* International Archives of Photogrammetry and Remote Sensing, vol. 27, part B8, pp. 20-31.

McKeown, D.M. and J.L. Denlinger (1988): *Cooperative Methods for road tracing in aerial imagery.* IEEE Proceedings on Computer Vision and Pattern Recognition, Ann Arbor, MI, pp. 662-672.

Quam, L.H. and T.M. Strat (1991): *SRI Image Understanding Research in Cartographic Feature Extraction.* In: Ebner et al. (ed), Digital Photogrammetric Systems, Wichmann Verlag, Karlsruhe, pp. 111-122.

Model-Based Road Extraction from Images

Carsten Steger, Clemens Glock, Wolfgang Eckstein, Helmut Mayer, Bernd Radig

Lehrstuhl für Bildanalyse und wissensbasierte Systeme
Technische Universität München
Orleansstraße 34, 81667 München, Germany
Ph.: +49-89-48095-122, Fax: +49-89-48095-203
e-mail: {stegerc|eckstein|radig}@informatik.tu-muenchen.de

Lehrstuhl für Photogrammetrie und Fernerkundung
Technische Universität München
Arcisstraße 21, 80290 München, Germany
Ph.: +49-89-2105-2671, Fax: +49-89-280-9573
e-mail: {clemens|helmut}@photo.verm.tu-muenchen.de

Abstract

In this paper we present an approach to the automatic extraction of roads from aerial images. We argue that a model for road extraction is needed in every step of the image interpretation process. The model needs to include knowledge about different aspects of roads, like geometry, radiometry, topology, and context. The main part of this paper discusses the parts of that knowledge that we have implemented so far. It is shown that roads can be successfully detected at various resolution levels of the same image. Furthermore, we show that combining the results obtained in each level helps to eliminate false hypotheses typical for each level. The approach has been successfully applied to a variety of images.

1 Introduction

The extraction of roads from images has received considerable attention in the past. Several schemes have been proposed to solve this problem at resolutions that range from satellite images to low altitude aerial images. The strategies proposed fall into two broad categories. The work described in Gruen et al. (1994), Heipke et al. (1994), and McKeown et al. (1988) deals with the semi-automatic extraction of roads. The human operator has to select a certain number of points of the road which is then extracted. On the other hand, the work presented in Barzohar et al. (1993), Ruskoné et al. (1994), and Zerubia et al. (1993) is concerned with the automatic extraction of roads.

The aim of the work presented here is the automatic extraction of roads from aerial images for input into geographic information systems (GIS), in particular the German Authoritative Cartographic Topographic Information System ATKIS DLM25, which uses a scale of 1 : 25 000. Orthophotos with a ground resolution of 25 cm per pixel are used because they contain the necessary detail for a successful interpretation and are sufficient to obtain the geometric precision required.

Section 2 gives a short outline of the kinds of knowledge necessary for the extraction process. In section 3 the implemented parts of the model are presented in detail. Section 4 shows results. Finally, section 5 concludes the paper.

2 A Model for the Extraction of Roads

In order to be able to extract roads from aerial images a vision system must incorporate various kinds of knowledge. The proposed model groups knowledge into the categories geometry, radiometry, topology, and context. For example, the knowledge that roads are linear objects that have parallel roadsides, and that their width is more or less constant belongs to the geometric part of the model. The fact that roads have a surface that is relatively homogeneous in the direction of the road, and that they can have bright markings is radiometric knowledge. Furthermore, roads build up a hierarchical network which is used to connect towns on a coarse level and buildings on a fine level. This means, for example, that usually roads do not end without there being an object to connect to nearby. This aspect belongs to the knowledge about the topology of roads. It can be used to reject road hypotheses that cannot be connected with other roads or to start a process that looks for a reason why no connection can be made, e.g., shadows cast by trees. Finally, the context in which roads can appear has to be modeled as well in order to be able to explain failures to extract roads and correct them, and to discriminate roads from objects that look like roads, e.g., drive ways or roofs.

The extraction process has to receive attention as well. We argue to use knowledge as early as possible. This leads us to use a process that uses the parts of the road network which can be extracted quickly and with high reliability as the starting point for all further computations. Furthermore, an automatically generated digital elevation model (DEM) can be used to exclude objects that are higher than their surroundings, e.g., buildings or trees, from processing in later steps. Additionaly, color can be used to exclude regions that cannot be roads, e.g., lawns or parks. Utilisation of this kind of knowledge in very early stages helps to avoid generating false road hypotheses in later stages and speeds up processing.

Furthermore, knowledge about the areas of application of operators and knowledge about their strengths and weaknesses have to be incorporated in to the system. This can be used to select the appropriate extraction strategy in different contexts. For example, within cities a DEM can give most of the relevant hints for finding roads, if "ravines" between buildings are extracted. On the other hand, in rural or suburban areas the DEM serves mainly to eliminate false hypotheses, while the main clues are provided by radiomety and geometry.

3 Implemented Parts of the Model

3.1 Detection of Lines at Low Resolutions

In this section we will describe how roads can be detected in aerial images of reduced resolution. Because of the implicit smoothing performed in the scaling process, several kinds of problems that make the detection of roads at high resolutions more difficult are alleviated, e.g., cars on the road or shadows cast onto the road by adjacent trees.

To start the extraction process the original orthophoto is reduced by a factor chosen such that roads in the reduced image are at most five to six pixels wide. At this level of detail roads can be modeled as lines that usually are brighter than their surroundings. To extract the lines from this image each pixel is compared to a locally computed Gaussian mean. Pixels that are brighter than the local mean by a certain threshold are selected for further processing (R_1). In a second step the image is thresholded such that only regions which have an intensity that lies within a certain range are selected (R_2). For the image in figure 1 only relatively bright regions were selected. Then the intersection of these two regions is computed, resulting in $R_3 = R_1 \cap R_2$. This result corresponds to the regions in the image that are bright and brighter than their surroundings. After this step the skeleton S of R_3 is computed since we are only interested in the center lines of the roads at this resolution.

Unfortunately, the process so far selects lines as well as certain kinds of edges in the image. To select the lines, contours are computed from S. A contour is defined as a sequence of 8-connected points that starts and ends either in a junction point or a point that has exactly one neighbour in S. The contour-points are then examined whether they are local maxima on a line perpendicular to the direction of the contour. Only contours that have more than a certain percentage of points that are local maxima are selected. The output of this step are the road hypotheses at this resolution level. Figure 1b) shows the result of this process.

3.2 Edge-Extraction and Polygonal Approximation

The next three sections will give details about the road extraction process at the highest resolution level. At this level we assume that roads are relatively homogeneous regions in the image that have a significantly different brightness than their sourrounding areas. Hence it follows that roadsides can be detected by an edge extraction algorithm. For this task we use a modified version of a Deriche edge detector that has been described in Lanser et al. (1992). The advantages of this operator are very good detection quality, accurate edge localisation, few multiple responses, and isotropic response. The resulting edges are then thinned by a non-maximum-suppression algorithm, yielding one pixel wide edges. From these edges contours are computed.

To reduce the amount of data which has to be handled, and to faciliate the perceptual grouping of parallel lines, a polygonal approximation of each contour is computed by the algorithm given in Ramer (1972). This algorithm splits contours into polygon segments which have a limited distance to the approximated contour. The advantage of this algorithm is that it yields rather long line segments with an acceptably small approximation error.

a) b)

Fig. 1: a) Reduced version of the original orthophoto (reduced by a factor of 8) b) Result of the line detection algorithm

Figure 2 shows the result of the edge detection and polygonal approximation step with the maximum deviation set to 2 pixels.

3.3 Perceptual Grouping of Parallel Edges

One feature which characterises roads is the parallelism of opposite roadsides. Therefore the next step in the road-extraction process is to construct relations of parallel polygons. To be included into this relation, line segments have to fulfill several criteria. First of all, the line segments should be parallel. Because of the discrete nature of the edge extraction process and because the edges are approximated by polygons, opposite roadsides are never perfectly parallel. Furthermore, the direction of a long line segment is determined more accurately than that of a short segment. Therefore the criterion to determine whether two line segments are parallel has to allow for a certain angular difference that depends on the length of the lines involved. The following formula is used as a threshold for the angle α that two lines are allowed to enclose:

$$\alpha \leq \tau + \frac{w\tau}{\max(l_1, l_2)} \qquad (1)$$

Here, l_1 and l_2 are the lengths of the line segments involved, and τ is a user-selectable threshold. This formula allows short lines to enclose a greater angle than long lines. The weight w can be used to control how far the threshold grows for short lines. In the current implementation it is set to 1.

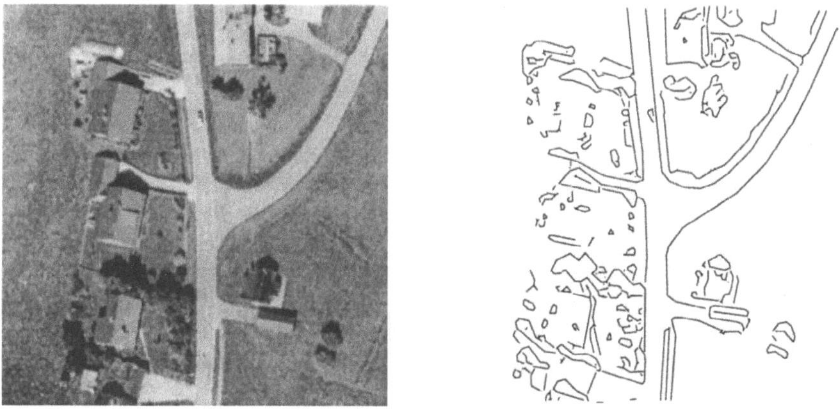

a) b)

Fig. 2: a) Input image for the edge detection process b) Detected edges approximated by polygons

The second criterion is that parallel line segments have to overlap. This is determined by projecting the end points of both lines onto a line that has a direction determined by the bisection of the angle that the lines enclose. If the projected lines overlap, then the original lines overlap as well. The last criterion is that the lines involved have to be closer than a certain threshold. This is motivated by the fact that roads cannot be wider than a certain distance. Figure 3 shows the results of this step with $\tau = 10°$ and a maximum distance of 7.5 m (30 pixels).

3.4 Selection of Homogeneous Areas between Parallels

The next step in our approach is concerned with the radiometric part of the road model. In this step we examine whether the area between the parallel line segments found in section 3.3 is homogeneous in the direction of the center line between them. This corresponds to the assumption that the surface intensity of a road is relatively constant in the direction of the road, whereas it can vary considerably across the road due to road markings or tire tracks.

To determine whether the region between two parallel line segments is homogeneous, slices parallel to the center line of the two lines are generated. These slices are 1 pixel apart, and the intensities within each slice are computed by bilinear interpolation. Figure 4a) shows how these slices are placed between two parallels. The solid line indicates the center line, while the dashed lines show the parallel slices. Note that only the minimal area between the parallels is used for computation.

The procedure then computes the mean intensity within each slice. If the mean falls outside a certain user-settable range, the region is immediately rejected. This is due to the

Fig. 3: Parallel line segments detected

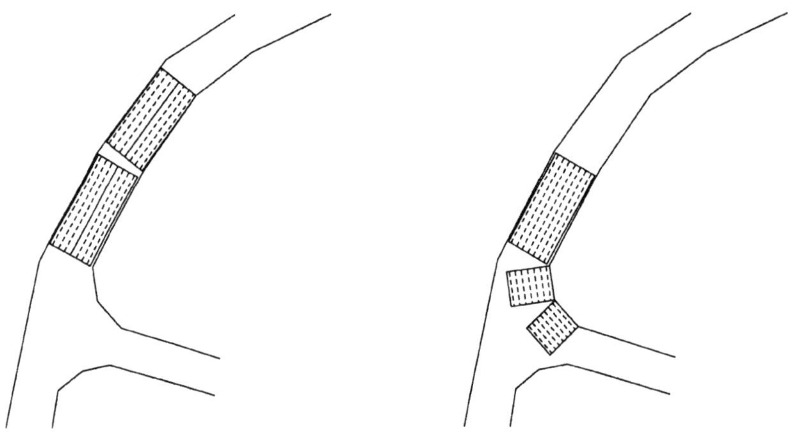

a) b)

Fig. 4: a) Slices computed between parallel line segments b) Slices computed while following the original edges

model assumption that a road is normally a relatively bright object, but it also allows one to select dark roads if present. After this the intensity variance in each slice is determined. This is the main criterion for homogeneity of the region in question. If the variance in each slice is smaller than a threshold, the parallel lines that enclose the region are accepted as hypothetical roadsides. The results of this step are shown in figure 5a).

As is evident from figure 5a), the algorithm is quite successful in finding the parts of the road where parallel roadsides exist. It fails in regions of the road where no parallelism can be detected, e.g., at intersections. Therefore all edges which neighbour the last parallel edge selected are examined to see if they still border homogeneous regions. To this end, rectangles are constructed that have the width of the last rectangle processed in the previous step. These rectangles are then sliced as above and examined for homogeneity. Figure 4b) shows the last segment processed and the two following segments that are examined. The result of this extension process can be seen in figure 5b).

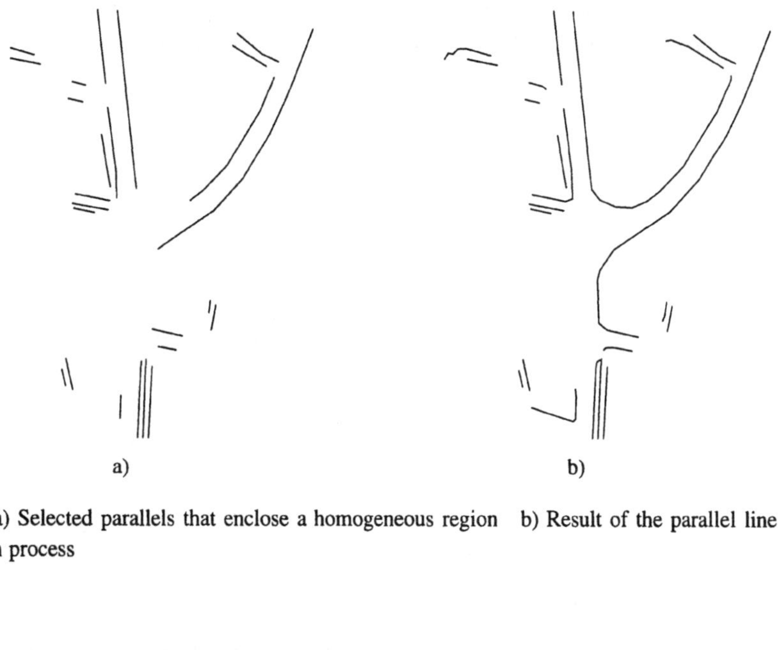

a) b)

Fig. 5: a) Selected parallels that enclose a homogeneous region b) Result of the parallel line extension process

3.5 Combining the Different Resolution Levels

As can be seen from the examples given so far, the extraction process in the two different resolution levels have their advantages and deficiencies. Therefore in this section we show how to combine the results of both levels to eliminate incorrectly detected road-segments and to extend results to regions where the extraction process so far has failed for some reason. The basic strategy in this step is to take the results of both levels which support each other. To start the process, parallel lines that enclose a homogeneous area of the high resolution level are selected if a center line of the reduced resolution level is found between the two parallels. Starting from these strong hypotheses, gradually all the roadsides are extracted.

This is done using a number of rules. One rule is, e.g.: Take roadside R_1 and its parallel roadside R_2. If there exists an edge R_3 that is colinear to R_1, and R_2 overlaps both R_1 and R_3, and there is a center line C that lies between R_2 and R_3, then accept R_3 as a new roadside. As can be seen from this example, the rules are rather conservative, selecting only edges as roadsides that have support in all resolution levels. Figure 6a) depicts the input to the combination process. The results obtained in section 3.4 are shown as solid lines, the original edges as dotted lines, and the center lines obtained in section 3.1 as dashed lines.

Figure 6b) shows the final result of the road extraction process. It can be seen that the algorithm was able to bridge the gap on the left hand side of the road at the intersection that resulted from lacking parallelism. Furthermore, some erroneously detected road hypotheses next to the houses on the left hand side of the image and a drive way on the upper right hand side were eliminated. The only errors that need to be eliminated are the drive ways on the left hand side of the image. The ATKIS model states that these roads must not be included in the GIS dataset.

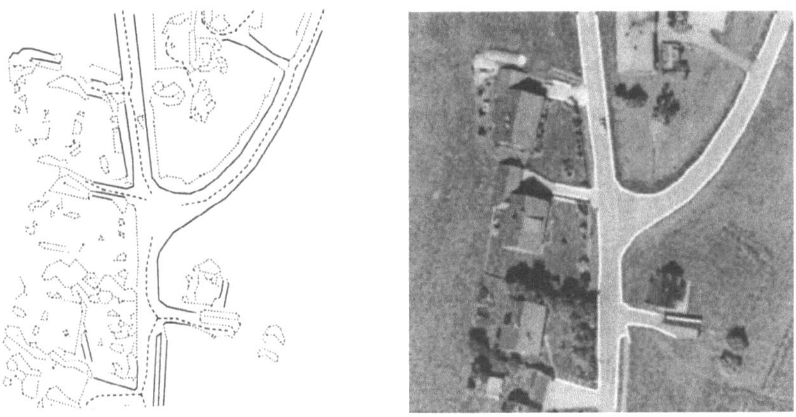

a) b)

Fig. 6: a) Input to the combination step b) Final result of the road extraction process

4 Results

In this section we will present two more examples of the results obtained with the approach we have described. The first example is a different part of the image shown in figure 1a). Figure 7 shows that most parts of the road network are extracted correctly. However, due to shadows cast onto the road by trees, some parts of the road in the upper part of the image could not be detected. Also, adjacent to that road, an erroneous road segment has been detected because the shadow of a tree has pushed the center line outside the road, thereby making it support that segment. Finally, not all parts of the track that leads into the field on the right hand side of the image have been eliminated because a short center

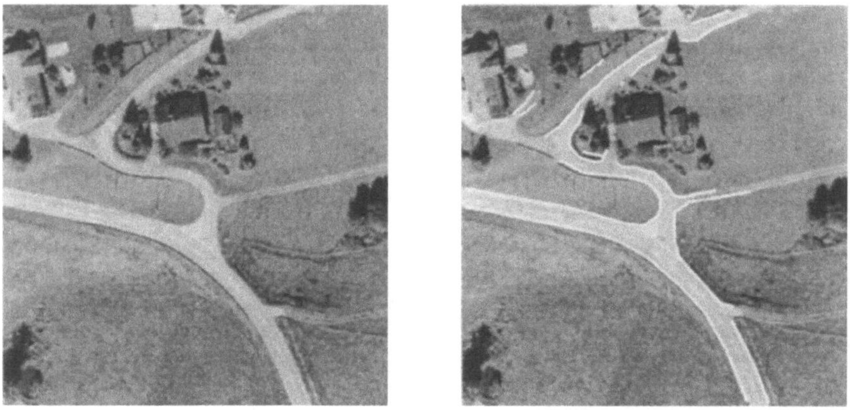

a) b)

Fig. 7: a) Input image to the algorithm b) Final result of the road extraction process

line was detected that gave support to the initial part of the track.

Figure 8 shows the result obtained with the image 5987 of the Ascona testset. Since this image has a ground resolution of 7.5 cm, it was reduced by a factor of 3 to fit our model. Again, most of the road network is detected correctly. However, several buildings are extracted as well. These errors could be eliminated very easily through the use of a DEM by excluding areas that are higher than their sourroundings from being input to the algorithm.

5 Conclusions

The results given in sections 3.5 and 4 indicate that the strategy we have chosen is quite general and leads to good results. However, more parts of the knowledge mentioned in section 2 have to be implemented for a successful automatic extraction of roads. Especially the topology and the context are crucial, as can be seen from the examples.

Therefore, we will focus on implementing knowledge about the topology of roads and the context in which roads appear in the next phase of the project. To faciliate this we plan to use color images and automatically computed digital elevation models as additional input data.

Acknowledgements

We would like to acknowledge the vital contributions of Heinrich Ebner to this project. This work was funded by the Deutsche Forschungsgemeinschaft under grant no. Eb 74-8/1.

a)

b)

Fig. 8: a) Image 5897 of the Ascona testset b) Final result of the road extraction process

References

Barzohar M., D.B. Cooper (1993) *Automatic Finding Of Main Roads In Aerial Images By Using Geometric - Stochastic Models and Estimation*, CVPR 93, pp. 459–464.

Gruen A., H. Li (1994) *Semi-Automatic Road Extraction by Dynamic Programming*, IntArchPhRS, Vol. 30, Part 3/1, pp. 324–332.

Heipke C., A. Englisch, T. Speer, S. Stier, R. Kutka (1994) *Semi-Automatic Extraction of Roads from Aerial Images*, IntArchPhRS, Vol. 30, Part 3/1, pp. 353–360.

Lanser S., W. Eckstein (1992) *A Modification of Deriche's Approach to Edge Detection*, ICPR 92, Vol. III, pp. 633–637.

McKeown D.M., J.L. Denlinger (1988) *Cooperative Methods For Road Tracking In Aerial Imagery*, CVPR 88, pp. 662–672.

Ramer U. (1972) *An Iterative Procedure for the Polygonal Approximation of Plane Curves*, Computer Graphics and Image Processing, Vol. 1, pp. 244–256.

Ruskoné R., S. Airault, O. Jamet (1994) *Road Network Interpretation: A Topological Hypothesis Driven System*, IntArchPhRS, Vol. 30, Part 3/2, pp. 711–717.

Zerubia J., N. Merlet (1993) *Classical Mechanics and Roads Detection in SPOT Images*, INRIA Rapport de Recherche No. 1889, INRIA Sophia Antipolis, France.

Map-Based Extraction

Automatic Extraction and Structuring of Objects from Scanned Topographical Maps - An Alternative to the Extraction from Aerial and Space Images ?

Stephan Nebiker, Alessandro Carosio

Institute of Geodesy and Photogrammetry
Swiss Federal Institute of Technology
ETH-Hönggerberg, CH-8093 Zurich, Switzerland
Phone: ++41 1 633 3049 (or 55), Fax: ++41 1 633 1101
e-mail: (sn or carosio)@ris.igp.ethz.ch

Abstract

This paper presents investigations and developments in the area of automatic extraction of cartographic features from scanned topographic maps. It focuses on the recognition and extraction of buildings for the establishment of topographic and cartographic information systems or for the provision of information supporting the automatic extraction of man-made objects from aerial or space images. The described solution combines knowledge-based pattern recognition techniques, raster data processing operations and raster-vector conversion procedures based on robust estimation and constrained adjustment techniques. Following a stringent quality control process the structured results can be exported in a selection of formats. The implemented solution allows largely automatic processing of entire map sheets with very high success rates within a few hours using standard UNIX workstations. Finally, the paper presents results from applications of the technology such as the recognition of buildings for the simulation and design of mobile communication networks for Swiss PTT.

1. Introduction

1.1. Motivation and Aim

Conventional topographical maps contain a vast amount of spatial and thematic information, which covers large areas. The scanned information from these maps serves as the basis for the transition to digital map production procedures and for the establishment of digital cartographic information systems. The same information could also play an important role in supporting the automatic extraction of man-made objects from aerial or space images or as a readily available and relatively cheap data source for numerous GIS applications. In the past, however, most attempts to extract and structure information from scanned topographic maps using traditional data acquisition methods such as manual digitising have proved not to be economically viable. One of the main problems is the enormous amount of information - e.g. several ten thousands of cartographic objects per map sheet - which requires largely automated solutions.

The main emphasis of our investigations was placed on the extraction of information from medium to small scale topographical maps (1: 25'000 and smaller). Some of the characteristics of these map types are: generally high graphical quality standards, a high information density and, as a result, a large amount of information and - most important - a geometric accuracy which is sufficient for numerous GIS applications. On the one hand, the available quality standards favour the use of standard automatic pattern recognition and vectorisation techniques. On the other hand, the high information density and the quality requirements for the final product demand for specific approaches and techniques. One goal of the automatic vectorisation approach is that the quality of the treated objects should be equal or superior to that obtainable from manual digitising. The main reason for choosing maps at a scale of 1:25'000 was the fact that they provide a good balance between a high information content and an acceptable geometric quality loss caused by geometric displacement and generalisation effects.

1.2. Chosen Approach: Overview

The method for extracting and structuring areal objects developed at the Institute of Geodesy and Photogrammetry (IGP), ETH Zurich can be divided into two different processing phases. The main purpose of each phase and the processing tasks involved are summarised below:

The Object Recognition Phase is aimed at the recognition of areal objects and their separation from other cartographic features. The investigated approach is utilising a *knowledge-based pattern recognition technique* and *raster processing operations* especially *adapted to cartographic characteristics* (Stengele, 1993). The entire process is carried out in the raster domain, without a prior raster-vector conversion. The object recognition phase is described in detail in chapter 2.

The Raster-Vector Conversion Phase is aimed at structuring the recognised objects and determining optimal object contours. The chosen approach is based on *robust estimation techniques, adjustment computations using geometric constraints* and a *quality control system*. The implemented solution operates irrespective of scale or scanning resolution. The raster-vector conversion phase is described in chapter 3.

The implemented solution has been successfully tested and installed for operational use. In an operational environment, the performance is in the range of one to two map sheets 1:25'000 per day, including visual inspection and manual editing.

1.3. Data Source and Characteristics

The investigations are based on raster data from scanned topographical maps (1:25'000) obtained from the Swiss Federal Office of Topography. Raster data for the different colour layers at a scanning resolution of 20 lines/mm (approx. 500 dpi) is commercially available for the entire country. The initial investigations were based on this type of data. Recent tests have been extended to data sets obtained with a scanning resolution of 40 lines/mm (~1000 dpi), a resolution which allows to exploit (and preserve) the full geometric quality of the original maps. The original assignment of objects to individual colour layers was primarily directed by the cartographic production processes. As a result each colour layer usually contains multiple thematic topics. The object extraction investigations were carried out using the *black layer* only. In addition to the thematic topic 'buildings', this layer also contains roads, railway lines, grid lines, text features, symbols, etc.

2. Object Recognition

The first step in the object extraction process is the actual recognition of areal objects, i.e. their separation from other cartographic features. The optimal recognition is a key factor for the quality of the final vectorisation results.

The object recognition procedure is directly carried out in the raster domain, which avoids problems related to raster-to-vector pre-processing operations *(Carosio/Stengele, 1993)*. The chosen approach for the object recognition consists of two steps. In a first step, undesired information (such as lettering, map symbols, etc.) is eliminated, and in a second processing step, areal features are separated from linear features.

2.1. Elimination of Undesired Information

The automatic recognition and extraction of undesired information (map symbols, grid lines, etc.) is performed by using a knowledge-based, template-matching approach. This method allows the recognition and extraction or -- as in the case of this application -- elimination of letters, figures, cartographic symbols, grid lines etc.

Knowledge-Based Template Matching. The key features of the method are outlined in *Stengele(1993)* and *Stengele/Carosio(1994)*. The characteristics of the method comprise:

- The use of multi-colour hierarchical templates with differentiation between situation and back-ground pixels at different significance levels. (Fig. 1)
- The use of only partially populated templates. This allows to filter out the negative influence on the similarity measure from nearby pixels belonging to other objects.
- The similarity measure between the template and the original raster image is computed by means of a modified Laplace-Distance, which limits the comparison to selected pixels. This increases the recognition rate and at the same time significantly reduces the computational effort.
- The matching process itself is carried out in a step-by-step approach starting with pixels in significant areas of the template. Thus, the comparison will only be computed if the original raster image contains sufficient situation pixels at the current location.

The described method provides a robust procedure particularly suited for cartographic applications. It provides high recognition rates as well as low error rates *(Carosio/Stengele, 1993)*.

Fig. 1: Multi-colour hierarchical template.

2.2. Extraction of Areal Objects

The actual extraction of areal objects is carried out by applying a series of raster-processing operations especially adapted to the characteristics of cartographic data. The main operations applied are: dilation, erosion and noise suppression filtering *(Göpfert, 1991)*. Additional possibilities to enhance the extraction of areal objects are again offered by template-matching operations using square or cross-shaped templates in line with the theory of mathematical morphology. The software package "KAMU" developed at the IGP utilises a macro language system for controlling the raster processing procedure. This allows to write customised script files in order to cater for different cartographic conventions and/or graphical standards.

The implemented solution yields high success rates (usually > 95%) but for ideal results it will still require a manual verification and editing process. A certain number of ambiguities - especially in areas with overlapping cartographic symbols - cannot be resolved with the exclusive use of algorithms. Common problems are: rotated map text features not aligned horizontally and linear cartographic features containing areal components which cannot easily be distinguished from small buildings.

2.3. Inspection of Extraction Results

A visual inspection of the object extraction results is generally recommended, but for certain applications it could also be postponed until the verification of the final results. At this stage, two main types of errors are to be dealt with: a) incorrectly recognised non-areal features and b) not recognised small areal features. The high success rate of the chosen object recognition method limits the manual editing task to an absolute minimum. However, the large number of areal objects per map sheet (approx. 10'000 to 30'000) requires the provision of efficient editing and correction tools for the incorrectly identified elements.

The verification and editing tasks are carried out using a raster visualisation program (RaVis) developed at the IGP. It allows to visualise entire map sheets in 2D or 3D and provides numerous tools, such as head-up digitising. Currently, the editing process is carried out by digitising one point within each incorrectly recognised object. This information is stored in a seed file, which is then used in a subsequent re-colouring process. With this approach, the editing effort for an entire map sheet (size A1) ranges between 1 and max. 4 hours.

Fig. 2: Results from object recognition process (black: recognised objects, grey: eliminated features).

3. Raster-Vector Conversion

3.1. Area-Preserving Contour Extraction and Object Assignment

Due to the stringent accuracy requirements, area loss during the raster-vector conversion step - as it results from most standard contour vectorisation methods - had to be considered in particular. Most standard methods determine the border pixels of areal objects and subsequently form closed polygons connecting the centres of adjacent border pixels. The described approach results in a reduction of the object dimensions in the order of the scanner resolution (e.g. scanner resolution 20 lines/mm ⇨ dimension reduction = 2 × 50% of scanner resolution ≡ 1/20 mm). This area reduction could reach a significant size, especially in cases with limited scanner resolution and in cases where such a raster-vector conversion step might have to be repeated at a later stage.

The implemented solution consists of two steps. In a first step, the exterior and (where present) interior border pixels of the recognised objects are determined. In a second step, the 'raw' object contours are determined and assigned to the appropriate object. The result consists of a closed polygon for each exterior and interior contour of an object connecting the outside corners of the contour pixels (Fig. 3).

3.2. Robust Vectorisation Technique

The chosen vectorisation technique is based on a 'robust straight-line fitting' approach using data points of the raw contour polygons as the observation input. The motivation for using a robust, i.e. a non-least-squares estimator, was its reduced susceptibility to outliers. This allows to simultaneously use the method for the detection of object corners (i.e. discontinuities in the data sets) and for the estimation of the parameters for the contour segments.

Least Absolute Deviation Estimator. The chosen least absolute deviation estimator is a so-called 'robust' estimator and belongs to the class of M-estimates which follow from maximum-likelihood estimates *(Wicki, 1992)*. M-estimates represent the most relevant class for the estimation of parameters and are related to the least-squares estimation method. A 'robust' estimator can be characterised by its low susceptibility to fractionally large outliers for a small number of data points *(Huber, 1981 & Rousseeuw/Leroy, 1987)*.

The least absolute deviation estimator is not the fanciest robust estimator available, but it provides a good combination of reliability, high breakdown point and reasonable computational demands. For the special case of fitting a straight line (3.1.) to a set of data points with this estimator, the following merit function (3.2.) has to be minimised *(Press, et al., 1992)*.

$$y(x;a,b) = a + bx \qquad \text{(a = intercept, b = slope)} \qquad (3.1.)$$

$$\sum_{i=1}^{N} \left| y_i - a - bx_i \right| \qquad (3.2.)$$

Due to its robust behaviour, the least absolute deviation estimator can simultaneously be used to detect object corners, which is equivalent to the detection of discontinuities in the raw data set. The advantage over least-squares procedures is the possibility to add several data points at once (e.g. 20 % of current number of points) without running the risk of affecting the result of the parameter estimation or of reaching the breakdown point of the estimation process.

In parallel to the contour estimation process, the intersections of adjacent line segments are computed, tested and any generalisation operations are applied according to the parameter settings of the vectorisation software "AutoVec". The generalisation functionality allows to clean-up ragged corners, remove too short segments and small 'ledges'.

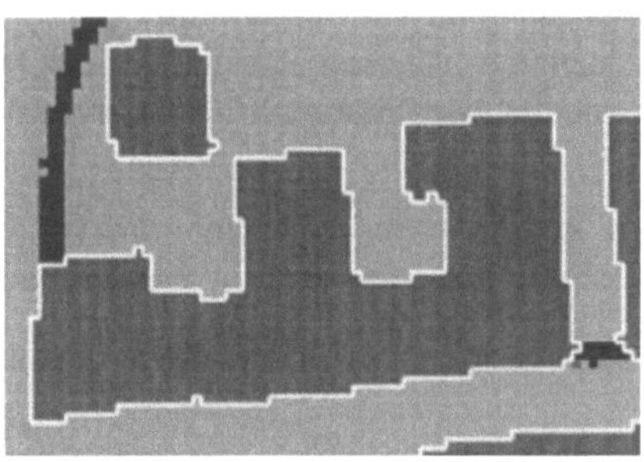

Fig. 3: Raw contour polygons (connecting the individual raw data points).

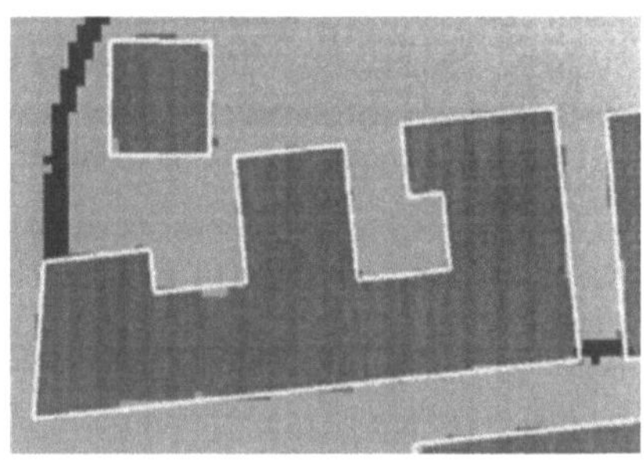

Fig. 4: Automatically adjusted object contour segments (using geometric constraints).

3.3. Adjustment of Contour Segments with Geometric Constraints

Following the parameter estimation, line intersection tests and generalisation operations, an option is provided to adjust the segments of each object using geometric constraints. The supported geometric constraints are: *parallelism* and *perpendicularity*.

In a first step, all segments of an object are tested for parallelism or perpendicularity against other segments and are then grouped accordingly. The tolerance criteria for assigning a segment to a group is based on the maximum across-line distance at the extreme points of the segment. In a second step, the orientation of each such group of parallel and perpendicular segments is estimated using a weighted mean. The weight of each segment is determined in function of its length.

3.4. Quality Control

In the final step of the vectorisation process, the quality of the results is assessed by determining the agreement between the raw data and the adjusted contour segments. This is achieved by: a) computing the offset of each raw data point from the nearest segment and b) computing the offset of each adjusted object corner from the closest raw data point. If any of these offsets within an object exceed a user-controlled tolerance, then the adjusted solution is rejected and replaced by the raw contour solution.

3.5. Results Output

Normally the results of the vectorisation process are provided as closed polygons for each object, each with an object identifier. Depending on the results of the QC procedure, these polygons either consist of the adjusted contour intersection points (Fig. 4) or of the raw contour data points (Fig. 3). Figure 4 shows an adjusted object with right-angle constraints. Alternatively, object reference points - either the centres of gravity or inner points - with object identifiers can be generated. The results can be exported to various data formats, such as DXF. Adjusted and raw objects can be exported on different layers or in different colours in order to facilitate any manual post-processing, head-up-digitising etc.

Objects touching the border of the processed map sheet are treated separately. They are excluded from the vectorisation process and are provided as specially tagged raw contour polygons. In an additional processing step, border objects from different map sheets can be extracted, combined and then be treated as above.

4. Implemented System

The described object extraction functionality was implemented at the Institute of Geodesy and Photogrammetry as part of the following program packages:

- "KAMU" (KArtographische MUstererkennung): Raster data processing software with special emphasis on cartographic pattern recognition. (
- "AutoVec" (AUTomatic VECtorisation): Software package for the automatic vectorisation and structuring of areal objects *(Nebiker, 1994).*
- „RaVis“ (Raster Visualisation): Cartographic 2D and 3D visualisation program supporting raster and vector data, digital terrain models and combinations thereof.

The packages were developed on workstations of the type IBM RS/6000 Model 530 (96-128 MB memory, ca. 32 MIPS) running under AIX. AutoVec and KAMU are also running on the SUN architecture.

The productivity of the automatic object extraction and structuring process largely depends on the density of information on an individual map sheet and on the scanning resolution. Performance figures for processing one entire map sheet (1:25'000, size A1, resolution 20 l/mm, approx. 15'000 buildings) on a relatively slow platform are described in *Nebiker (1994)*. The ratio of correctly recognised objects is approx. 96-98 % and the success rate of the automatic vectorisation is typically better than 99.5 %.

5. Applications

5.1. Overview

The following list should give an indication of some of the possible applications of the automatic extraction of objects from topographic maps:

- digital cartography: establishment of cartographic databases with vector data suitable for (and compatible with) existing map revision procedures (e.g. photogrammetric restitution), establishment of geographic names databases
- photogrammetry and remote sensing: provision of 'a priori' information to support automatic object extraction processes
- planning / architecture: 3D-views of built-up areas, visualisation of major construction projects in combination with existing buildings in the area
- telecom: simulation of signal propagation and decision-aiding tool for selection of transmitter locations (especially in urban areas)
- military: various simulation applications
- statistics: areal statistics, building statistics, construction density

In most of these applications the greatest benefit is obtained by combining the automatically extracted information with digital terrain model information and - most important - with powerful 3D visualisation tools.

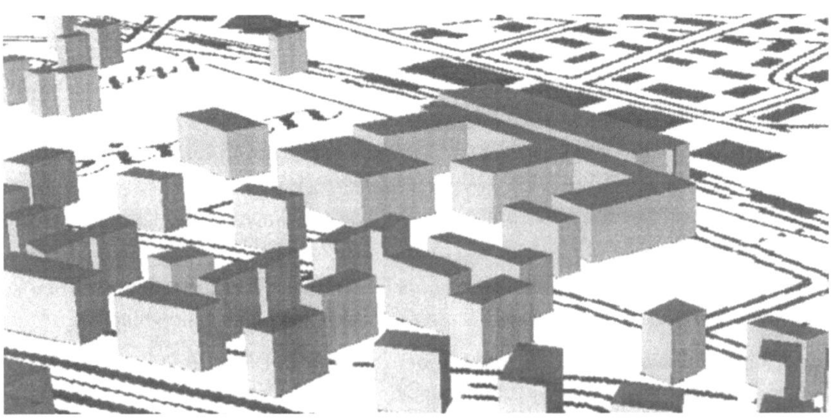

Fig. 5: 3D-map-visualisation with automatically recognised buildings in the foreground.
(Data source: digital topographic map (PK25) and digital height model (DHM25) © Swiss Federal Office of Topography, Bern.)

5.2. Generation of Cartographic 3D-Views

Figure 5 shows a perspective 3D-view on the basis of a scanned topographic map 1:25'000 and the national digital terrain model of Switzerland (DHM25). The 3D-view was generated by the visualisation package RaVis including information from automatically extracted buildings in the foreground.

5.3. Extraction of Buildings for Swiss PTT

The extraction of objects from topographical maps has been funded by the Mobile Communications Department of Swiss PTT in view of a specific application. The requirement was to develop a solution for the automatic extraction of buildings from scanned topographic maps (1:25'000). The resulting vector data is used for the simulation and design of mobile communication networks. A separate software package at the Mobile Telecoms Department allows to model signal propagation based on digital height data and on the results from the automatic building extraction process. Figure 6 shows the automatically extracted buildings for an area of 3.5 km × 3.0 km in the city centre of Bern, Switzerland.

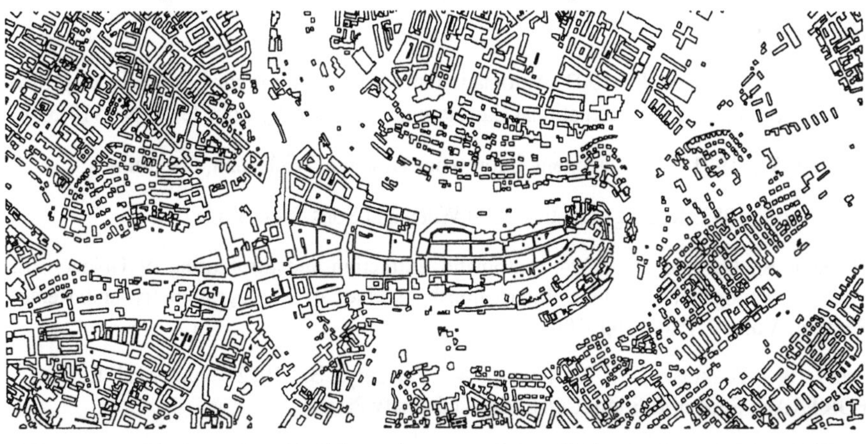

Fig. 6: Automatically extracted buildings, city centre of Bern (Switzerland), vector data (DXF).

6. Conclusions

First production runs have shown that the described method provides an economical, largely automated solution with high recognition rates for reliable and accurate vectorisation results. Thus, the goal of developing an economical solution - with an accuracy superior to that of manual digitising - was achieved. Currently, the object recognition and vectorisation processing steps are best performed in batch-mode, for example overnight. With state-of-the-art workstation technology, however, it is soon becoming possible to perform these tasks on-line or at least in the back-ground, while, for example, performing editing and data handling tasks. The current solution could be further enhanced by a preceding extraction of linear features and of rotated map symbols. Due to the nature of topographic maps with revision cycles of several years, with geometric displacement and cartographic generalisation effects, object extraction from this data source is no equal alternative to future extraction procedures from aerial images. However, object extraction from topographic maps is probably one of the most efficient GIS data acquisition methods available in the short and medium term future and it could also provide a valuable a priori input for the object extraction from aerial and space images.

7. Outlook

The current investigations at the Institute of Geodesy and Photogrammetry in the field of extraction of objects from topographical maps focus on the following main issues:

- Recognition and elimination of linear cartographic features with a complex symbology (e.g. highways, railway lines).
- Recognition and elimination of rotated map symbols, text elements and figures.
- Development of additional editing functionality specifically suited for object extraction applications.
- Integration of map based object extraction functionality into future raster DBMS.

References

Carosio, A., Stengele, R., 1993. Automatic Pattern Recognition for Economic Map Revision. Proceedings, 16th International Cartographic Conference, Köln 1993, pp. 518-526.

Göpfert, W., 1991. Raumbezogene Informationssysteme. Wichmann, Karlsruhe.

Huber, P.J., 1981. Robust Statistics. John Wiley and Sons, New York.

Nebiker, S., Carosio, A., 1994. Automatic Extraction and Structuring of Objects from Scanned Topographical Maps. International Archives of Photogrammetry and Remote Sensing, Vol. 30, Part 1, pp. 180-187.

Press, W., Teukolsky, S., Vetterling, W., Flannery, B., 1992. Numerical Recipes in FORTRAN, Second Edition. Cambridge University Press.

Rousseeuw, P.J., Leroy, A.M., 1987. Robust Regression & Outlier Detection. John Wiley and Sons, New York.

Stengele, R., 1993. Kartographische Mustererkennung durch Template Matching. Bericht Nr. 230, Institute of Geodesy and Photogrammetry, Swiss Federal Institute of Technology, Zurich.

Stengele, R., Carosio, A., 1994. Cartographic Raster Data and their Use by GIS. Proceedings, FIG, XX. International Congress Melbourne, Australia, pp. 550ff.

Wicki, F., 1992. Robuste M-Schätzer und Zuverlässigkeit. Bericht Nr. 190, , Institute of Geodesy and Photogrammetry, Swiss Federal Institute of Technology, Zurich.

Cooperative use of aerial images and maps for the interpretation of urban scenes

**Henri Maître, Isabelle Bloch,
Henri Moissinac, Christophe Gouinaud**

Ecole Nationale Supérieure des Télécommunications, Département Images,
46 rue Barrault, 75634 Paris Cedex 13, France.
Ph.: +33-1-45 81 76 55, Fax: +33-1-45 81 37 94
e-mail: maitre@ima.enst.fr

Abstract

We present a methodology for the interpretation of aerial images on urban areas. It is based on a careful examination of the most relevant elements used by a human person to understand such complex scenes. These elements are well reflected in a geographical map. For this reason it seems of prior interest to manage simultaneously cartographic and pictural information.

1 Towards another image interpretation

Conventional image interpretation aims at providing a partition of the image into objects with a label attached to every object. The more semantic the label content is, the deeper the interpretation. Conventional image processing techniques associated with conventional image interpretation are based on a bottom up detection of primitives which are gradually associated or combined to build higher level primitives. Possible feed-back loops are used to solve dead-end situations when no upward issue is foreseeable. In more complex systems a complex tree of hypotheses may be developed to keep track of competitive hypotheses.

When quite complex situations arise, the limits of the conventional approach are soon reached with the present-time state of software capabilities. Two reasons may explain this failure.

- At first the complexity of the image message is such that there is not enough information in it to "explain" everything that is, for instance, deduced by a human observer. This is typically what happens with a urban scene seen from above (see figure 1). With a reasonable resolution (say 1 meter per pixel), the main features which constitute the basis of the interpretation (the roads, the buildings) are hidden by many elements which make them very difficult to detect with a good reliability: as far as streets are concerned for instance, they are often hidden by high buildings or trees, and their appearance may be confused because of cars, road marks, or urban furniture.
- On the other hand, the multitude of representations of the scene provides several different interpretations for a same object, or for the association of several objects, which make one interpretation not only multiform but also questionable. Let us take again the image of a city: when an aerial image suggests a vaste field, a map may indicate it as a school yard; on the contrary, when the map says that a block is an admistrative building, a stereo pair will differentiate several different houses with different heights. This clearly illustrates that the interpretation of a complex scene strongly depends on the information we have on it. It should take into account the origin and the nature of the documents, but also the date of their acquisition, which will be necessary to solve possible conflicts.

The work which is presented here is proposing a solution to the interpretation of urban scenes from several different image sources: aerial images or geographic maps. The purpose is to provide the user with a synthetic interpretation of the different sources, which not only collects the different elements which may be derived from every source, but also takes benefit of an information obtained from a source to help deriving new results from another source, thus performing a complex fusion of different pieces of information. We will show that a coherent description of the scene is of prime importance to guarantee the efficiency of the process, requiring a consistent uncertainty management context. In the second part, the logical framework of the representation and the methodological guideline for the interpretation will be introduced. In the third part, the knowledge management is described. The fourth part is devoted to the picture processing and the map processing levels, the last part presents some results.

2 A methodological framework for urban scene interpretation

2.1 How is a urban area image interpreted

Despite the scientific interest of conventional aerial image interpretation techniques [6], [3], [4], [2], it is clear that a bottom-up approach to the interpretation of a dense urban area is not yet able to provide useful results for most of the present demands of automatic cartography. The reason of this failure may be seen at different levels: the insufficient development of low level processing softwares, the complexity of the task and the variety of its formulations depending on geographic locations, the poor experience we have on this problem, etc. But we are convinced that much has to be done in proposing a clear and efficient strategy for the interpretation which makes the best of the main feature which characterizes the urban scene: its global coherence. We will base our recognition strategy on this coherence which may be stated in the following words [8]:

- the structure of the town is reflected by the network of the streets;
- this structure may often present a geometrical distribution which is typical from geographical and historical conditions: regular square grid, radially distributed structure, concentric structure, etc.;
- the street network has to verify certain functional properties which may vary from city to city: connectedness of the graph, orthogonal crossings, regular distances, etc.;
- the dual graph of the street graph is the block graph which carries most of the "complementary" information on the town;
- within an elementary cycle of the street graph the city block is itself a substructure with placement rules of its own, which may give rise to a within-block graph, where structures like individual houses, gardens, alleys may be represented;
- the within block structures are ruled by their own functional properties which garantee the internal coherence of the block.

2.2 The structural description

These rules are certainly not immediately reflected by an aerial photo, they proceed from the social and economical role of the town and could be said as "internal" coherence rules. They emerge obviously from a careful examination of a map which is a simplification of reality preserving (and enhancing) the main functional attributes of the city (figure 1). Thus we commend to base the description of the urban area on the description issued from a map and we will derive our analysis scheme from the structural scheme of a map which provides exactly the skeleton of the analysis which is needed to preserve the coherence of the urban web [8], [9].

The main information comes from the street and the road networks. Therefore we base our town description on a road-graph $\mathcal{G}_{road}$. The dual of this graph is called the block-graph and is denoted $\mathcal{G}_{block}$. Each node of the block-graph may be developed as a graph, the within-block-graph $\mathcal{G}_{w-b}$, describing the way the components of the block are distributed. For one image source only, the global coherence of the description may be checked by a careful examination of the set of graphs: $\mathcal{S} = \{\mathcal{G}_{road}, \mathcal{G}_{block}, \mathcal{G}_{w-b}\}$ under the light of the structural rules which have been previously listed.

2.3 The multisource information

Another important property of the urban scene description is the variety of the description arising from different sources of information: geographical maps or photos taken at different times. These sources are at the same moment independent (since they may reflect different aspects of the scene, some being possibly conflictual), and dependent (since one information in one image may only take sense because of the simultaneous presence of another information in a second source). The global interpretation of the scene cannot easily be reduced to a single description, it is made of the collection of the many graph-sets $\mathcal{S}^i = \{\mathcal{G}^i_{road}, \mathcal{G}^i_{block}, \mathcal{G}^i_{w-b}\}$ built for every source i. In order to allow a practicable comparison of these many descriptions, some conditions are introduced which constrain all the graphs of a similar nature (for instance all the road graphs $\mathcal{G}^i_{road}$ for different i) to be homotopic [8].

3 Information management

The information which is used to interpret the scene is coded in two ways :

- the functional (or operational) information is stored in the structure of the analysis (the graph representation), and in the way the different low-level primitives are combined to provide higher level ones;
- the factual information comes from the sources and is extracted with conventional picture processing tools.

In order to allow the management of information, each feature x which is manipulated by the user has a measure of confidence attached to it which is called reliability and is denoted by $Re(x)$. When the feature is obtained from the combination of lower level features, its reliability is made of the composition of the low-level primitive reliabilities and a coherence term. A special attention is paid to the combination of the reliabilities between the different graph elements [9].

4 Picture processing

4.1 Map processing

Most of the structural information which is used for the scene interpretation comes from the map. In other cases it is possible to have it directly from a Geographic

Information System or a data-base, but such systems are not everywhere available. We make use of scanned maps. Many different features are extracted from it:

- the road-network,
- the nature of the blocks: high or low buildings, cottages, parks, sports-fields, etc.
- the internal structure of the blocks: individual houses, yards, alleys,
- the names and the written information.

This information is mostly obtained by an image segmentation using a color clustering, thresholdings and combined mathematical morphology operations. This stage feeds the map graphs: $\mathcal{S}^{map} = \{\mathcal{G}^{map}_{road}, \mathcal{G}^{map}_{block}, \mathcal{G}^{map}_{w-b}\}$. We see on figure 1 the graphs resulting from the map segmentation [7].

4.2 Image processing

Several picture processing stages are combined to extract features to build the image set $\mathcal{S}^{image}$.

Our attention is first focused on the road network. Several processes may be used in order to extract the streets. None is powerful enough to provide a reliable result on our complex areas. Thus we constrain the research to be made close from the network which has been proposed by the map. This may be done in several concurrent ways: one makes use of snake-like techniques [7], another one uses a Kalman filter [11], a third one is based mostly on classifications [8]. These processes result in several different road graphs which may be fused according to the respective reliabilities of each feature.

The blocks are analyzed using cooccurrence matrices in order to take advantage of the different textural information due to different block contents[10]. We see on figure 2 the comparison of several cooccurence matrices for different natures of blocks.

When a pair of stereo images is available, a matching is made in each block to get 3D information on the buildings according to the method proposed in [5]. This allows to extract several types of information: the average height of the block (this will secure the previous classification), the height of individual houses (this information, although poorly reliable, may be used for the within-block graph) [12]. We see on figure 3 the result of a stereo matching proposing the height of buildings.

5 Results

We experimented the method in a suburb of Paris, with 2 aerial images (IGN photos, scale 1:20000), and one geographic map (IGN scale 1:25000). We added the ground truth collected by visiting each building and noting the main features of interest (number of levels, type of roofs, building sites, etc.). We have only developed the road- and block-graphs up to now, the within-block-graph having not yet been filled with data.

After a coarse registration of the photos on the map, the road-graph of the map is transferred to the photos. Taking into account the nominal accuracy of maps [1], we build the image road-graph, i.e. we distort this graph to adapt it to the image signal [8]. For each street, after a competition of the three possible road detectors, we keep the best solution (according to the reliability criterion). On figure 4 we see the result of the detection where the final reliability of the best candidate is also shown.

The inter-source coherence on the ground occupancy has not yet been completely tested between maps and images, because the processing of the image is not yet fully automated. However, first results are presented on figure 5, and show a few differences between image block classification and map classification.

Comparison between the map and the ground truth collected shows the evolution of the ground occupation (since there is virtually no error in the processing of the map or in the evaluation by the human expert on the ground).

6 Conclusion

We have presented a global scheme to process several sources of information on a urban area. This scheme is proposing a coherent framework for the management of information. A great attention is paid to the propagation of the confidence from low-level to high-level, and to the maintenance of coherence between different sources.

Acknowledgements: we would like to thank Jaime Lopez-Krahe, Michel Roux, Jean-Pierre Crettez, Philippe Guérin, Sophie Paquerault and Natacha Zamani for their participation to this project. This project has received a support from Thomson LCR, the Ecole Normale Supérieure and the CNET.

References

[1] **P. Guérin, O. Jamet, H. Maître :** *Distorsion Model in Road Networks from Topographic Maps: Identification and Assessment*, SPIE Aerospace Sensing, Orlando, April 1995.

[2] **A. Huertas, W. Cole, R. Neviata:** *Using Generic Knowledge in Analysis of Aerial Images: A Case Study*, 11th International Joint Conference on Artificial Intelligence, 1642-1648, 1989.

[3] **D. M. Mc Keown. W. A. Harvey, J. Mc Dermott:** *Rule based Interpretation of Aerial Imagery*, IEEE Trans. on Pattern Analysis and Machine Intelligence, Vol. 7, No-5, 570-585, 1985.

[4] **D. M. Mc Keown:** *The Role of Artificial Intelligence in the Integration of Remotely Sensed Data with Geographic Information Systems*, IEEE Trans. on Geoscience and Remote Sensing, Vol. 25, No-3, 330-348, 1987.

[5] **H. Maître, W. Luo:** *Using Models to Improve Stereo Reconstruction*, IEEE Trans. on Pattern Analysis and Machine Intelligence, Vol. PAMI-14, No-2, 269-277, 1992.

[6] **T. Matsuyama:** *Expert Systems for Image Processing, Knowledge-based Composition of Image Analysis Processes*, 9th Int. Conf. on Pattern Recognition, 125-133, Rome, Nov. 1988.

[7] **H. Moissinac, P. Gurin, I. Bloch, H. Matre** : *Image Interpretation Using Symbolic Data*, EUSIPCO 94, Edinburgh, 431-434, 13-16 Sept. 94.

[8] **H. Moissinac, H. Maître, I. Bloch:** *Urban Aerial Image Understanding Using Symbolic Data*, EUROPTO Conference on Image and Signal Processing for Remote Sensing, Rome, SPIE Vol. 2315, 310-321, 26-30 Sep. 1994.

[9] **H. Moissinac, H. Maître, I. Bloch:** *Graph Based Urban Scene Analysis Using Symbolic Data*, SPIE Aerospace Sensing, Orlando, April 1995.

[10] **S. Paquerault:** *Reconnaissance de zones urbaines sur des critères de textures en imagerie aérienne*, DEA Technical Report, ENST, 1994.

[11] **J.P. Veran:** *Suivi de routes dans une image aérienne par filtrage de Kalman*, Technical Report ENST, 93D007, 1993.

[12] **N. Zamani:** *Traitement d'images aériennes : calculer les hauteurs dea bâtiments en milieu urbain*, DEA Technical Report, ENST, 1994.

Figure 1: Geographic map, aerial picture and the final interpretation built, based on the joint examination of the road network (in black) and the block adjacency graph (in grey, with different colored squares for the different kinds of city blocks: parks, factories, suburban houses, large high buildings, dense low constructions ...).

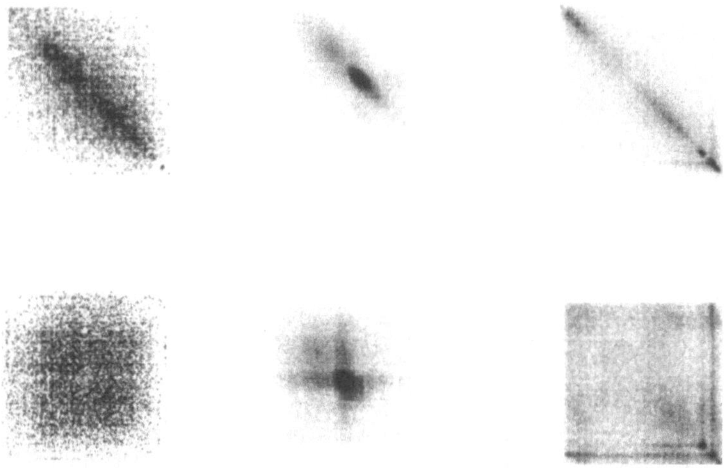

Figure 2: Cooccurrence matrix of different blocks: from left to right, suburban houses, park, large and high buildings, vector displacement of one (first line) and of eight pixels (second line)

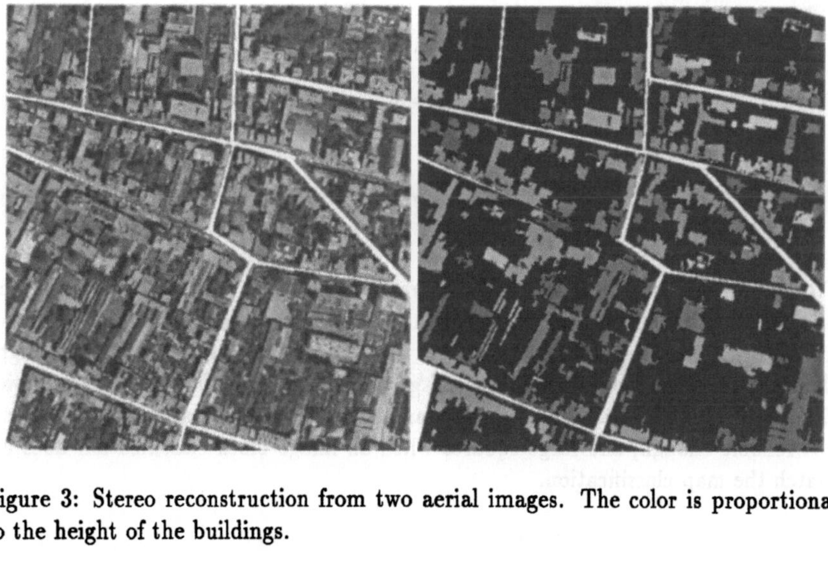

Figure 3: Stereo reconstruction from two aerial images. The color is proportional to the height of the buildings.

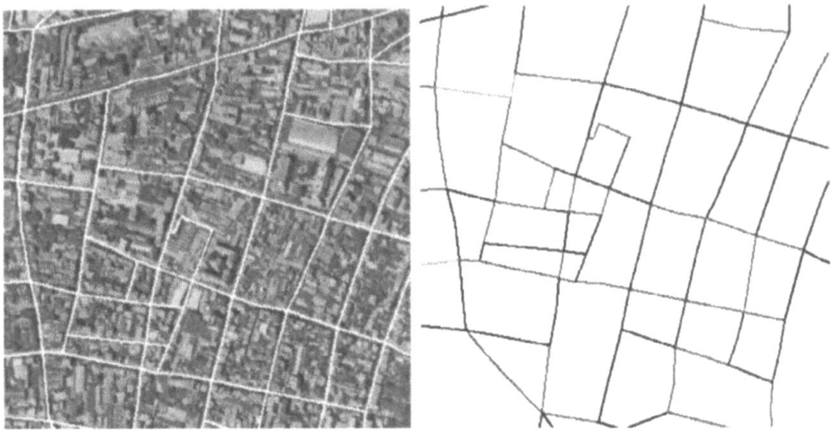

Figure 4: Road network detection on the aerial picture. A measure of confidence is evaluated automatically for each road arc in order to illustrate the final reliability of the presented results (high in dark, low in bright).

Figure 5: Aerial picture (left) and its analysis (middle): the road graph detected, printed with the reliabilities of the arcs and the block classification (suburban houses in black, large buildings or parks in dark grey, dense buildings in light grey).Right: the final analysis: merging all the road graphs detected provides a better solution (improved reliabilities). Small squares are printed when a final image road arc is not reliable enough, and large squares when an image block classification does not match the map classification.

Map–based semantic modeling for the extraction of objects from aerial images

Franz Quint, Manfred Sties

Institute for Photogrammetry and Remote Sensing
University of Karlsruhe
D – 76128 Karlsruhe, Germany
Ph.: +49-721-6083092, Fax: +49-721-694568
email: quint@ipf.bau-verm.uni-karlsruhe.de

Abstract

Images taken from satellite or airborne platforms usually do not represent isolated information of man's environment. In most countries, valuable context data are available which may be integrated successfully in the image interpretation procedure. This paper presents the verification phase of a *M*ap *O*riented *SE*mantic image under*S*tanding process[1] (MOSES). It is implemented as a model driven process, where semantic networks are used as modeling tools. In a three stage scheme, the models are successively refined and for image analysis an automatically generated semantic network, specialized on the analysis of the underlying scene is used. Digitized topographic maps serve as a principal knowledge source.

1 Introduction

The analysis of aerial and satellite images for exploration and monitoring of our environment gains rising importance. In many cases, it is the only applicable method for solving given problems. As soon as the reliability of the automated processing schemes increases, they are advantageous under cost and time aspects. Therefore, it is not surprising, that the international community supports big efforts for developing efficient and reliable image understanding procedures.

These efforts are encouraged by the fact, that in near future Geo-Data will be available in digital form; however, this is not yet generally the case except for Satellite Remote Sensing Imagery. At the moment many national surveying and cartographic administrations are establishing digital data bases of conventional topographic maps in vector form (see e.g. Harbeck, 1994).

[1]The research is funded by the Deutsche Forschungsgemeinschaft (DFG)

Understanding of aerial images is one of the challenging problems in computer vision. Due to its complexity, knowledge based systems have been found to be mandatory since the mid seventies: Agin (1979) developed a knowledge based system for detection of roads and vehicles in aerial images.

A rule-based aerial image understanding system (SPAM) for interpretation of aerial scenes has been presented by McKeown et al. (1985). Several production rules are applied to generate partial interpretations, and then a set of rules for consistency examination is applied.

A blackboard based production system (BPI) is used by Stilla and Hajdu (1994) for a map-aided analysis of structures in aerial images. Based on the knowledge obtained by map analysis, estimations for attribute values of objects are defined. Objects found in the image analysis process are assessed relative to the hypotheses generated by map and object model.

The knowledge based system for aerial image understanding presented by Nicolin and Gabler (1987) is composed of four components: long term memory, short term memory, procedures and control module. In a first, data-driven stage, initial interpretations and assessments are assigned to striking structures in the image. In the second, model-driven stage, hypotheses for not yet classified objects are build using a-priori knowledge and the results of the first stage.

The system SIGMA developed by Matsuyama and Hwang (1990) consists of four loosely coupled reasoning modules. The *Geometric Reasoning Expert* as the central module of the system stores knowledge in frames. Object classes are described with attributes and relations and instantiated objects can generate hypotheses for neighboring objects.

In the system MESSIE (Sandakly, Giraudon, 1994) knowledge is also stored in frames and the communication between the *specialists* takes place using a blackboard mechanism. A specialist starts processing after a query by another specialist (model-driven phase) or when it is triggered by events (data-driven phase).

In this paper the verification part of MOSES (*M*ap *O*riented *SE*mantic image under*S*tanding) is presented. We incorporate knowledge from the map domain for image analysis. Using a semantic network for analysis of the map, a description of the scene in map domain is build. Another semantic network specialized in the analysis of the underlying scene is automatically generated using the scene description in map domain and a generic model in image domain. This later network is used for image analysis.

2 Image and map data

The methodologies for map-based verification and recognition of objects are developed for urban scenes. As a test area, a sector of the urban environment of the city of Karlsruhe was selected, which contains both typical metropolitan densely populated areas as well as extended park and forest areas.

The digital image data were acquired by scanning aerial color photographies of 230 by 230 mm. The aerial images were taken from a flying height of 2 km, at a scale of 1:6300 and scanning of the photography was done with 50 μm spot size and a grey value resolution of 8 bit.

The context information was acquired from a topographic map, the German Topographic Base Map 1:5000. The corresponding sector to the aerial images was acquired by manual digitization of the contours and is stored in the data base of GIS ARC/INFO. To each digitized contour a class is assigned.

The transformation parameters between the coordinate system of the images and the world coordinate system are determined by an adjustment from the ground control coordinates. The transformation parameters allow to project the map information onto the image. Figure 1 shows the digitized contours from the map as an overlay (white lines) on the image of a scene.

Fig. 1: Extract from the aerial image with map information as overlay (white lines)

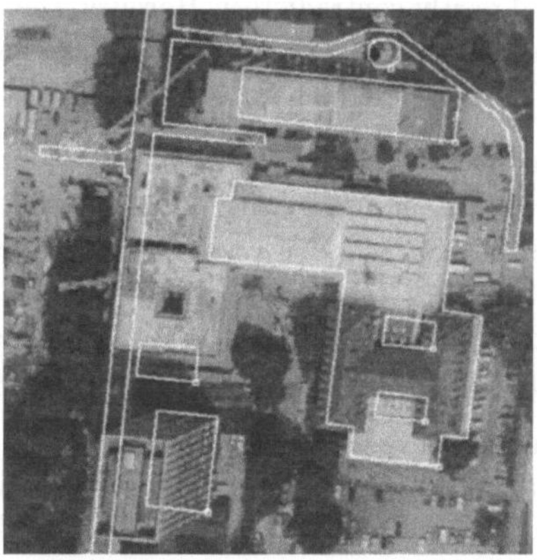

Fig. 2: Differences between the aerial image and the map information (white lines as overlay)

This examples shows, that the object borders acquired out of the map are not at all identical with the contours visible in the image. There are multiple reasons for this effect:

- The map does not include height information. Therefore transformation parameters were only modeled on a 2D-base.
- The perspective distortions in the image, which are extreme especially for tall buildings at the image borders.
- The map shows the ground plans of the buildings, whereas in the images the border lines of the roofs are visible. They necessarily do not fully match.

- Occlusions between objects occur.
- Inaccuracies during the digitization and ground control point determination.

Due to the different acquisition dates of map and image, there are also differences because of the changed reality. These differences are frequent in build-up areas, especially due to activities in construction (Fig. 2). A system for image analysis has to tolerate the differences in representation of the same object, but has to detect the differences due to changed reality.

3 Verification and classification

In general, aerial photography is more economic and faster than the production of new maps. Automatic updating of maps on the basis of aerial images is desirable. As previously shown, the representations of a scene in map and image are different and partly contradictory. However, the map contains knowledge about the scene and can be used as a (incomplete) model for image interpretation. In achieving our goal of image understanding, we define two phases: verification and classification phase.

In the verification phase we rely on the contents of the map: we use the map as a model for image interpretation and verify the map contents. The result of the verification phase is a description of the image, in which for each object of the map, features extracted from the image are measured and an assessment is computed, how good these features match with the model. This description is still incomplete, since it contains only the objects represented in the map.

In the subsequent classification phase, the verification result is used as context for classifying the objects, for which a satisfactory verification could not succeed, and for classifying image primitives not addressed in the first phase.

In this paper we will discuss only the verification phase. The image analysis process for verification of the map contents is conceived as a model driven algorithm. A main component of the model is the map contents to be verified. As a modeling environment, the semantic network system ERNEST (Niemann et al., 1990) is used. For a general discussion of semantic networks, see e.g. Findler (1979). During the analysis process, several models are build and each model is represented in its own semantic network. Because of this unique correspondence between a model and a semantic network, these terms are used synonymously in the following.

4 Usage of map knowledge

The mechanism of inheritance and the hierarchies in semantic networks allow an efficient modeling and use of domain knowledge for pattern recognition tasks.

Using only general, common sense knowledge will lead to simple semantic networks, which are comparatively easy to build. However, we have to accept the drawback,

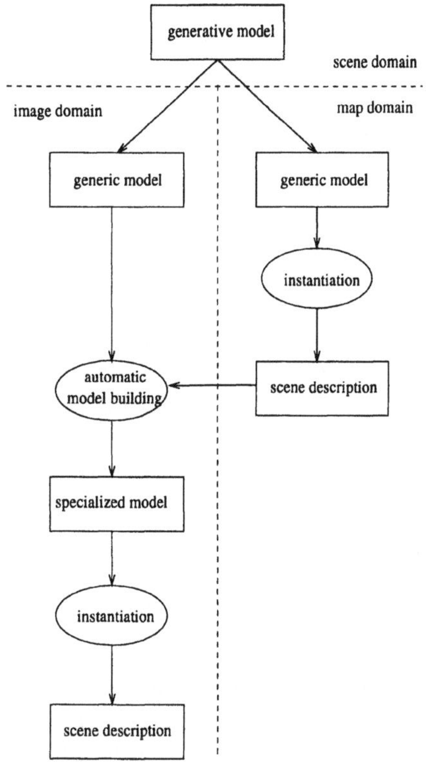

Fig. 3: Structure of the analysis process

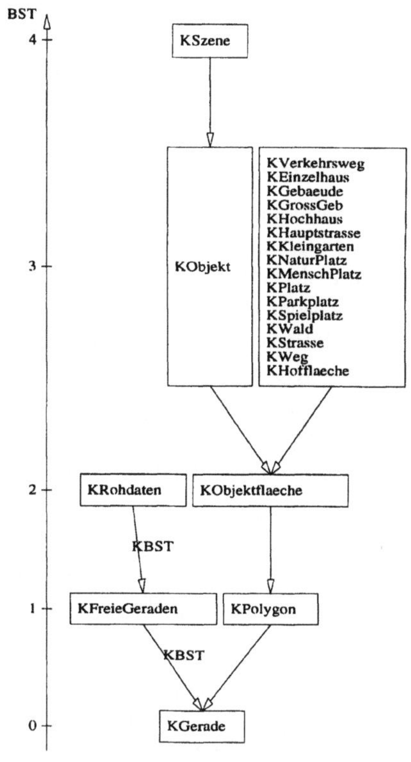

Fig. 4: Part-of hierarchy of the generic model in map domain

that these models in general are not as powerful for the image analysis task as specialized models. Building these requires bigger efforts and they will perform well on the specialized problem they were build for, but generally they will have a poorer ability of generalization.

We present a three stage approach for modeling and use of knowledge: relatively simple, general models are specified by the system designer, whereas specialized models for image analysis are automatically build using the results of map analysis.

The structure of the analysis process is shown in figure 3. The models and the results of an analysis process (scene descriptions) are represented as rectangles, whereas processes are represented as ellipses. The direction of the arrows indicates the direction of data flow.

We distinguish between three domains: the scene domain, the image domain and the map domain. The scene domain is our environment, the real world. The map domain and the image domain comprise representations of the real world in map and image, respectively.

The generative model is a semantic network, which describes our environment in scene domain. The knowledge contained in it is general, common sense knowledge we have about our environment. Objects of the real world are described, as they are perceived by the system designer, regardless of their representation in a map or an image. Since analysis will not be performed in scene domain, there is no need for procedural knowledge in this semantic network.

The generic models in the map and in the image domain describe the mapping of the scene in map and image domain, respectively. Therefore, they roughly contain the same declarative knowledge as the generative model. Both generic models are specializations of the generative model, they reflect, however, particularities of the representations in map and image domain: Since in the map, for example, vehicles are not represented, during the specialization of the generative model to the generic model in map domain, the *part-of* link in the semantic network to the concept representing a vehicle is marked as non existent. In the image, vehicles can be observed and the link to the vehicle-concept is preserved in the generic model in image domain.

Being specializations of the generative model, both generic models describe a general, typical scene and therefore they can be used for analysis of various kinds of scenes. The part-of hierarchy of the generic model in map domain, as drawn by the network analysis tool of ERNEST, is given in figure 4.

Since analysis in image and in map domain is desired, procedural knowledge is present for both generic models. Part of the procedural knowledge are functions for feature extraction, for the calculation of attribute values from image or map data and functions for calculation of confidence values. The procedural knowledge is specific for map domain and for image domain, but it is not scene-specific.

Although both generic models can be used for analyzing the scene in the respective domain, we use only the generic model in map domain for analysis. Because map contours are available in digital form, fault tolerant feature extraction procedures are not necessary. It is expected, that building of the scene description in map domain can be performed with higher reliability than the one in image domain using the unspecific generic model.

The instantiation process in map domain gives us a complete description of the scene in the limits of the map contents. Using this description and the generic model in image domain, we automatically build a new semantic network, which is a specialized model of the concrete scene to be analyzed. Whilst the generic model contains statements of the form:

- buildings are optional parts of the scene, or
- a building may have inner courts,

the specialized model of the actual scene to be analyzed contains precise information about the objects of the scene, e.g.:

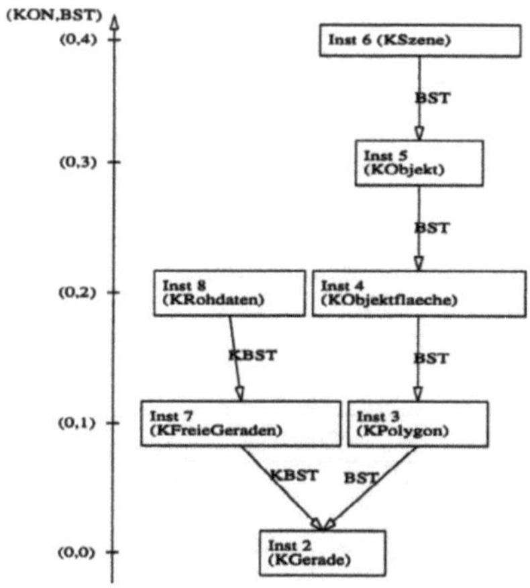

Fig. 5: Intermediate scene description graph

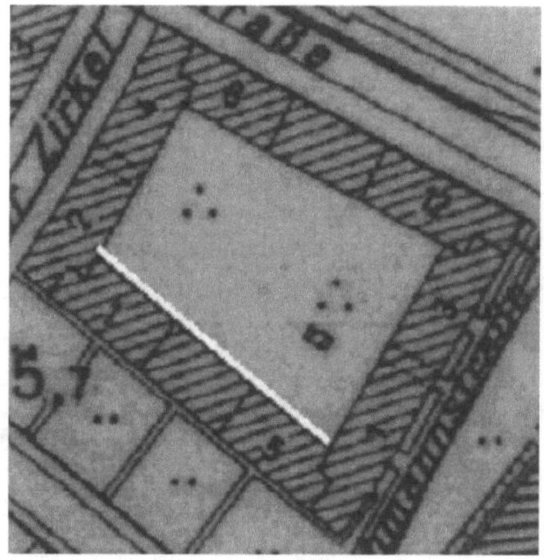

Fig. 6: First instantiated line as overlay in the map

- building number 5 is rectanglarly shaped and the coordinates of corner points are such and such.

The specialized model is now used for image analysis. Because of the concrete modeling of the expected objects in the scene, using this model for image analysis will give more reliable results than using the generic model. Being automatically generated, expense for the system designer is not increased.

For verification of the map objects in the image, we do not compare the results of an analysis process in map domain with those of an analysis process in image domain, where both processes are performed independently, but we use the knowledge obtained from map analysis for the image analysis process.

5 Map analysis

We present some fundamental aspects of the analysis process (instantiation) in our semantic network by an example of map analysis.

Instantiation is an alternate top-down and bottom-up processing. It starts top-down: The target concept, which in our case is *KSzene* (Fig. 4) and represents the scene to be analyzed, is submitted for instantiation. Since *KSzene* depends on other concepts, its parts, a final instance of it can be created only when all its parts are instantiated. The *part-of* hierarchy is tracked down until a basic concept is found, which can be instantiated by directly computing its attributes from available data. In the following bottom-up phase the concepts, for which previously it was not possible, are now instantiated. An intermediate scene description graph showing the situation after the first cycle is given in figure 5. In figure 6, the line corresponding to instance no. 2 (*KGerade*) of figure 5 is overlayed in white onto the map.

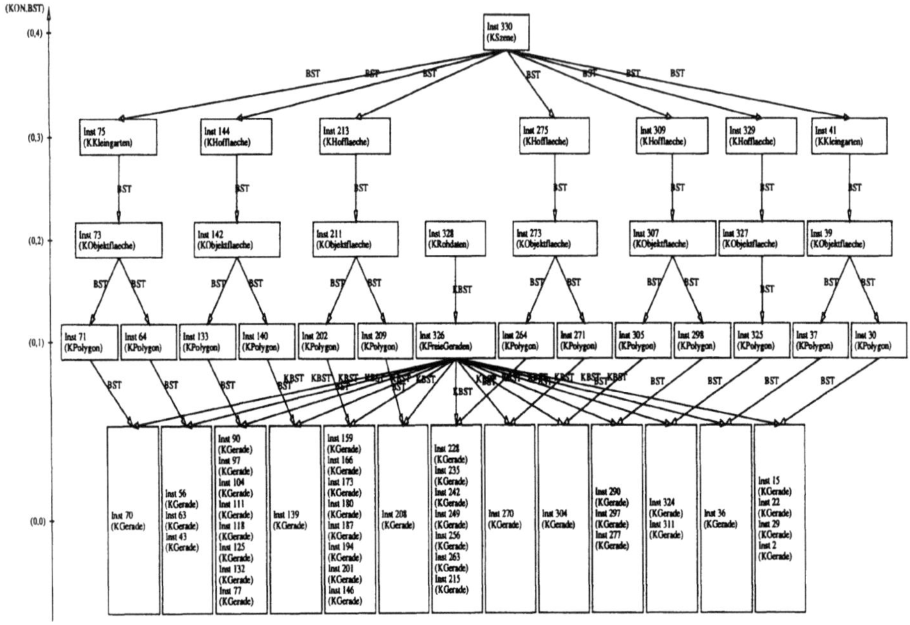

Fig. 7: Scene description graph after map analysis

In top-down phase, constraints are propagated. For example, the coordinates of the next line to complete the contour of the courtyard (Fig. 6) must satisfy the constraints imposed by instance no. 2.

Instantiation continues until no more digitized contours are available. The resulting scene description graph contains the knowledge extracted from the map in a structured fashion. Figure 7 shows an example of such a scene description graph at the end of a map analysis process. This scene description graph is used for automatically building the specialized model for image interpretation.

6 Specialized model for image analysis

The program for automatically building the specialized model for scene analysis relies only on the supposition, that the generic model in map domain and the generic model in image domain are two semantic networks, which originate from the same parent, the generative model. Beginning with the target concept of the scene analysis process, for each concept in generic model in image domain, instances of corresponding concepts in the generic model in map domain are searched in the scene description and for each correspondence found, a new concept in the specialized model is created. Using the part-of hierarchy of the network and of the scene description graph, this is done recursively.

The attributes and procedural knowledge attached to the newly created concepts

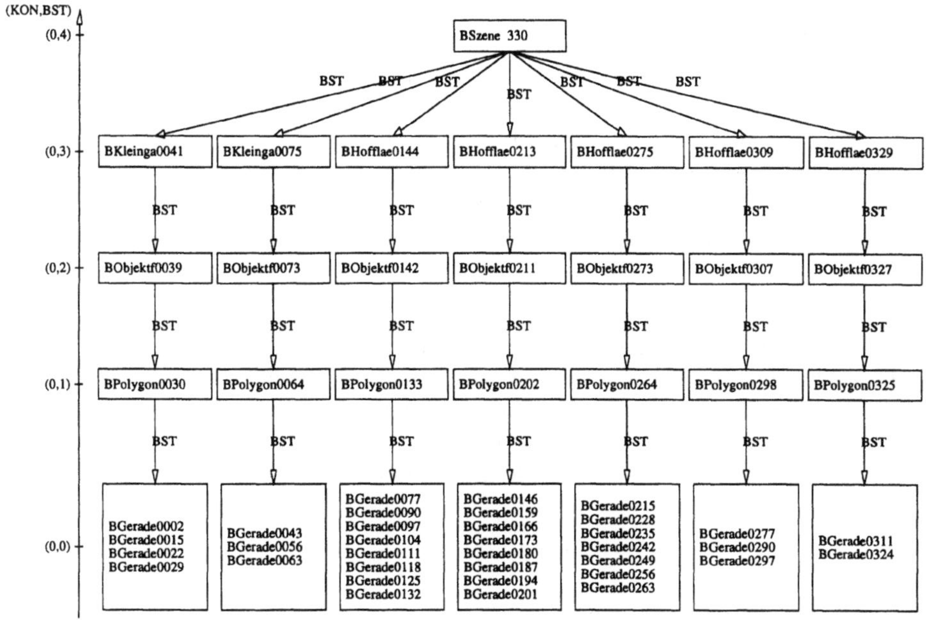

Fig. 8: Part-of hierarchy of the specialized model for image analysis

are gained from the generic model in image domain. The values of the (mainly geometric) attributes of the instances build during map analysis are transformed from map domain to image domain and stored as constraints of the corresponding attributes in the concepts of the specialized model. These constraints lead to a substantial reduction of search space during image analysis.

The part-of hierarchy of the specialized model for image analysis of the scene with the description graph in figure 7 is given in figure 8. The names of the concepts are derived from the names of the concepts in the generic model, to which the number of the instance in scene description was appended. One can observe the correspondence between the part-of hierarchy of the newly created semantic network and the scene description graph after map analysis.

Because of an implementation detail, not all instances in the scene description have a correspondence in the specialized model. It is a characteristic of the semantic network ERNRST, that calculation of attributes is done during instantiation. And even if no proper attribute values satisfying the constraints are found, the instance remains in the resulting description graph. However, these instances are marked as not corresponding. This is the case e.g. for instance no. 70 *KGerade* or no. 71 *KPolygon* in figure 7. For these instances no corresponding concepts in the specialized model are build.

7 Summary and conclusions

We presented a new scheme for image analysis using contextual knowledge. The interpretation task is divided in two phases: verification and classification phase. As a modeling tool, we use semantic networks. Starting with a very general model (generative model), we build two models describing the scene in map and image domain (generic models). The model in map domain is used for analysis of the map. The resulting scene description graph is combined with the generic model in image domain and a new semantic network, specialized in the analysis of the concrete scene is built.

This procedure is very effective, since map knowledge is automatically incorporated in the analysis process. The resulting specialized model is tailored for image analysis of the underlying scene. Presently, only line segments are used in the image analysis process. We plan to integrate region-based features for better grouping of line segments and experiment with different assessment procedures in the generic model in image domain.

References

Agin, J. (1979) *Knowledge based detection and classification of vehicles and other objects in aerial images*, Proceedings of the DARPA Image Understanding Workshop, Palo Alto, CA, pp. 66-71.

Findler, N.V. (ed.) (1979) *Associative Networks*, Academic Press, Orlando.

Harbeck, R. (1994) *Das Geoinformationssystem* ATKIS *und seine Nutzung in Wirtschaft und Verwaltung*, Landesvermessungsamt Nordrhein–Westfalen, Bonn.

Matsuyama, T., Hwang, V. (1990) SIGMA*: A knowledge based aerial image understanding system*, Plenum Press, New York, London.

McKeown, D., Harvey, W. and McDermott, J. (1985) *Rule based interpretation of aerial imagery*, IEEE–PAMI, Vol. 7, No. 5, pp. 570–585.

Nicolin, B., Gabler, R. (1987) *A knowledge based system for the analysis of aerial images*, IEEE Transactions on Geoscience and Remote Sensing, Vol. 25, No. 3, pp. 317–329.

Niemann, H., Sagerer, G., Schröder, S., Kummert, F. (1990) ERNEST*: A semantic network system for pattern understanding*, IEEE–PAMI, Vol. 12, No. 9, pp. 883–905.

Sandakly, F., Giraudon, G. (1994) *Multispecialist system for 3D scene analysis*, Proceedings of ECAI 94, J. Wiley & Sons Ltd., pp. 771–775.

Stilla, U., Hajdu, A. (1994) *Map-aided structural analysis of aerial images*, Proceedings of the ISPRS Commission III Symposium, München, Vol. 30, Part 3/2, pp. 769–775.

List of Workshop Participants

Peggy Agouris, Institute of Geodesy & Photogrammetry, Swiss Federal Institute of Technology, ETH - Hoenggerrberg, CH-8093 Zurich, Switzerland

Emmanuel Baltsavias, Institute of Geodesy & Photogrammetry, Swiss Federal Institute of Technology, ETH - Hoenggerrberg, CH-8093 Zurich, Switzerland

Meir Barzohar, RAFAEL, 12 Lincoln Street, Haifa 34369, Israel

Frank Bignone, Institute of Communications Technology, Swiss Federal Institute of Technology, ETH Zentrum, Gloriastrasse 35, CH-8092 Zurich, Switzerland

David Burtin, INRIA Lorraine & CRIN/CNRS, 615 rue du Jardin Botanique, 54602 Villers-les-Nancy Cedex, France

Jose Colomer, Institut Cartographic de Catalunya, Balmes 209, 08006 Barcelona, Spain

Heinrich Ebner, Technical University Munich, Chair for Photogrammetry and Remote Sensing, Arcisstr. 21, D-80290 Munich, Germany

Christoph Eidenbenz, Bundesamt für Landestopographie, Seftigenstrasse 264, CH-3084 Wabern, Switzerland

Edward Falkov, State Research Institute of Aviation Systems (Gos NIIAS), 7 Victorenko Str., Moskow 125319, Russia

Wolfgang Foerstner, Dept. of Photogrammetry, University of Bonn, Nussallee 15, D-5300 Bonn 1, Germany

Gianfranco Forlani, Politecnico di Milano, Dip. IIAR - Sezione Rilevamento, Piazza Leonardo Da Vinci 32, I-20133 Milano, Italy

Pascal Fua, SRI International, Artificial Intelligence Center, 333 Ravenswood Ave., Menlo Park, CA 94025, USA

Claudia Fuchs, Institute for Photogrammetry, University of Bonn, Nussalle 15, D-53115 Bonn, Germany

Armin Gruen, Institute of Geodesy & Photogrammetry, Swiss Federal Institute of Technology, ETH - Hoenggerrberg, CH-8093 Zurich, Switzerland

Eberhard Guelch, Dept. of Geodesy & Photogrammetry, Royal Institute of Technology (KTH), S-100 44 Stockholm 70, Sweden

Norbert Haala, Institute for Photogrammetry, University of Stuttgart, Keplerstr. 11, 70174 Stuttgart, Germany

Allen Hanson, Dept. of Computer Science, Box 34610, Lederle Graduate Research Center, University of Massachusetts, Amherst, MA 01003-4610, USA

Jan Heikkilä, Helsinki University of Technology, Institute of Photogrammetry & Remote Sensing, Otakaari 1F, FIN-02150 Espoo, Finland

Olof Henricsson, Institute of Communications Technology, Swiss Federal Institute of Technology, ETH - Zentrum, Gloriastrasse 35, CH-8092 Zurich, Switzerland

Andres Huertas, Institute for Robotics & Intelligent Systems, Powell Hall of Engineering 204, Univ. of Southern California, Los Angeles, CA 90089-0273, USA

Olivier Jamet, Institut Géographique National (IGN), Laboratoire MATIS, Service de la Recherche, 2 avenue Pasteur, 94160 Saint-Mandé, France

Bruno Jedynak, INRIA, Domaine de Voluceau - Rocquencourt, B.P. 105, 78153 Le Chesnay Cedex, France

Taejung Kim, University College London, Dept. of Photogrammetry and Surveying, Gower Str., London WC1E 6BT, UK

Thomas Kolbe, Inst. für Informatik, Universität Bonn, Römerstr. 164, 53117 Bonn, Germany

Olaf Kuebler, Institute of Communications Technology, Swiss Federal Institute of Technology, ETH - Zentrum, Gloriastrasse 35, CH-8092 Zurich, Switzerland

Felicitas Lang, Institute for Photogrammetry, University of Bonn, Nussalle 15, D-53115 Bonn, Germany

Stéphane Laveau, INRIA, 2004 Route des Lucioles, B.P. 93, 06902 Sophia-Antipolis Cedex, France

Franz Leberl, Institut für Computergraphik, Technische Universität Graz, Münzgrabenstr. 11, A-8010 Graz, Austria

Haihong Li, Institute of Geodesy & Photogrammetry, Swiss Federal Institute of Technology, ETH - Hoenggerrberg, CH-8093 Zurich, Switzerland

Jean-Luc Lotti, INRIA, 2004 Route des Lucioles, B.P. 93, 06902 Sophia-Antipolis, France

Henri Maître, Ecole Nationale Supérieure des Télécommunications, Département Images, 46 Rue Barrault, F-75634 Paris Cedex 13, France

Scott Mason, Institute of Geodesy & Photogrammetry, Swiss Federal Institute of Technology ETH - Hoenggerrberg, CH-8093 Zurich, Switzerland

Helmut Mayer, Technical University Munich, Chair for Photogrammetry and Remote Sensing, Arcisstr. 21, D-80290 Munich, Germany

Alfons Meid, Leica AG PMU, CH-5035 Unterentfelden, Switzerland

Eckhard Michaelsen, FIM, Eisenstockstr. 12, 76275 Ettlingen, Germany

Walter Mueller, GDE Systems Inc., P.O. Box 509008, San Diego, CA 92150-9008, USA

Jan-Peter Muller, Dept. of Photogrammetry & Surveying, University College London , Gower Str., London WC1E 6BT, UK

Carla Nardinocchi, Università degli Studi di Ancona, Dipartimento di Scienza delle Terre, Via Brecce Bianche, 63100 Ancona, Italy

Stephan Nebiker, Institute of Geodesy & Photogrammetry, ETH Hoenggerberg, CH-8093 Zurich, Switzerland

Walter Neuenschwander, Institute of Communications Technology, ETH - Zentrum, Gloriastrasse 35, 8092 Zurich, Switzerland

Ramakant Nevatia, Institute for Robotics & Intelligent Systems, Powell Hall of Engineering

204, Univ. of Southern California, Los Angeles, CA 90089-0273, USA

Marko Pasko, Institute for Computer Graphics, Technical University Graz, Muenzgrobenstr. 11, A-8010 Graz, Austria

Petros Patias, The Aristotle University of Thessaloniki, Dept. of Cadastre, Photogrammetry & Cartography, Univ. Box 473, GR-54006 Thessaloniki, Greece

Berthold Plietker, Universität Stuttgart, Institut für Photogrammetrie, Keplerstr. 11, 70174 Stuttgart, Germany

Franz Quint, Institut für Photogrammetrie und Fernerkundung, Universität Karlsruhe, Postfach 6980, D-76128 Karlsruhe, Germany

Luc Robert, INRIA, 2004 Route des Lucioles, B.P. 93, 06902 Sophia-Antipolis Cedex, France

Guy Ruckebusch, Dept. S2IG, Syseca, 66 Avenue Pierre Brossolette, Malakoff 92, France

Tapani Sarjakoski, Finnish Geodetic Institute, Department of Cartography and Geoinformatics, Ilmalankatu 1A, FIN-00240 Helsinki, Finland

Toni Schenk, Dept. of Geodetic Science and Surveying, The Ohio State University, 1958 Neil Avenue # 404, Columbus, Ohio 43210, USA

Monika Sester, Institute of Photogrammetry, Stuttgart University, Keplerstrasse 11, D-70174 Stuttgart, Germany

Vittala Shettigara, Defense Science and Technology Organization, P.O. Box 1500, Bldg. 79 Labs, Salisbury, S.A., Australia 5108

Anthony Stefanidis, Institute of Geodesy & Photogrammetry, Swiss Federal Institute of Technology, ETH - Hoenggerrberg, CH-8093 Zurich, Switzerland

Carsten Steger, Technische Universität München, Lehrstuhl für Informatik IX, Orleanstrasse 34, 81667 München, Germany

Manfred Sties, Universität Karlsruhe, Institut für Photogrammetrie & Fernerkundung, Postfach 6980, D-76128 Karlsruhe, Germany

Uwe Stilla, FIM, Eisenstockstr. 12, D-76275 Ettlingen, Germany

Thomas Strat, SRI International, 333 Ravenswood Ave., Menlo Park, CA 94025, USA

Rolf Thiemann, Abteilungsleiter Raster-GIS, Strässle Informationssysteme GmbH, Talangerstr. 5-7, 82152 Krailling, Germany

Margarita Torre, Institut Cartographic de Catalunya, Balmes 209, 08006 Barcelona, Spain

John Trinder, School of Geomatic Engineering, The University of NSW, Sydney, NSW 2052, Australia

George Vosselman, Laboratory of Photogrammetry & Remote Sensing, Faculty of Geodetic Engineering, Delft University of Technology, Thijsseweg 11, 2629 JA Delft, The Netherlands

Wolfram Willuhn, Institute of Communications Technology, Swiss Federal Institute of Technology, ETH- Zentrum, Gloriastrasse 35, 8092 Zurich, Switzerland

Sergei Zheltov, State Research Institute of Aviation Systems (Gos NIIAS), 7 Victorenko str., Moskow 125319, Russia

Author Index

If you have any concerns about our product,
you can contact us at
ProductSafety@springernature.com

In case Publisher is established outside the EU,
the EU authorized representative is:
Springer Nature Customer Service Center GmbH
Europaplatz 3, 69115 Heidelberg, Germany

Printed by Libri Plureos GmbH
in Hamburg, Germany

MIX
Papier aus verantwortungsvollen Quellen
Paper from responsible sources
FSC® C105338

If you have any concerns about our products,
you can contact us on
ProductSafety@springernature.com

In case Publisher is established outside the EU,
the EU authorized representative is:
Springer Nature Customer Service Center GmbH
Europaplatz 3, 69115 Heidelberg, Germany

Printed by Libri Plureos GmbH
in Hamburg, Germany